TREATISE ON
ADHESION AND ADHESIVES
volume 5

TREATISE ON ADHESION AND ADHESIVES

volume 5

edited by
Robert L. Patrick

MARCEL DEKKER, INC. New York and Basel

MARCEL DEKKER, INC.
270 Madison Avenue, New York, New York 10016

LIBRARY OF CONGRESS CATALOG CARD NUMBER: 66-11285
ISBN: 0-8247-1399-0

Current printing (last digit):
10 9 8 7 6 5 4 3 2 1

PRINTED IN THE UNITED STATES OF AMERICA

Preface

The *Treatise on Adhesion and Adhesives* was originally set at three volumes: the first volume to be theory, the second, technology, and the third, materials. As with all good intentions, the concept was carried out in the first case, was modified in the second case, and was substantially changed in the third. In addition, *Volume 4* provides a state-of-the-art review of structural adhesives in the form of a Materials Advisory Board report of the National Research Council of the National Academy of Sciences. Several respected colleagues debated the idea of including the MAB report; however, it would seem that the general public is often not aware of this type of document, and therefore some worthwhile reviews have been made available. Also, *Volume 4* provides a generalized roadmap for future work in the field.

Finally, *Volume 5* of the *Treatise* consists of a selection of special topics, starting with a critical review by J. R. Huntsberger of the relationships among interfacial energies, contact angle equilibria, and the work of adhesion. Since the early work of Zisman and coworkers, the contact angle has been used and misused by numberless workers in the field. In a concise treatment, Huntsberger points out the direct relation of the contact angle to estimating interfacial energies and the work of adhesion when solids are involved. The importance of a thorough understanding of fundamentals is stressed.

A new area of interest is covered in a chapter by Drzal concerning the advanced material, high modulus graphite. In the next few years, much will be heard concerning graphite fiber reinforced composites, and the requirement to understand and improve the graphite surface and surface interactions is paramount. The chapter reviews recent exciting work that was carried out by the author when examining the surface composition and surface energetics of graphite fibers derived from polyacrylonitrile. The third chapter is essentially a continuation of the work reported in Volume 3 by J. Dean Minford. In an exhaustive treatment of durability studies carried out for as long as eight years, the author points out the important interacting relationships of substrate composition, physical state, and geometry; the adhesion composition, as well as curing temperature; and the environmental considerations with and without stress. The chapter is essentially state-of-the-art as far as durability of bond with aluminum substrates is concerned. The next two chapters represent a monumental effort by Smith and Kaelble to examine the results of a detailed study of the effect of surface treatment on aluminum and titanium from a failure point of view. Using elegant techniques developed during the work, they have generated an outstanding contribution to the understanding of failure mechanisms.

Good has been concerned for several years about the semantics of adhesion; he offers a provocative discussion involving this subject. Lewis and Gounder have been considering the extension of the Prot equation to predict the performance and, therefore, the failure of adhesive joints. The approach is controversial and is included to stimulate thought and, hopefully, further experimentation in this area.

Volume 5 falls into the general category of special topics; however, it is obvious that the interface is preeminent in five of the seven chapters. This is not surprising inasmuch as there is considerable understanding of the substrates and the adhesives, respectively, while the interface is poorly understood. It is to be hoped that this volume will shed some additional light on this poorly illuminated area.

Robert L. Patrick

Contributors

Lawrence T. Drzal, Air Force Materials Laboratory, Wright Patterson Air Force Base, Ohio

Robert J. Good, Department of Chemical Engineering, State University of New York, Buffalo, New York

James R. Huntsberger, E. I. du Pont de Nemours & Company Wilmington, Delaware

David H. Kaelble, Rockwell International, Thousand Oaks, California

Armand F. Lewis, Lord Corporation, Erie, Pennsylvania

J. Dean Minford, Alcoa Laboratories, Alcoa Center, Pennsylvania

R. Natarajan Gounder,* Lord Corporation, Erie, Pennsylvania

Tennyson Smith, Rockwell International, Thousand Oaks, California

*Present affiliation: RCA Laboratories, Princeton, New Jersey

Contents

Treatise on Adhesion and Adhesives
Contents of Other Volumes

editor
Robert L. Patrick

TREATISE ON
ADHESION AND ADHESIVES
volume 5

1

Interfacial Energies, Contact Angles, and Adhesion

James R. Huntsberger

E. I. du Pont de Nemours & Company
Wilmington, Delaware

1. ATTRACTIVE FORCES AND ENERGIES

This chapter comprises a review of the relationships among interfacial energies, contact angle equilibria, and the reversible work of adhesion. The review is not intended to be comprehensive, but hopefully is sufficiently current and critical to be of some value.

Let us start by defining "adhesion" (as it will be used throughout this work) as the force or energy of attraction between two materials or phases which are in contact with each other. These attractive forces and energies are due mainly to dispersion and dipolar and induction interactions. In some cases hydrogen bonding or acid-base-type interactions may also be involved.

Various workers [1-6] have proposed models and calculated these forces using polarizabilities, electron vibrational frequencies, ionization potentials, and dipole moments as the source of attractive interactions. For the purpose of this review the important results are the apparent validity of the geometric mean as a means of calculating interactions between two materials and recognition that at interfacial separations less than 10 nm attractive forces are inversely proportional to the third power of the effective mean distance of intermolecular separation and attractive energies inversely proportional to the second power of the separation.

Another important result of these calculations is that they lead to a direct relationship between interfacial energies and the maximum attractive force or the reversible work of adhesion per unit area between phases. From Gardon [7]:

$$\sigma_{max} = \frac{2.05\phi(\gamma_1\gamma_2)^{\frac{1}{2}}}{\bar{r}_{12}} \tag{1}$$

and from Good and Girifalco [1]:

$$W_{adh} = 2\phi(\gamma_1\gamma_2)^{\frac{1}{2}} \tag{2}$$

In both of these equations ϕ is Good's interaction parameter, which corrects for any defects in the models used for calculation and for departures from ideality or regularity in interfacial interactions.

The parameter ϕ can be broken down into two factors: ϕ_a, which accounts for the departures from ideality in the intermolecular attractions; and ϕ_r, which accounts for departures from regularity in interfacial separations.

The importance of variations of ϕ_r has been dismissed frequently (e.g., Ref. 6) on the premise that only when differences between the average intermolecular separations of the phases is large will ϕ_r be significantly less than unity. This is the result of assuming that the interfacial separation $\bar{r}_{12}$ is the arithmetic mean of the average intermolecular separations of the individual phases $\bar{r}_{11}$ and $\bar{r}_{22}$.

This assumption seems unwarranted, especially when polymers comprise at least one of the phases. The configurations and restrictions on conformations of polymer molecules at an interface make it highly probable that the effective mean interfacial separations will be appreciably greater than the arithmetic mean based on segment volumes.

As a conservative estimate, if $\bar{r}_{11}$ for one phase were 20% larger than $\bar{r}_{22}$ for the other, and the effective mean interfacial separation were equal to $\bar{r}_{11}$, ϕ would be only 0.83 and the work of adhesion would be 16% less than that calculated using $\phi_r = 0.99$ (based on the arithmetic mean assumption for the value of $\bar{r}_{12}$). For this same example the maximum or critical stress would be 28% less than that calculated using $\phi = 0.99$.

In reality, discrepancies may frequently be greater than these.

Wu [8] found better agreement between measured values of γ_1, γ_2, and γ_{12} and values for ϕ_a calculated using harmonic means rather than geometric means. There is no satisfactory theoretical basis for this. Harmonic means lead to smaller values for work of adhesion and larger values for γ_{12} when $\gamma_1 \neq \gamma_2$. These energies are dependent on the values of ϕ and it seems likely that the apparent better agreement using harmonic means was due in reality to unrecognized and unaccounted for decreases in ϕ_r associated with polymers.

The molecular density at interfaces and the influence of molecular density on mean interfacial separations is in fact the most important single factor in determining surface and interfacial tensions. This is suggested for example by the relationship between surface tension and parachor which shows γ is directly proportional to the fourth power of the density. Hoernshemeyer [9] showed that the low surface energies of fluoropolymers are mainly due to their low molecular surface densities rather than the character of the attractive forces.

It is clear that close packing of surface layers can diminish interfacial interactions <u>only</u> when increasing packing diminishes intermolecular penetration into a surface layer. A macroscopic analogy is provided by the influence of surface roughness on contact angle hysteresis. When $\theta > 90°$ increasing roughness increases advancing contact angles, but when $\theta < 90°$ increasing roughness decreases advancing angles.

In the following discussion the interaction parameter ϕ will be considered as making the work of adhesion an exact thermodynamic value for a given pair of materials. However, whenever a value for ϕ is *calculated* rather than determined experimentally the value is an *estimate* of ϕ.

Estimates of ϕ_a can be obtained using equations of the form:

$$\phi_a = (d_1 d_2)^{\frac{1}{2}} + (p_1 p_2)^{\frac{1}{2}} + 0.0721\,(d_2 p_1)^{\frac{1}{2}} + 0.0721(d_1 p_2)^{\frac{1}{2}} + \Delta H + \Delta ab \qquad (3)$$

where d_i is the fraction of total interaction due to dispersion forces, p_i the fraction due to polar interactions, ΔH interactions due to hydrogen bonding, and Δ_{ab} acid-base interactions.

The third and fourth terms are due to induction [3].

At the present time inclusion of all of the terms in Eq. (3) is probably not warranted and the simplified equation

$$\phi_a = (d_1 d_2)^{\frac{1}{2}} + (p_1 p_2)^{\frac{1}{2}} \qquad (4)$$

will be found useful and convenient.

Estimates for ϕ_r have been given [6] as:

$$\phi_r = \frac{r_{11} r_{12}}{[(r_{11} + r_{22})/2]^2} \qquad (5)$$

but as discussed earlier are probably poor estimates of little value.

There are several fundamental relationships which are the basis for using contact angle equilibria for making inferences concerning adhesion. These include:

$$\gamma_{SL} = \gamma_{SO} + \gamma_{LV} - 2\phi(\gamma_{SO}\gamma_{LV})^{\frac{1}{2}} \qquad (6)$$

$$\gamma_{SV} - \gamma_{SL} = \gamma_{LV} \cos\theta \qquad (7)$$

$$\gamma_{SO} - \pi_e = \gamma_{SV} \qquad (8)$$

$$\pi_{SL} - \pi_e = \gamma_{LV} \cos \theta \tag{9}$$

$$\pi_{SL} = \gamma_{SO} - \gamma_{SL} \tag{10}$$

$$\gamma_{SO} - \pi_e - \gamma_{SL} = \gamma_{LV} \quad \text{when } \theta = 0 \tag{11}$$

In these equations γ_{SO} and γ_{LV} are values for the surface tensions of the solid and liquid, respectively, in equilibrium with their own vapors. The equilibrium spreading pressure of the vapor on the solid is π_e, and π_{SL} is the equilibrium spreading pressure of the liquid on the solid.

Equation (11) is a statement of Anatoff's rule *at equilibrium*. This was shown [10] to be valid.

In general, the initial spreading coefficients in three-phase systems are defined by expressiond of the form:

$$S_{2/13} = \gamma_{13} - \gamma_{12} - \gamma_{23} \tag{12}$$

$$= W_{adh\,12} + W_{adh\,23} - W_{adh\,13} - 2\gamma_{2V} \tag{13}$$

where $S_{2/13}$ is the coefficient for phase 2 spreading along the 13 interface.

For the special case where phase 3 is vapor:

$$S_{2/13} = W_{adh\,12} - 2\gamma_{2V} \tag{14}$$

The following special values are obtained from the fundamental equations:

At $\theta = 0$; $S = 0$ (S = equilibrium spreading coeff.)

At $\theta = 90°$; $\pi_{SL} = \pi_e$

Plotting these relationships in a dimensionless form as ratios with respect to γ_{SO} provides some interesting insights.

The work of adhesion as W_{adh}/γ_{SO} is plotted vs. γ_{LV}/γ_{SO} in Fig. 1-1. Note that the work of adhesion increases monotonically with increasing γ_{LV}.

The interfacial tension γ_{SL}/γ_{SO} is plotted vs. γ_{LV}/γ_{SO} in Fig. 1-2. The values for γ_{SL}/γ_{SO} exhibit minima at $\gamma_{LV}/\gamma_{SO} = \phi^2$. The minimum values for $\gamma_{SL}/\gamma_{SO} = (1 - \phi^2)$. This was shown earlier by Good [6].

The spreading pressure of the liquid on the solid π_{SL}/γ_{SO} is plotted in Fig. 1-3. The initial spreading coefficient S_i/γ_{SO} and the equilibrium spreading coefficient S/γ_{SO} are plotted in Figs. 1-4 and 1-5.

From Figs. 1-3 to 1-5 and equations (2), (6), (8), and (10) the following relationships are clear:

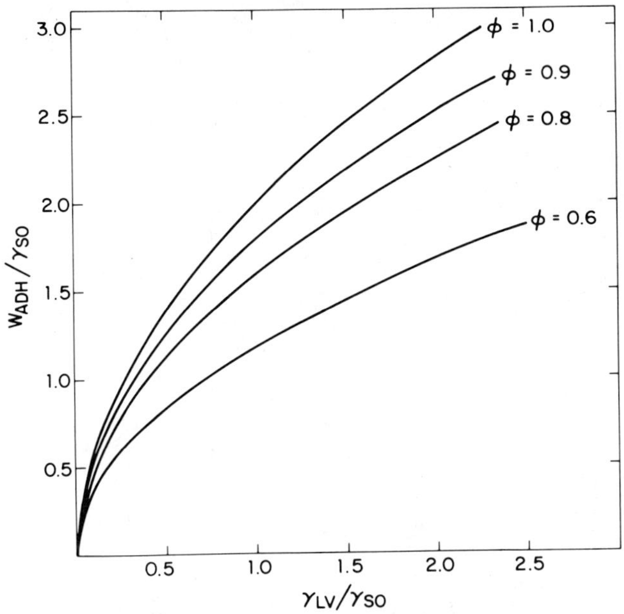

FIG. 1-1 Work of adhesion/γ_{SO} vs. γ_{LV}/γ_{SO} for various values of ϕ.

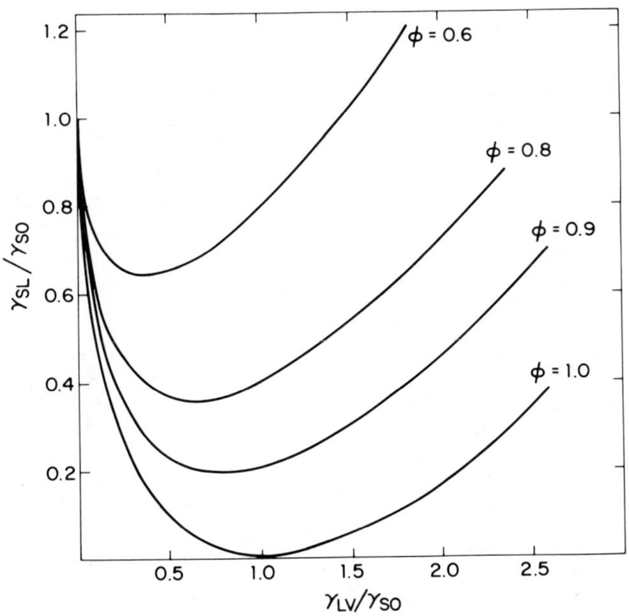

FIG. 1-2 Interfacial tension γ_{SL}/γ_{SO} vs. γ_{LV}/γ_{SO} for various values of ϕ.

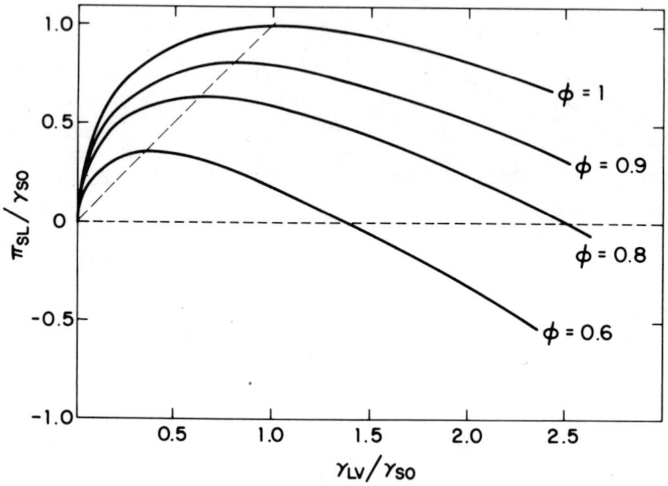

FIG. 1-3 Spreading pressure of liquid on the solid $(\pi_{SL})/\gamma_{SO}$ vs. γ_{LV}/γ_{SO} for various values of ϕ.

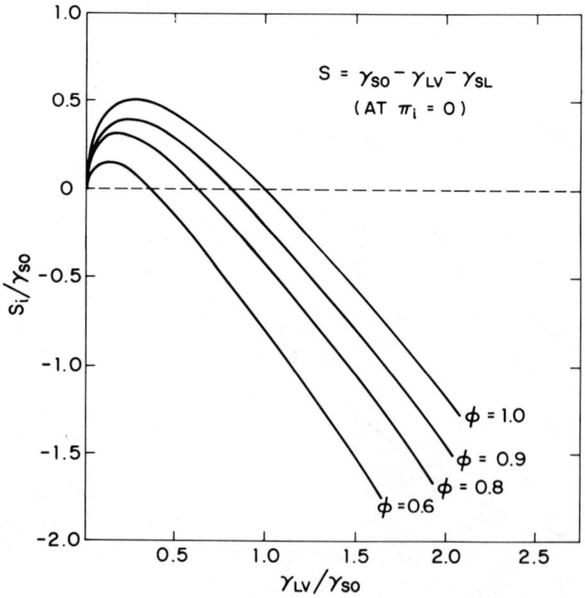

FIG. 1-4 Initial spreading coefficient S_i/γ_{SO} vs. γ_{LV}/γ_{SO} for various values of ϕ.

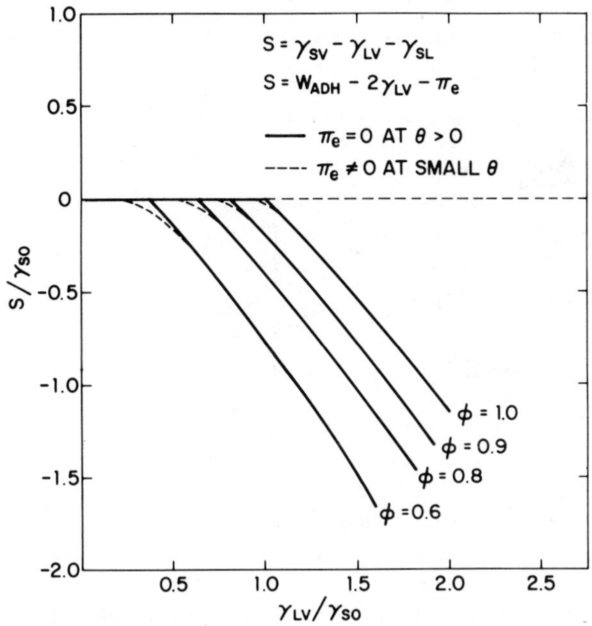

FIG. 1-5 Equilibrium spreading coefficient S/γ_{SO} vs. γ_{LV}/γ_{SO} for various values of ϕ.

$$W_{adh} - \gamma_{LV} = \pi_{SL}$$

$$W_{adh} - 2\gamma_{LV} = S_i$$

$$W_{adh} - 2\gamma_{LV} - \pi_e = S$$

when $\theta = 0$, π_{SL}/γ_{SO} exhibit maxima at $\gamma_{LV}/\gamma_{SO} = \phi^2$, $S = 0$, and $\pi_e = S_i$. Maxima for the S_i/γ_{SO} curves occur at $0.25\phi^2$. The maximum values for S_i/γ_{SO} and consequently also for π_e/γ_{SO} are $= 0.5\phi^2$.

The plots of $\phi(\gamma_{LV}/\gamma_{SO})^{\frac{1}{2}}$ and $\phi(\gamma_{SO}/\gamma_{LV})^{\frac{1}{2}}$ in Fig. 1-6 show the failure locus for *reversible* separation. When both values are <1 interfacial failure is predicted. While these criteria are only for reversible separation, recall that when ϕ becomes appreciably less than unity the critical stresses usually diminish more rapidly than the interfacial energies, and for samples with only small flaws or voids in the bulk phases the probability of interfacial separation is greatly increased. There is no reason to consider interfacial separation a rare or unlikely event.

In Fig. 1-7 cos θ is plotted vs. the dimensionless γ_{LV}/γ_{SO}. These curves represent the theoretical basis for Zisman plots of cos θ vs. γ_{LV}.

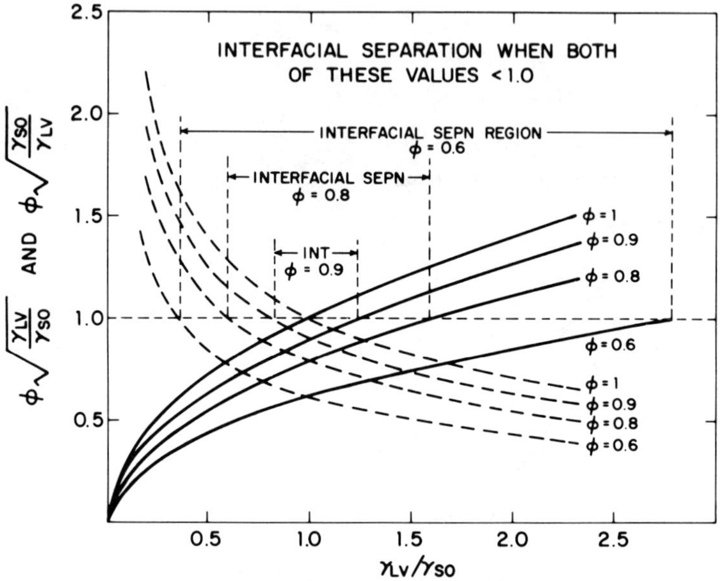

FIG. 1-6 Failure criteria for reversible separation at various values for ϕ: —— = $\phi\sqrt{\gamma_{LV}/\gamma_{SO}}$; - - - = $\phi\sqrt{\gamma_{SO}/\gamma_{LV}}$.

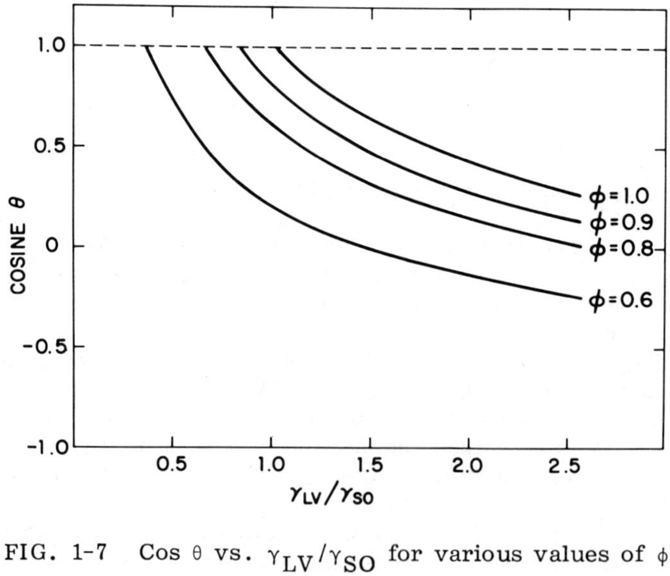

FIG. 1-7 Cos θ vs. γ_{LV}/γ_{SO} for various values of ϕ.

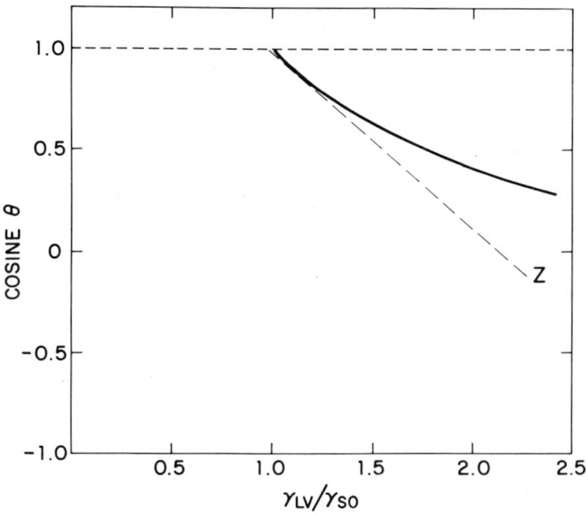

FIG. 1-8 Cos θ vs. γ_{LV}/γ_{SO}. The dashed line Z is a plot corresponding to $\cos\theta = 1 + b(\gamma_C - \gamma_{LV})$ and using Zisman's value for b.

Figure 1-8 shows a Zisman plot for various liquids on polyethylene. Finite spreading pressures of the test liquids at low θ cause the curve to approximate a straight line in this region and also lead to values for $\gamma_c \sim 3$ dynes less than the value for γ_{SO} reported by others [11-13]. The approximately linear form of the data at higher θ and higher γ_{LV}/γ_{SO} is due to changes in ϕ for interactions between the liquids and polyethylene. The line Z in Fig. 1-8 was drawn using Zisman's value for b [14] in the equation

$$\cos\theta = 1 + b(\gamma_c - \gamma_{LV}) \tag{15}$$

Zisman uses Eq. (15) and

$$W_{adh} = \gamma_{LV}(1 + \cos\theta) \tag{16}$$

to obtain:

$$W_{adh} = (2 + b\gamma_c)\gamma_{LV} - b\gamma_{LV}^2 \tag{17}$$

which is the equation for a parabola.

In Fig. 1-9 Zisman's data are plotted on the curves for W_{adh}/γ_{SO} vs. γ_{LV}/γ_{SO}. It is evident that the parabolic form simply reflects the diminishing values for the interaction parameter ϕ. Zisman recognized this as an empirical equation, and in the light of subsequent

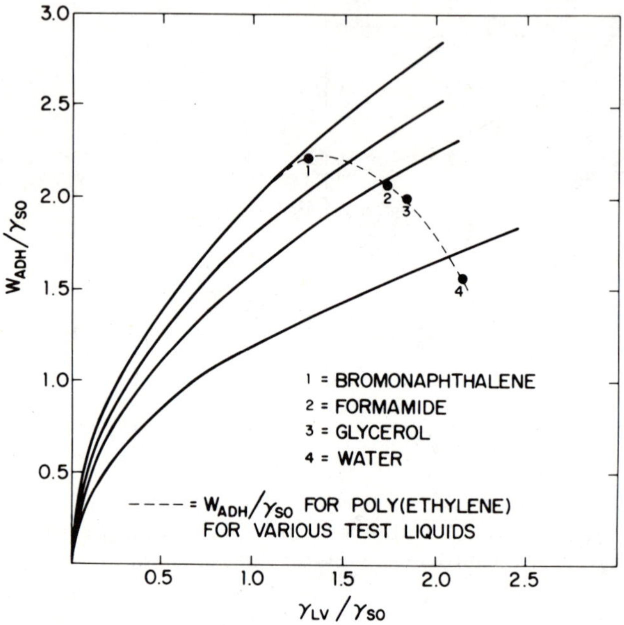

FIG. 1-9 W_{adh}/γ_{SO} vs. γ_{LV}/γ_{SO} for various values of ϕ.

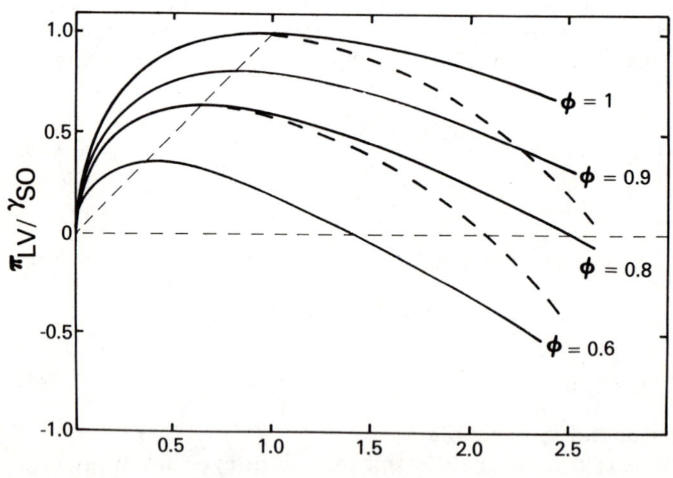

FIG. 1-10 Spreading pressure of liquid on the solid $(\pi_{SL})/\gamma_{SO}$ vs. γ_{LV}/γ_{SO} for various values of ϕ.

clarification Eq. (17) does not appear to be of any practical value and is subject to misinterpretation.

Another equation bearing a resemblance to Eq. (17) was proposed recently [15] by Neumann et al.:

$$\gamma_{LV} \cos \theta = a\gamma_{LV}^2 + b\gamma_{LV} + c \tag{18}$$

Curves based on Eq. (18) were plotted as $\gamma_{LV} \cos \theta$ vs. γ_{LV}. From Eq. (9) we see that these curves bear the same relationship to the π_{SL}/γ_{SO} curves of Fig. 1-3 that the Zisman curves had with Figs. 1-1 and 1-7. That is the π_{SL}/γ_{SO} (minus π_e/γ_{SO}) curves are the theoretical basis for the empirical N, G, H, and S curves. This is shown in Fig. 1-10. Here again the departure of the empirical curves reflects diminishing values of ϕ with increasing γ_{LV}/γ_{SO} and suggests that Eq. (18) has no special value.

N, G, H, and S used the intercept of the curves with the 45° line to arrive at a value $(\gamma_{LV})^*$ which is in fact identical to Zisman's long-established γ_c. These authors made unwarranted and highly restrictive assumptions ($\pi_e = 0$ at $\theta \geq 0$, and $(\gamma_{SL})^* = 0$, where $(\gamma_{SL})^*$ is the interfacial tension between the solid and liquid at $\gamma_{LV} = (\gamma_{LV})^*$). This latter assumption is valid only when $\phi = 1$ *and* $\gamma_{LV} = \gamma_{SO}$. An error fatal to their further argument was caused by substituting $(\gamma_{SV})^* = \gamma_c$ into Eq. (6), obtaining

$$\phi = \frac{\gamma_c + \gamma_{LV} - \gamma_{SL}}{2(\gamma_c\gamma_{LV})^{\frac{1}{2}}} \tag{19}$$

instead of using the correct:

$$\phi^2 = \frac{\gamma_c}{2(\gamma_c\gamma_{LV})^{\frac{1}{2}} + \gamma_{SL} - \gamma_{LV}} \tag{20}$$

They used Eq. (19) and values for γ_{SL} obtained using Eq. (7) to determine ϕ as a function of γ_{SL} and reported

$$\phi = 1 - 0.0075\gamma_{SL} \tag{21}$$

Since γ_c is itself a function of $\phi(\gamma_c = \phi^2\gamma_{SO})$ it is clear that Eq. (19) cannot lead to correct values.

It can be easily shown from the fundamental relationships that $d\phi/d\gamma_{SL}$ is linear only at constant γ_{LV}/γ_{SO}, and that the slope varies with changes in γ_{LV}/γ_{SO}. The slopes which would correspond to the incorrect value 0.0075 in Eq. (21) range over a broad region of γ_{LV}/γ_{SO} values from 2× to 10× the value from Eq. (21).

It is unfortunate that the value for θ given in Eq. (21) was used as the basis for the Sell-Neumann equation [16]:

$$\cos\,\theta = \frac{(0.015\,\gamma_{SO} - 2)(\gamma_{SO}\gamma_{LV})^{\frac{1}{2}} + \gamma_{LV}}{\gamma_{LV}[0.015(\gamma_{SO}\gamma_{LV})^{\frac{1}{2}} - 1]} \tag{22}$$

It follows that Eq. (22) also is not valid.

2. CONTACT ANGLE EQUILIBRIA

With the preceding discussion as a basis, let us examine some examples of contact angle equilibria. Figure 1-11 shows the contact angles for (a) water on methylene iodide (using experimental values but with methylene iodide represented as an undeformed plane surface), and (b) water on mercury using experimental values which have been reported for γ_{SO} (mercury) γ_{LV}, γ_{SL}, and π_e and a calculated value for θ. These figures show the various interfacial tensions and spreading pressures at equilibrium in each example.

Examples of the uncertainty involved in experimental values when π_e is unknown and not necessarily = 0 are illustrated in Fig. 1-12. Note that when γ_{LV} is appreciably less than γ_{SO} significant values for π_e are not unlikely.

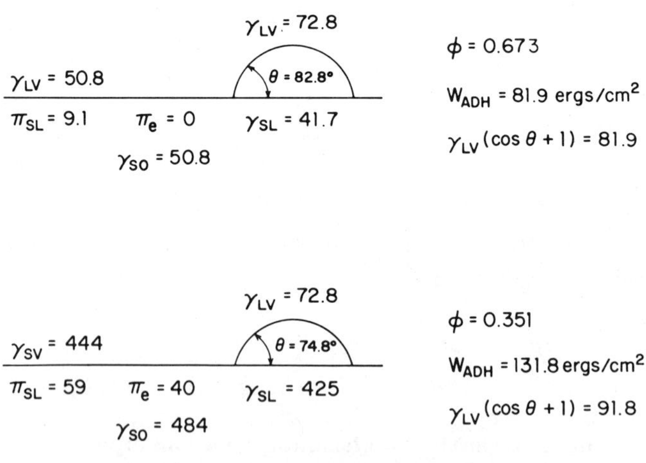

FIG. 1-11 Some examples of contact angle equilibria: (top) for water-methylene iodide (with experimental values—but modified to show the methylene iodide surface as undeformed); (bottom) for water-mercury (θ calculated using a reported value of π_e = 40).

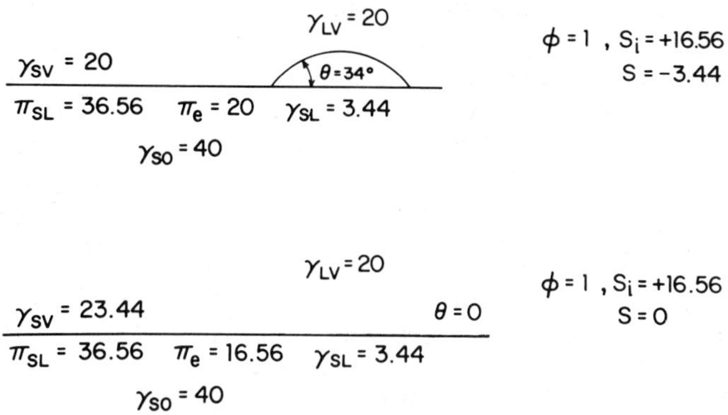

FIG. 1-12 Examples of uncertainty among relationships in contact angle equilibria when π_e is unknown and perhaps does not = 0.

Despite some uncertainties and limitations the contact angle used in conjunction with the basic relationships can be a useful tool for estimating the surface energies of solids. This approach was used by Owens and Wendt [17] for determining the dispersion and polar contributions of solid-liquid interactions. They measured contact angles of two test liquids of different, known surface tensions and fractional polarities—water ($\gamma = 72.8$, $d = 0.3$, $p = 0.7$) and methylene iodide ($\gamma = 50.8$, $d = 0.97$, $p = 0.03$)—on a common surface. The surface energies of the solids and their dispersion and polar contributions were calculated using simultaneous equations of the form:

$$1 + \cos\theta = 2\sqrt{\gamma^d_{SO}}\left(\frac{\sqrt{\gamma^d_{LV}}}{\gamma_{LV}}\right) + 2\sqrt{\gamma^p_{SO}}\left(\frac{\sqrt{\gamma^p_{LV}}}{\gamma_{LV}}\right) \qquad (23)$$

Estimating the surface tensions and fractional polarities of solids using the Owens and Wendt method allow prediction of the work of adhesion between two such solids when they have been brought into equilibrium interfacial contact. The interaction parameter ϕ can be estimated from the fractional polarities using Eq. (4) and the work of adhesion using Eq. (2).

Dann [18] used an approach similar to that of Owens and Wendt but used ethanol-water mixtures for the test liquids. Interesting observations can certainly be obtained using such liquid mixtures, but it is evident that selective adsorption may occur at each of the three interfaces involved in contact angle equilibria. This would alter the interfacial tensions and contact angles. Satisfactory or reliable interpretation of such data requires additional information.

Adsorption and its influence on interfacial tensions is a very real problem when using contact angles. Very small amounts of impurities or additives contained in practical samples can lead to data very different from that of similar systems of high purity. This can be troublesome, but also provides a sensitive way of detecting, following, and interpreting the behavior of complex systems.

Johnson and Dettre [19] showed that material contained within the liquid phase can often spread and adsorb on the solid surface in advance of the line of contact between the bulk phases.

Relative adsorption at the various interfaces is a function of the concentration of adsorbing species. The thermodynamics of adsorption were given by Smolders [20]. Figure 1-13 from Johnson and Dettre [19] shows the results of Smolders treatment. The contact angle and liquid surface tension are plotted against the log of the concentration of surfactant. At low concentration adsorption was predominantly at the solid interface causing increasing θ. At intermediate concentrations adsorptions were such that changes at the solid-vapor interface were counterbalanced by the changes in γ_{LV} and γ_{SL} and θ remained constant. At higher surfactant concentrations no further changes in γ_{SV} occurred but decreasing γ_{LV} and γ_{SL} caused θ to decrease.

Contact angle equilibria are further complicated by composite surfaces. When two materials forming a composite surface are not uniformly

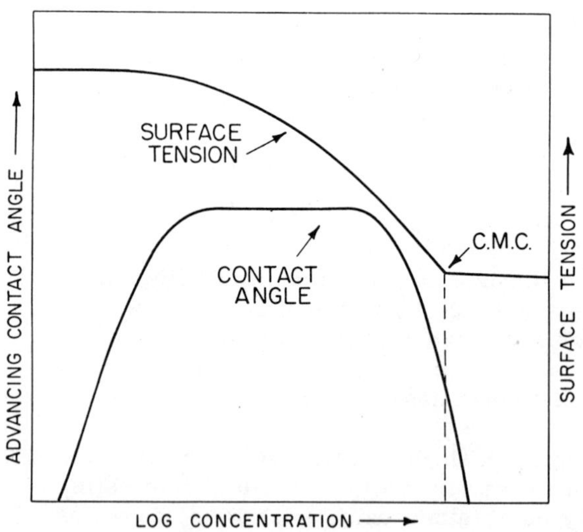

FIG. 1-13 Contact angle and surface tension vs. log surfactant concentration.

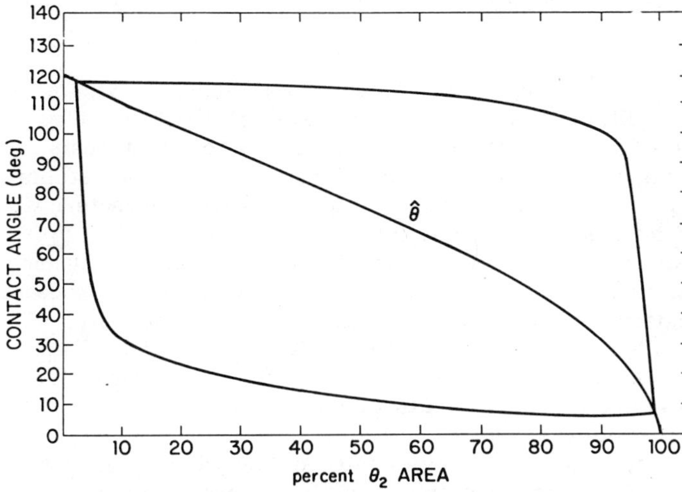

FIG. 1-14 Contact angle hysteresis: upper curve for advancing angles; lower curve for receding angles; $\hat{\theta}$ = estimated equilibrium angles.

distributed on a molecular scale but rather exist as domains with linear dimensions of ∿100 nm or greater or as patches separated by ∿100 nm, contact angle hysteresis may be found.

Hysteresis resulting from heterogeneity has been thoroughly treated by Johnson and Dettre [19]. This form of hysteresis is shown in Fig. 1-14. Considerable uncertainty about the composition of such surfaces is caused by the large influence of relatively small amounts of material in setting both advancing and receding angles. Some indications about the relative amounts of the two materials can sometimes be discerned by the reproducibility of the advancing and receding angles. When only a few percent of low-energy material is present in the solid surface, advancing angles exhibit poor reproducibility. When small amounts of high-surface-energy material are present, receding angles are poorly reproducible.

There is good reason to believe that surfaces of block copolymers may sometimes exhibit isolated areas comprising molecular segments of a single component of the copolymer which are large enough to lead to hysteresis. It is unfortunate that some workers have failed to recognize that valuable information may be lost when receding angles are not measured.

3. CONTACT ANGLES, INTERFACIAL EQUILIBRIUM, AND WETTING

This discussion has been based on samples at equilibrium. Stable thermodynamic equilibrium is required for meaningful relationships between contact angles and interfacial energies. When hysteresis occurs contact angles correspond to one of many possible metastable states which depend on both the physical and chemical characteristics of the solid and on measuring techniques.

Stable thermodynamic equilibrium implies a condition of minimum free energy and steady states for any adsorption-desorption processes. No changes in adsorption or spreading will occur spontaneously.

To relate contact angle equilibria to practical adhesion problems it is necessary to establish that both experimental systems used for measurements and the adhesive bonds which are their practical counterparts are in fact at equilibrium.

When liquid adhesives are brought into contact with rough solid surfaces, the question arises whether equilibrium corresponds to completely wetted or partially wetted interfaces.

Analysis of the system depicted in Fig. 1-15 [21] leads to the conclusion that virtually all practical bonds will be completely wetted at equilibrium. Only very rare cases would fail to meet the thermodynamic requirement that for spreading and spontaneous wetting the product $\cos \theta(\Omega/A)$ must be more positive than -1 (where Ω is the actual interfacial area and A is the apparent or projected area.) Since equilibrium corresponds to completely wetted interfaces, the reversible work of adhesion per unit of interface is precisely $2\phi(\gamma_{SO}\gamma_{LV})^{\frac{1}{2}}$.

This result and the important distinction between the spreading process and the state of wetting and their proper relationship to contact angle equilibria have apparently been overlooked or misunderstood by many. Zisman, for example, believes this difference is a matter of

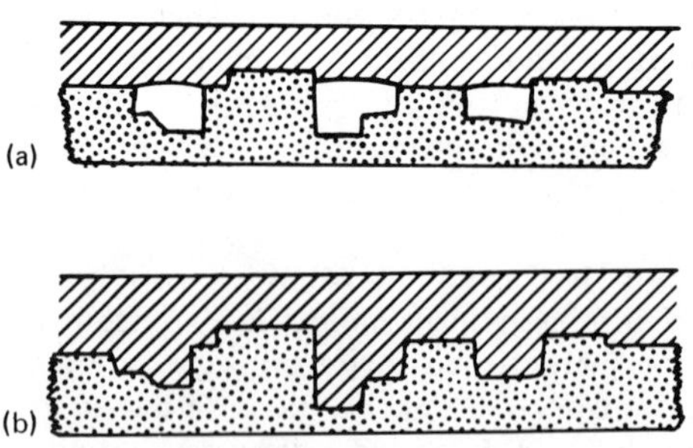

FIG. 1-15 Wetting of a solid substrate by a fluid adhesive: (a) substrate incompletely wetted; (b) substrate completely wetted.

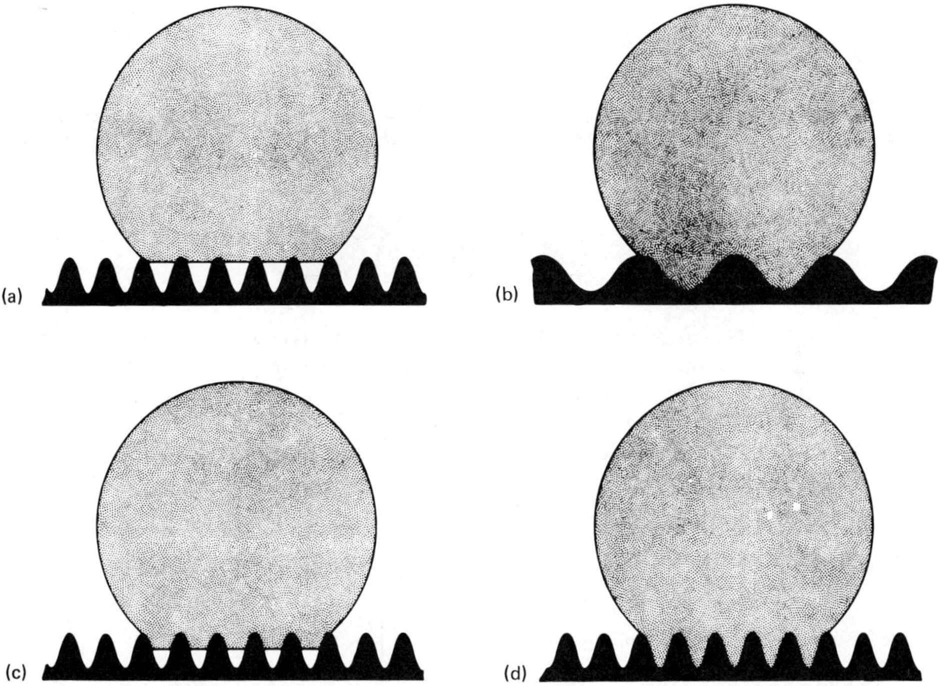

FIG. 1-16 Wetting of idealized rough homogeneous surfaces: (a) intrinsic contact angle = 120°—partially wetted at equilibrium, (b) completely wetted at equilibrium; (c) intrinsic contact angle = 105°—partially wetted in metastable state; (d) completely wetted at equilibrium.

semantics [22]. However, he concludes that maximum wetting and optimum adhesion require that $\gamma_{LV} \leq \gamma_c$. This conclusion is not supported by the preceding argument nor by the fundamental definition of the work of adhesion.

The fallacy of relating the state of wetting to contact angle measurements is further illustrated by Figs. 1-16 and 1-17.

All of the examples of Fig. 16 would be considered nonwetted by those who relate nonwetting to systems exhibiting $\theta > 90°$. Systems (a) and (b) of Fig. 1-17 would be designated as nonwetted and partially wetted, respectively, while the incompletely wetted heterogeneous surface would be considered wetted.

Viscous polymer solutions or melts require long times for establishing equilibrium. This is very important in experimental measurements of contact angles, and is of paramount importance in creating satisfactory adhesive bonds.

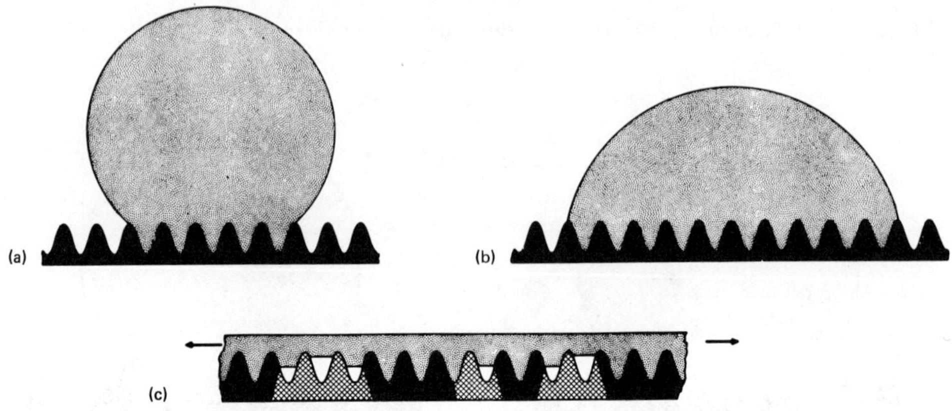

FIG. 1-17 The fallacy of relating "wetting" to the spreading behavior: (top) for homogeneous surfaces: (a) nonwetting or poor wetting (depending on definition); (b) partial wetting (better than a); (bottom) for heterogeneous surfaces: (c) contact angle phase 1 = 0; contact angle phase 2 = 120°; completely wetted (if based on receding contact angle).

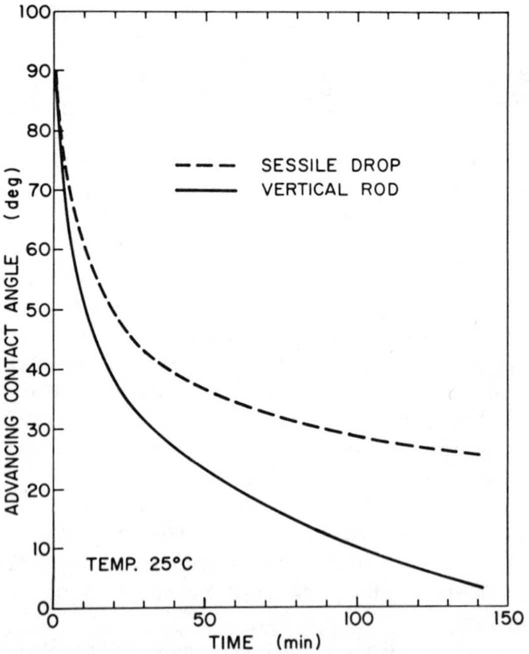

FIG. 1-18 Relative rates of approach to equilibrium for sessile drop vs. wetting a verticle rod.

TABLE 1-1 Approximate Times Required for Polymers to Attain Their Equilibrium Advancing Contact Angles Against Glass Wilhelmy Slides

Polymer	Viscosity	Temp.	Time to reach equilibrium
Epoxidized poly (butadiene)	200 Pa·s	25°C	∿200 minutes
Aromatic epoxy resin	10 Pa·s	25°C	∿15 minutes
Silicone oil	60 Pa·s	25°C	∿30 minutes
Silicone oil	.04 Pa·s	25°C	∿.4 minutes
Polyisobutylene	—	120°C	∿20 minutes
Poly(vinylacetate)	—	123°C	∿35 minutes
Poly(vinylacetate)	—	154°C	∿10 minutes
Poly(n-butyl methacrylate)	—	159°C	∿90 minutes

With viscous materials the Wilhelmy plate and vertical rod methods offer the advantage of requiring much shorter times for equilibrium than experiments using sessile droplets. This is shown in Fig. 1-18 for a liquid whose viscosity was 130 Pa·s. Times to reach equilibrium using a Wilhelmy plate are given for liquids of various viscosities in Table 1-1. A recording Wilhelmy plate also permits observation of sorption-desorption and shows clearly when equilibrium is reached.

4. CONCLUSIONS

Contact angle equilibria are directly related to interfacial energies. They can be used to estimate interfacial energies and work of adhesion when solids are involved and can provide the basis for useful information not readily accessible by other methods.

Their application, however, requires a thorough understanding of the fundamentals involved, an appreciation of the consequences of some frequently employed assumptions, and strict attention to cleanliness and purity.

5. SUMMARY

The dimensionless plots of the several important quantities—work of adhesion, interfacial tension between adhesive and adherend, initial and equilibrium spreading coefficients, and the spreading pressures π_{SL} and π_e—allow comparing some empirical equations with these fundamental quantities in a way which shows that deviations result from changes in the interaction parameter ϕ.

These comparisons also reemphasize that frequently used criteria relating contact angle equilibria and adhesion are not correct.

Attention is focused on interfacial contact density as the primary factor in determining the work of adhesion. Low contact density which is normally unaccounted for in estimates for ϕ is highlighted as the probable source of relatively high interfacial tensions between polymers.

Contact angle equilibria can provide useful information not readily accessible by other means. Precise application of the fundamental concepts to experimental situations is necessary. This means including π_e when required, insuring that equilibrium has been established, and distinguishing between metastable equilibria associated with hysteresis and stable thermodynamic equilibrium.

REFERENCES

1. L. A. Girifalco and R. J. Good, J. Phys. Chem. 61:904 (1957).
2. F. M. Fowkes, Ind. Eng. Chem. 56, No. 12:40-52 (1964).
3. J. L. Gardon, *Encyclopedia of Polymer Science & Technology*, Vol. 3, 1965.
4. P. J. Debye, *Adhesion and Cohesion* (A. A. Weiss, ed.), Elsevier, New York, 1962.
5. J. F. Padday and N. D. Uffindell, J. Phys. Chem. 72:1407 (1968).
6. R. J. Good, Ind. Eng. Chem. 62:54-78 (1970).
7. J. L. Gardon, *Treatise on Adhesion and Adhesives* (R. L. Patrick, ed.), Marcel Dekker, New York, 1967.
8. S. Wu, J. Pol. Sci, Part C, 34:19 (1971).
9. D. Hoernshemeyer, J. Phys. Chem. 70:2628 (1966).
10. R. E. Johnson and R. B. Dettre, J. Coll. Int. Sci. 21:610 (1966).
11. H. W. Starkweather, Jr., SPE Trans. 5:1 (1965).
12. R. J. Roe, J. Phys. Chem. 69:2809 (1965).
13. S. Wu, J. Coll. Int. Sci. 31, No. 2:153 (1969).
14. W. A. Zisman, Ind. Eng. Chem. 55, No. 10:19 (1963).
15. A. W. Neumann, R. J. Good, C. J. Hope, and M. Sejpal, J. Coll. Int. Sci. 49:291 (1974).
16. P. J. Sell and A. W. Neumann, Z. Phys. Chem. 47:20 (1965).
17. D. K. Owens and R. C. Wendt, J. App. Pol. Sci. 13:1241 (1969).
18. J. R. Dann, J. Coll. Int. Sci. 32:302 (1970).
19. R. E. Johnson, Jr., and R. H. Dettre, *Surface and Colloid Science*, Vol. 2 (Matijevic, ed.), Wiley-Interscience, New York, 1969.
20. C. A. Smolders, Rec. Trav. Chim. 80:650 (1961).
21. J. R. Huntsberger, C & E News 42:82 (1964).
22. J. R. Huntsberger (discussion), *Adhesive Restorative Dental Materials*—II, U.S. Dept. HEW, Public Health Service Publication No. 1494, 1966.

2

The Surface Composition and Energetics of Some PAN-Based Graphite Fibers

Lawrence T. Drzal

Wright Aeronautical Laboratory
Wright-Patterson Air Force Base, Ohio

1. INTRODUCTION

Significant advantages can be made in performance of aircraft structural components through the expanded use of graphite-fiber-reinforced composite materials. These composites, using graphite fibers that are light, stiff, and high-strength, provide a structural material that is not only lighter and stronger than many equivalent metals, but also provide a material that is potentially easier to fabricate and more durable [1].

A necessary question that must be answered, however, is what interactions occur at the interface between reinforcing fiber and polymer matrix; to what degree are these interactions responsible for composite properties and durability; and can this interaction be significantly improved?

Graphite fibers can be fabricated from any organic material, but the most common precursors in use are rayon and polyacrylonitrile (PAN) fibers. In addition, pitch is being developed as a possible low-cost source for the future. These precursor materials are stabilized, carbonized, and graphitized with increasing heat treatment temperature to produce the graphitized fiber used in composites.

Variations in the fiber processing are responsible for variations in fiber mechanical properties. Research [2,3] has shown that orientation of the graphitic basal planes in the fiber are responsible for fiber properties. For example, low-modulus fibers that have undergone a moderate graphitization step have both axial and radial structure (Fig. 2-1).

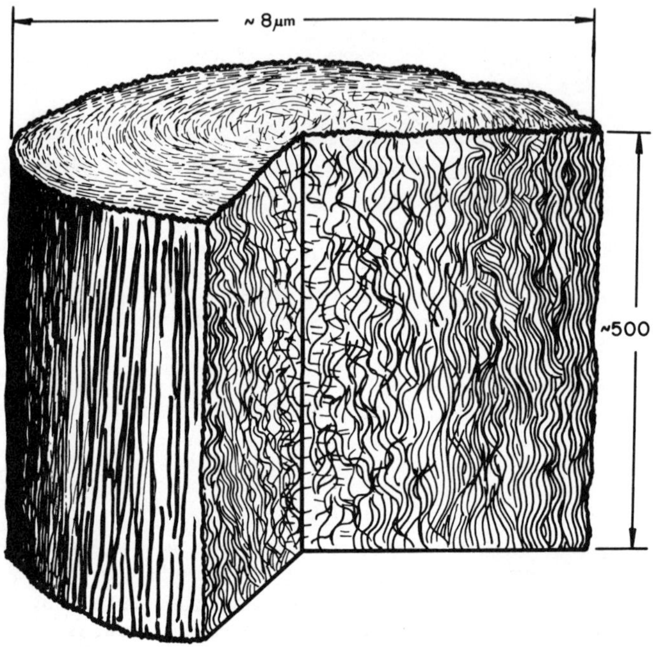

FIG. 2-1 Schematic 3-D structural model of polyacrylonitrile-based fiber with properties similar to type AU. (Source: Ref. 2.)

The graphitic basal planes form undulating ribbons which align along the length of the fiber. As the degree of graphitization increases and the fiber modulus increases (Fig. 2-2), the thickness and width of these ribbons increases and their undulation decreases so that the ribbons become more closely aligned with the fiber axis. At the same time, the basal plane surface of the ribbon becomes more closely aligned with the outer fiber surface and the ribbon intertangling decreases.

Coincident with the increase in mechanical properties of the fiber with increased graphitization is a decrease in interaction between fiber and matrix in composites as measured by interlaminar shear strength. To improve the interlaminar shear properties, surface treatments have been developed. Improvements in shear properties like those listed in Table 2-1 can be achieved. It appears that the function of the surface treatments is to make the surface more carbonlike to increase fiber-matrix interaction. This can be achieved through direct oxidation of the surface by a variety of techniques or coating of the surface with

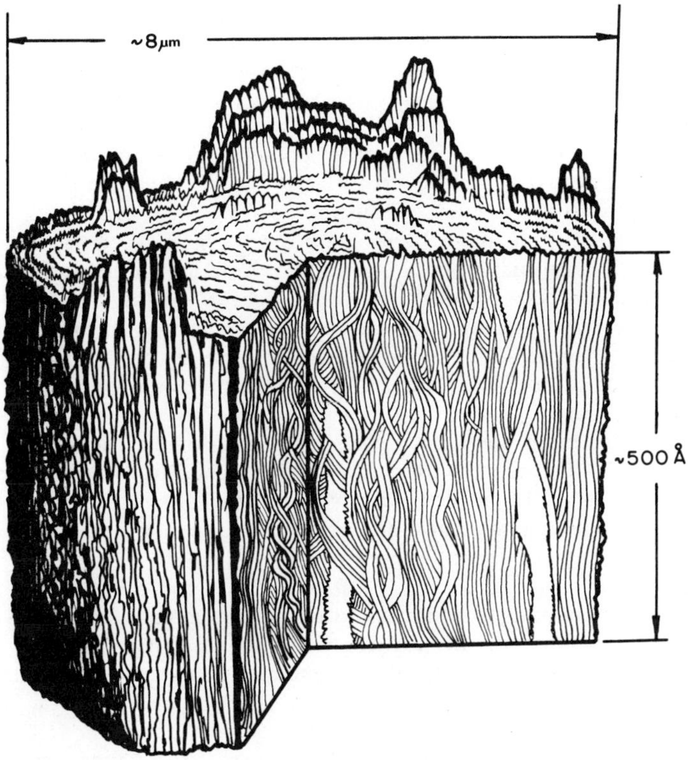

FIG. 2-2 Schematic 3-D structural model of polyacrylonitrile-based fiber with properties similar to type HMU. (Source: Ref. 2.)

TABLE 2-1 Effect of Surface Treatments on Composite Shear Strengths

Untreated fiber	4000 psi
Gaseous oxidation (air, ozone, RF plasma)	+10%
Liquid oxidation (HNO_3, NaClo, Elec. NaOH)	+100-200%
Vapor deposition (SiC, FeC, CH_4)	+60-100%
Solution reduction [$FeCl_3$, $Fe(C_5H_5)_2$, PPQ]	+100-200%
Whiskerization (Si_3N_4, TiO_2, SiC)	+200-300%
Carbon fiber	13,000 psi

a carbonized layer. In any case, one still sees a drop in shear properties of surface treated fibers with increasing modulus although the improvement brought about by surface treatment makes these high-modulus fibers acceptable for structural applications.

Many experimental studies have been completed which have established a dependence of composite properties on interfacial bonding between the fiber and matrix. Kaelble et al. [4], and Larsen et al. [5], have shown a relationship between interlaminar shear strength and fiber-matrix bonding. Hertz [6] has measured enhanced transport of water into composites in directions along the fiber axis over transport perpendicular to the fibers.

While attempts have been made to deduce the type of surface sites most responsible for the interfacial properties [7], actual determination of the existence of the type of polymer-fiber bonding (chemical vs. physical) and the amount of each type of bonding occurring between fiber and polymer has not occurred.

The purpose of this work is to lay the foundations for a study directed at obtaining a thorough understanding of the molecular interactions occurring at the polymer-fiber interface and relating these to the engineering parameters that completely describe composite material behavior.

To accomplish this, a combination of techniques has been chosen to characterize the graphite fiber surface. Determination of isosteric heats and differential entropies of krypton adsorption was chosen as a means of detecting and quantifying small surface changes in the graphite fiber surface. This technique has the advantages of not disturbing the surface as energy- or particle-excited surface spectroscopic techniques do and the advantage that inert gas molecules will adsorb on the outermost surface, i.e., even on preadsorbed gases [8]. Since only dispersion forces are involved in the interaction between krypton and the graphite fiber and since the interaction between krypton adatoms has been evaluated on homogeneous surfaces [9,10], slight variations in heats of adsorption with coverage should reflect changes in the surface state of a graphite fiber.

Mass spectral analysis of fiber thermal desorption products was also used in this investigation. The amount and type of adsorbed contaminant could be detected by thermal desorption analysis after exposure of the surface to the environment and after thermal treatment.

Finally, ion-scattering spectroscopy (ISS) and positive secondary ion mass spectroscopy (+SIMS) were conducted on these fiber surfaces to provide detailed information about the atomic structure of the outermost layer.

The combination of these techniques provides the information to quantify the graphite fiber surface energetically and physically and ultimately to determine the exact interactions that occur at the fiber-matrix interface which are responsible for interfacial strength and durability.

2. EXPERIMENTAL

2.1. System

A volumetric determination of the adsorption of krypton near its boiling point was chosen as the method of measuring surface energetic changes. This technique was chosen because of the proper combination of sensitivity, versatility, and adaptability to other experiments planned for this study.

An apparatus similar to the one used here was described in detail elsewhere [11]. Added to this system was a UTI-100C quadrupole mass spectrometer for analysis of thermal desorption products. This system was optimized to measure adsorption over the range of coverage of 0.01 to 2 monolayers on the low-surface-area graphite fibers (~0.5 M^2/g) of interest for use in advanced composites. A schematic drawing of the system is shown in Fig. 2-3. The details of this system have been recently documented [12].

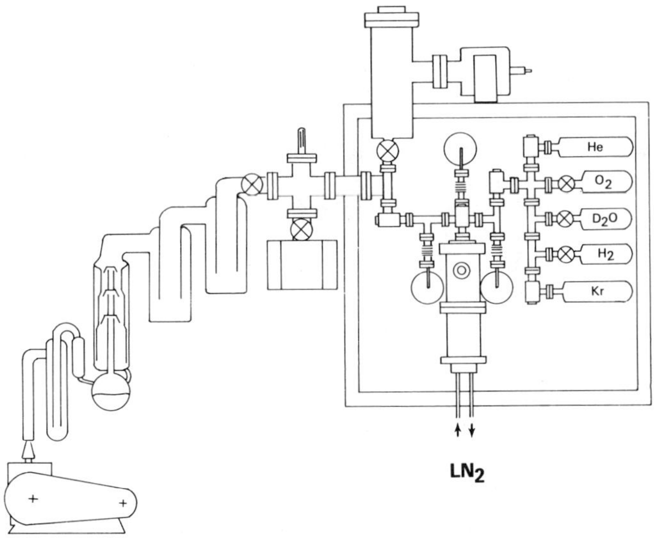

FIG. 2-3 Schematic representation of the volumetric adsorption system.

2.2. *Fibers*

PAN-based fibers were obtained from a single batch from Hercules Inc. [13]. Untreated low-modulus fibers were designated AU and those which had undergone the manufacturer's proprietary surface treatment were designated AS. An untreated high-modulus fiber without surface treatment was designated HMU. No additional sizing was used. All fibers were loosely wound on a 6-mm Pyrex rod to a total fiber weight of about 10 g. These fibers were then placed in a 19-mm Pyrex sample tube and evacuated at room temperature for 12 hr.

2.3. *Experimental Procedure*

2.3.1. Thermal Desorption Analysis

After this initial overnight evacuation, the sample was valved off from the system and a tube furnace was placed around it. The system temperature was raised to 70°C. Subsequently, the sample temperature was raised in approximately 50°C increments and held at each temperature for about 1 hr. At the end of this period the desorbed gases trapped in the sample tube were expanded into the calibrated dosing chamber where the pressure and temperatures were recorded.

The UTI-100C [14] operating at ~120°C was used to obtain a mass spectrum of the thermally desorbed gas trapped in the dosing chamber. A spectrum of the background gas, obtained before sampling, was subtracted from this spectrum before analysis.

After this mass spectral analysis was completed, the sample was evacuated to system base pressure, valved off, and the temperature was again incremented to the next level. The entire procedure was repeated until a sample temperature near 300°C was reached. At this point the entire system was baked out for 12 hr at 150°C with the sample remaining at 300°C. (Experiments have shown that an insignificant amount of additional material is removed during bakeout.) At the end of this period, the sample tube was valved off and the system cooled and configured for the krypton-adsorption experiments.

The amount desorbed was calculated using the ideal gas law relationship

$$n = P_1 \left(\frac{V_{DC}}{RT_C} + \frac{V_{ST}}{RT_B} \right) \tag{1}$$

where n is the number of moles of gas desorbed, R is the gas constant, V_{DC} is the volume of the dosing chamber, V_{ST} is the sample tube volume, and T_B and T_C are the temperatures of the sample tube and dosing chamber, respectively. Compositional analysis of the mass spectral data was accomplished through a zeroth-order peak height and background subtraction scheme. Major peaks were identified and cracking patterns [15] for these sample components were fit to the data to estimate the gas phase composition.

2.3.2. Krypton-adsorption Studies

After cooldown a cryostat was placed around the sample tube and evacuated. The cryostat temperature was lowered to the adsorption temperature and allowed to stabilize overnight. Research-grade krypton was expanded through valves and a manifold into the system, except the sample tube, and reevacuated once to flush out any possible residual bases. Purity of the adsorbate was checked with the mass spectrometer.

Once stabilized, an adsorption run was started by letting krypton into the dosing chamber until approximately two monolayers of krypton were adsorbed. Exact details of this procedure, the corrections which are applied to the raw data, and the entire computerized calculation procedure are listed elsewhere [16].

Generally, replicate adsorption isotherms were obtained on a given surface rather than one isotherm with many points to keep the accumulation of errors from the dosing procedure small. Superposition of these isotherms allows the best curve to be computer fitted to the data of a combination of isotherms. Artifacts introduced by computer curve fitting and interpolation have been checked for and are not present when a sufficient number of points are present.

The isosteric heat of adsorption (Q_{ST}) was calculated through use of the Clausius-Clapyron equation [17]

$$Q_{ST} = -R\left(\frac{\partial \ln p}{\partial (1/T)}\right)_n \qquad (2)$$

For these experiments, this relationship was used in the integrated form so that

$$Q_{ST} = -R\left(\frac{T_1 T_2}{T_1 - T_2}\right) \ln\left(\frac{P_2}{P_1}\right)_n \qquad (3)$$

The differential entropy of adsorption was calculated from Ref. 17:

$$\bar{S}_S - \underline{S}_g = \frac{Q_{ST}}{RT^2} \qquad (4)$$

with the reference state changed from the gas to the liquid [18] so that Eq. (4) becomes

$$\bar{S}_S - \underline{S}_L = -\frac{Q_{ST}}{T} + \frac{H_L}{T} - R\ln\left(\frac{P}{P_0}\right) \tag{5}$$

Here, H_L is the heat of vaporization of krypton. The error in the determination of these thermodynamic quantities is directly related to errors associated with the isotherm-measuring process. Replicate runs used to determine this error gave a value of ±200 J/mol for this system.

The reason for selection of the isosteric heat to monitor surface changes becomes apparent if an analysis of the various energies that contribute to the isosteric heat of adsorption is made. The ability of this quantity to detect surface energetic changes becomes apparent. Figure 2-4 is a representation of the energy-distance relationship for an adsorbing molecule. At some average distance from the surface, a krypton molecule from the gas phase is adsorbed on the surface. The isosteric heat measured during this adsorption process is composed of energy losses due to thermal energy, translational, and rotational changes as well as to the attractive potential of the adsorbent for the adsorbate. For an inert gas molecule like krypton, changes in the nature of the attractive potential of the surface for krypton, either through changes in surface chemistry or adsorbed material, will have an effect on the measured isosteric heat of adsorption. In a real situation, as the surface coverage increases, the lateral attraction between adsorbed krypton molecules must also be taken into account on energetically homogeneous surfaces.

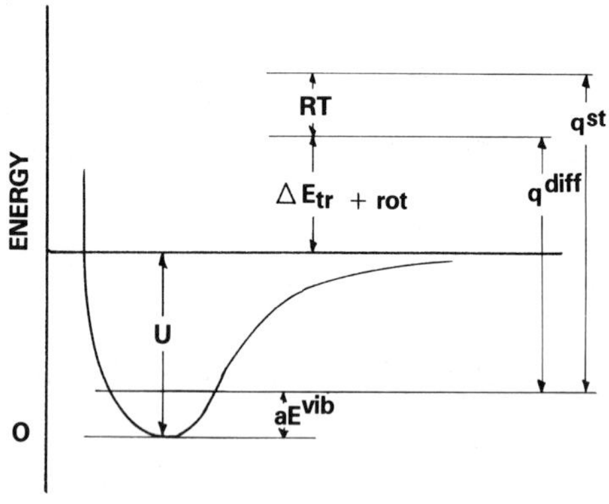

FIG. 2-4 Potential energy diagram for physical adsorption showing the relationship of the isosteric heat of adsorption to the surface potential.

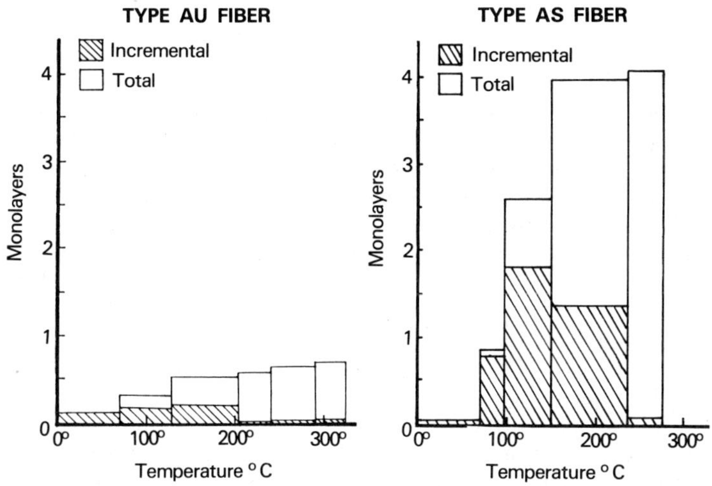

FIG. 2-5 Incremental and total amount desorbed from AU and AS fibers.

2.3.3. Ion Scattering Spectroscopy and Positive Secondary Ion Mass Spectroscopy

Samples of these same graphite fibers were cut into 10-mm lengths and packed in parallel into a 6 × 12 mm × 3 mm deep cavity which, when placed in position, aligns the fibers parallel with the scattering ion beam. The samples were handled in such a way as to preclude contamination of the surface.

The ISS and +SIMS spectra of these samples were obtained in a commercial Model 520 ion scattering spectrometer manufactured by the 3M Company [19], to which a modified UTI Model 100C [14] mass filter had been added. Helium 3 was used as the scattering ion. Details of this apparatus and its application to adherend surfaces have been published [20].

3. RESULTS AND DISCUSSION

3.1. Thermal Desorption

The incremental and total amounts of noncondensable materials desorbed from the surfaces of AU and AS fibers as a function of vacuum heat treatment temperature (VHT) are shown in Fig. 2-5.

The total amount of material desorbed differs greatly for the two fibers. While the AU fiber desorbs the equivalent of 0.6 monolayer at 275°C, the AS fiber has desorbed over 4.0 monolayers of material at that same temperature. The incremental quantities desorbed, shown as the shaded areas, are a maximum around 150°C for both fibers.

The compositional breakdown of the desorbed material shown in Fig. 2-6 shows two different types of material being removed. On both fibers below 150°C, mostly H_2O, CO, and CO_2, are desorbed. These components can be ascribed to physically adsorbed material on the fiber surface. Above 150°C, however, the composition reflects the nature of the fiber surface. The untreated PAN fiber (AU) begins to desorb some CO, NH_3, CH_4, and organic materials at higher temperatures, while more NH_3 and CO products are desorbed with increasing temperature on the chemically treated fiber surface (AS). Returning to Fig. 2-5, the actual amount of the material desorbed from both fibers is a very small portion of the surface at these high temperatures (>150°C). It is concluded that the bulk of this material represents removal of weakly bound surface species with little desorption of structural components. Thermal desorption work [21] has shown that much

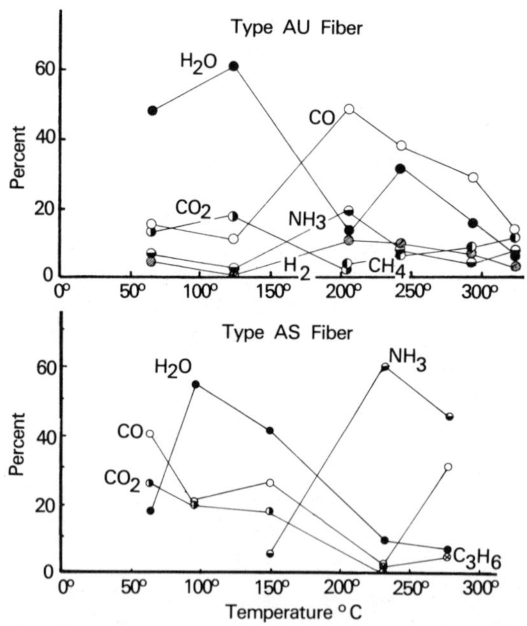

FIG. 2-6 Compositional analysis of material desorbed from the AU and AS fibers.

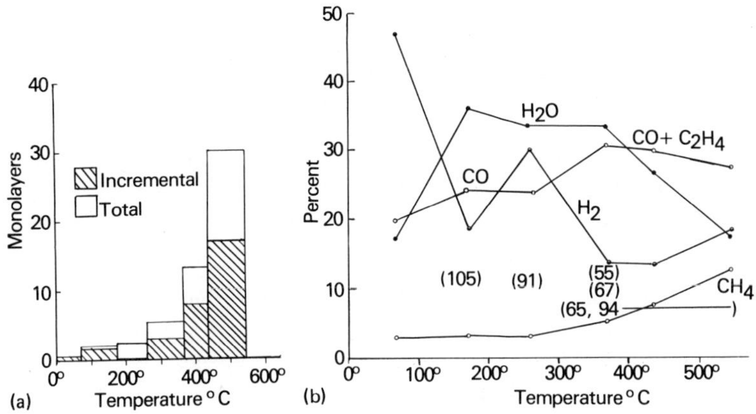

FIG. 2-7 Incremental and total amount desorbed and the compositional analysis of an epoxy compatible fiber surface finish.

higher temperatures (>500°C) are needed to remove surface functional groups from graphite surfaces.

Figure 2-7 is the thermal desorption data for a fiber coated with an epoxy-compatible finish as opposed to an oxidized surface. Here also, the equivalent of thirty monolayers of material can be thermally desorbed at normal composite fabrication conditions. The character of the desorbing material, however, is different and reflects the organic nature of this surface coating.

3.2. *Surface Area Analysis*

An important factor to consider in attempting to understand the interfacial behavior of graphite fiber composites is the amount of fiber surface area available for interaction with the matrix material. Surface areas of these fibers were calculated from the krypton adsorption isotherms for each of the fibers investigated using the Brunauer, Emmett and Teller (BET) method [22]. Although these values may not give absolute estimates of the surface area, they do produce consistent results reflecting variations with treatment on the same material. The fibers investigated here did not exhibit the uniform homogeneous adsorptive behavior of highly graphitized carbons which would preclude using this analysis as shown recently by Putnam and Ford [10].

Results of this BET analysis are shown in Table 2-2. The data shown are a summary of all the isotherms obtained. The first column lists the fiber preconditioning. An overnight 300°C vacuum heat treatment is represented by VHT. An isotherm obtained after a 4-hr room temperature atmospheric exposure is represented by "AIR."

TABLE 2-2 BET Analysis of Graphite Fibers

Fiber treatment	Kr adsorption temp. (K)	C	n_m (mol/g)	Σ (M^2/g)
Type AU				
VHT	103.48 ± 0.02	131	3.40 ± 0.03	0.436 ± 0.004
VHT	114.82	68	3.43	0.452
VHT/AIR	103.48	166	3.37	0.432
VHT/AIR	114.82	81	3.44	0.453
VHT/AIR/VHT	103.48	123	3.29	0.420
VHT/AIR/VHT	114.82	79	3.32	0.430
Type AS				
VHT	103.58	75	3.77	0.483
VHT	114.88	67	3.70	0.488
VHT/AIR	103.55	61	3.35	0.429
VHT/AIR	114.90	49	3.45	0.455

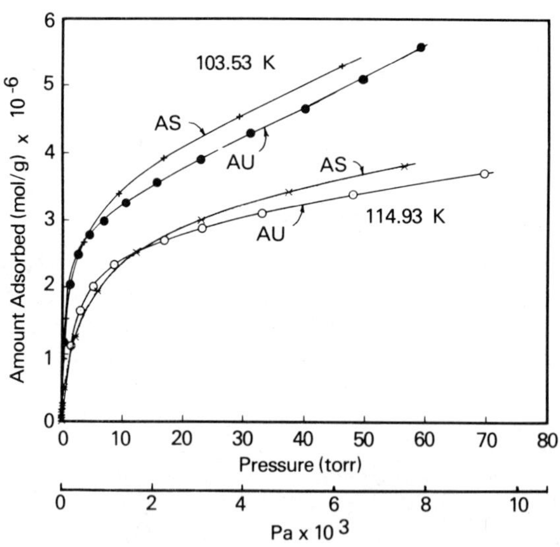

FIG. 2-8 Typical adsorption isotherms of krypton on AU and AS fiber surfaces.

Typical isotherms for the AU and AS fibers are shown in Fig. 2-8. These isotherms could be classified as type II according to Brunauer et al. [22] and show no vertical submonolayer portion characteristic of inert gas adsorption on a homogeneous adsorbent.

The BET analysis provided a straight line over the relative pressure range of 0.05 to 0.25. Results in terms of the BET "c" constant and the number of moles in a monolayer are displayed in columns 3 and 4. The value for the surface area, Σ, was calculated using a molecular area of 2.13 nm^2 and 2.19 nm^2 at 103.48 and 114.82 K respectively, as discussed by Drzal and Fort [23].

Comparing surface areas of the two fibers given the same treatment (VHT) shows a larger surface area for the surface-treated (AS) fiber than nontreated (AU) amounting to 10% which is attributable to the surface treatment procedure. However, when surface areas are compared for the same fibers after being subjected to a 4-hr laboratory air exposure (AIR) followed by room temperature evacuation, no discernible difference in areas is noted. The AU fiber surface area remains unchanged after AIR exposure while the AS fiber surface area is reduced by 10%. The thermal desorption study described earlier showed that the AS fiber thermally desorbed about 8 times as much material as the AU fiber which represented four monolayers and one-half monolayer, respectively. If this material was uniformly spread on the fiber surface, for the 7 to 8-μm fibers of this study, the amount adsorbed would have negligible influence on the surface area. It must, therefore, be concluded that laboratory air exposure effectively blocks about 10% of the surface from krypton adsorption. This surface area then is most probably present in surface cracks and/or pores created by the surface treatment.

3.3. *Isosteric Heat and Differential Entropies of Adsorption*

The isosteric heats (Q_{ST}) and differential entropies of adsorption ($\bar{S}_S - \underline{S}_L$) were calculated from the adsorption isotherms using Eqs. (3) and (5). (For a detailed derivation of adsorption dynamics, see Ref. 24, p. 609.) Replicate runs were used to provide a sufficient number of isotherm points while maintaining overall accuracy of the isotherms. A computer interpolation method was used to generate the data necessary for the calculations. Interpolated curves were checked for smoothness to preclude artifacts in the thermodynamics caused by the machine interpolation method. A sufficient number of points were used to insure that variations in the thermodynamic data were real. This procedure insures that comparisons made between data from different surfaces were consistent and free of operator error. The exact details of this scheme and the associated computer programs are published elsewhere [16].

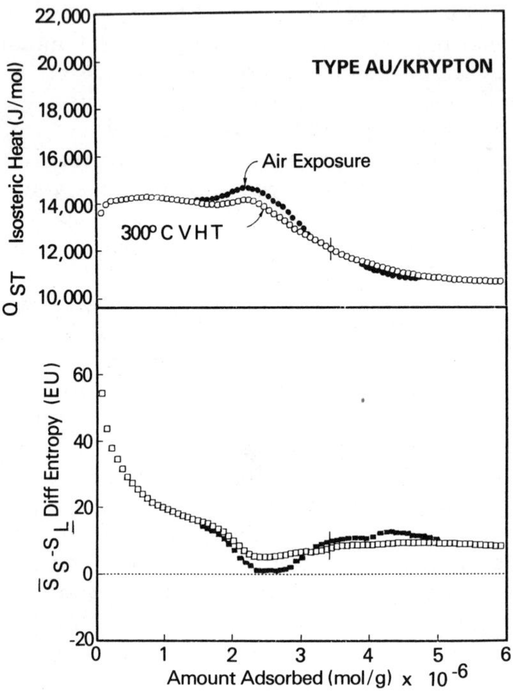

FIG. 2-9 Isosteric heat and differential entropy of adsorption for krypton on an air-exposed (•) and 300°C vacuum-heat-treated (o) AS fiber.

Isosteric heats and differential entropies of adsorption of krypton on the AU surface are shown in Fig. 2-9. Curve 300°C VHT represents the data for the vacuum-heat-treated surface. The monolayer point is indicated at 3.40 μmol/g as determined by BET analysis for this surface. Using this value for the monolayer, the isosteric heats are constant up to about 70% coverage at which time they begin to decrease toward the heat of liquifaction. If the fiber surface was energetically uniform, the krypton heats of adsorption would increase with coverage due to the lateral interaction of the adsorbate until a monolayer was completed. The heats of adsorption would then decrease to the heat of liquifaction. The type of energetic behavior exhibited here is seen, however, in systems exhibiting strong interaction between adsorbate and adsorbent such as the benzene-Graphon system reported by Kieslev and Isirikyan [25]. One would then conclude that the adsorbent surface is uniform in attraction for krypton up to 70% of a monolayer and that energy barriers exist which prevent 2-D lateral mobility across the surface. Comparing the limiting value of the isosteric heat extrapolated to zero coverage for this fiber (14000 J/mol) to that for krypton on an energetically uniform carbon such as Graphon under the same conditions (12600 J/mol) [23] does indeed indicate that there is a stronger attraction for the krypton on this fiber surface.

The differential entropies displayed in the lower part of Fig. 2-8 decrease in value with coverage to a point at n = 2.3 μmol/g. There is no pronounced minimum associated with this point characteristic of completion of a monolayer of inert gas on a homogeneous surface.

The data indicated by the solid points (AIR EXPOSURE) was obtained from isotherm data on this same AU surface after it had been exposed to laboratory air for 4 hr and then reevacuated at room temperature. Very little difference is observed in the thermodynamic data except near n = 2.3 μmol/g, where there is a significant rise in the isosteric heat. The remainder of the curve is the same as the AU/VHT data. The differential entropy data likewise is unchanged except for an enhancement in the depth of the knee at n = 2.3 μmol/g.

Mass spectral analysis of the gases removed from this surface in subsequent thermal desorption studies have shown it to be mostly water and carbon monoxide. The rise in the isosteric heats is most probably due to increased lateral interaction between adsorbate molecules caused by adsorption of water and carbon monoxide on surface high energy sites which reduces some of the barriers to 2-D mobility.

On repetition of the thermal vacuum heat treatment, the AU surface is returned to its original state as indicated by coincidence with the isosteric heat and differential entropy curves of the VHT surface.

3.3.2. Type AS Fiber

The thermodynamic data from krypton adsorption on the AS fiber reflects the results of the surface treatment the fiber has received. The isosteric heat values of Fig. 2.10 show an enhanced adsorption for krypton as manifested by the higher values in Q_{ST} for the first μmol of krypton adsorbed. Thereafter, the isosteric heat varies with coverage in about the same manner as on the untreated AU fiber.

The isosteric heats are constant in value up to a coverage of 2.1 μmol/g of krypton, whereupon the values begin to drop to the heat of liquifaction. It was shown earlier that the surface areas of the AU and AS fibers were within 10% of each other on a vacuum-heat-treated surface. Recalling this, the isosteric heat is virtually unchanged except for the first μmol adsorbed. Therefore, this initial large value must be attributed to the surface treatment process. The differential entropies likewise are the same as on the AU fiber except for the values which are reduced for the first μmol adsorbed. The entropy values are lower, indicating a decrease in adsorbate mobility.

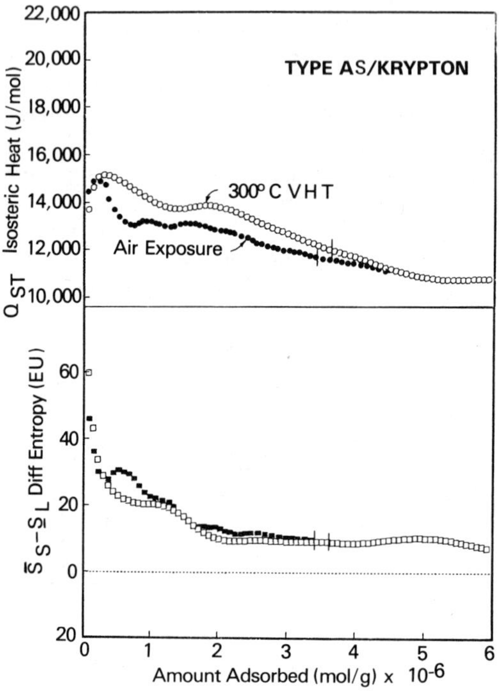

FIG. 2-10 Isosteric heat and differential entropy of adsorption for krypton on an air-exposed (●) and 300°C vacuum-heat-treated (o) AS fiber.

The curve represented with the solid symbols shows the degradation in attraction for krypton that results from 4 hr of exposure to laboratory air. Although the high-energy portion of the surface, as reflected in the high values of Q_{ST}, has decreased slightly from 1 to 0.6 μmol, it retains its high attraction for krypton indicating it is relatively unaffected by air exposure under these conditions. (The possibility remains, however, that surface degradation by air exposure is time-dependent.) The remainder of the surface, however, has decreased in attractiveness for krypton. The isosteric heat has dropped significantly below the values before air exposure. In view of the large quantities of material (approximately four monolayers) that this surface has been shown to desorb earlier, its degradation is not surprising. The values of the differential entropy are much higher in this condition and are indicative of enhanced adsorbate mobility.

3.3.3. HMU Fiber

Figure 2-11 is a plot of the isosteric heat and differential entropy of adsorption for the untreated low-modulus fiber (AU) and an untreated high-modulus fiber (HMU) made from the same PAN precursor. The adsorption thermodynamics are quite different for these two materials.

Initially at low coverages, there is very little distinction in the heats of adsorption. However, as the amount adsorbed increases, the heats of adsorption rise to a maximum and then rapidly fall off. This type of behavior is observed for adsorption on energetically homogeneous graphites consisting primarily of graphitic basal plane surfaces. The rise with coverage is attributed to the increasing lateral interaction between adsorbed krypton molecules having two dimensional freedom to move laterally over the surface. Likewise, the differential entropy has a pronounced minima at this point where one statistical monolayer

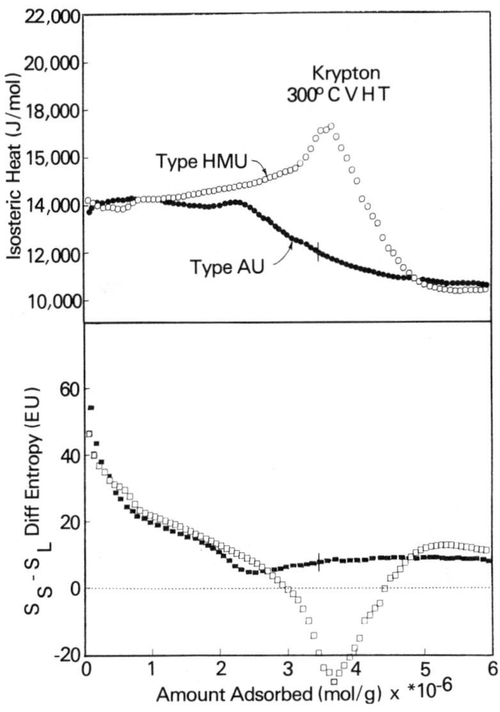

FIG. 2-11 Isosteric heat and differential entropy of adsorption for krypton on a type AU (•) and type HMU (o) fiber treated at 300°C in vacuum.

is completed. Molecules adsorbing in the second layer have greater entropy because of their increased 2-D mobility.

This behavior confirms the hypothesis that the graphitization conditions necessary to produce the high-modulus fibers produce a surface composed of graphitic basal planes. This fiber is a surface with potentially few sites for any strong interaction with polymer matrices and indicates why the low interlaminar strength values exist in composites made from untreated high-modulus graphite fibers.

3.4. Fiber Surface Composition by ISS and +SIMS

Figure 2-12 displays both the ISS and +SIMS spectra obtained on the AU fiber. Background noise is much higher on this graphite fiber bundle than are obtained on more planar surfaces, but clear spectral features are evident. The ISS spectra obtained with ^3He clearly shows major peaks corresponding to carbon and sodium and minor peaks appearing at positions ascribed to oxygen and potassium. The +SIMS spectra of the surface atoms desorbed by the collision of the ^3He ions with the surface shows one major peak at AMU = 23 which is identified as sodium. The grouping of peaks at AMU = 27 is attributable to organic fragments and the peak at AMU = 39 is attributable to potassium. However, the signal strength of the latter two peaks is at least a factor of 100 less than sodium.

Quantitative assignments of concentration to peak heights are difficult because the scattering yields vary with the chemical environment of the secondary ion. However, some semiqualitative comparisons can be made on this and the other surfaces. The sodium concentration at

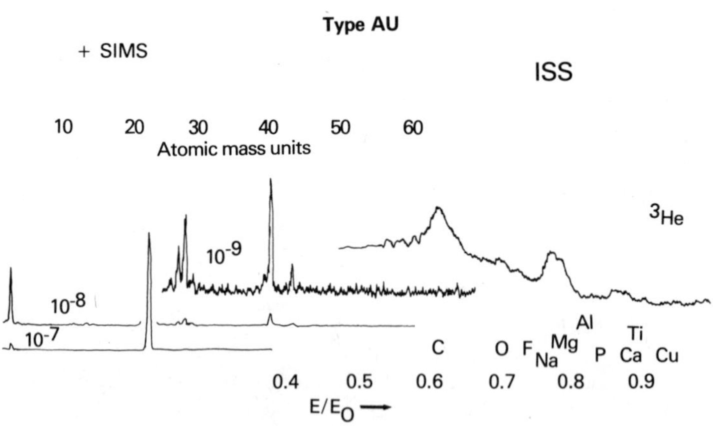

FIG. 2-12 ISS and +SIMS spectra of a type AU fiber.

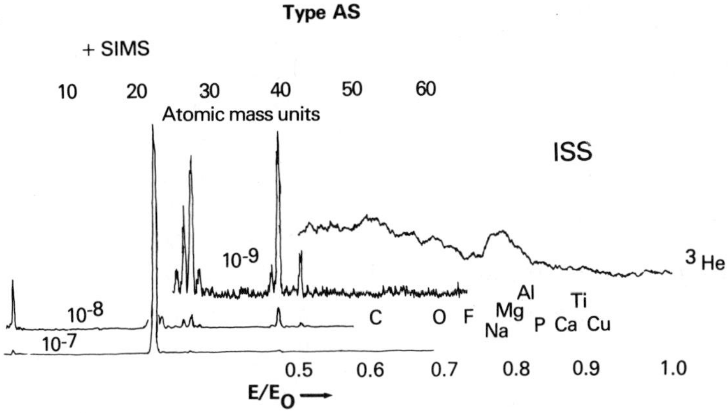

FIG. 2-13 ISS and +SIMS spectra of a type AS fiber.

the fiber surface is by far greater than the other components. This is not a system artifact since ISS and +SIMS spectra on other graphite fibers have been obtained in this same system where the sodium signal is reduced by a factor of 100.

Figure 2-13 is the ISS and +SIMS spectra for the AS fiber. Similar features are observable for sodium. Qualitatively, the AS fiber has a larger sodium signal than the AU fiber.

Other experiments conducted on these same fibers after thermal treatments, water washing, and acid washing have shown that the surface sodium content as measured by ISS and +SIMS cannot be significantly reduced. Indeed, the thermal treatments enhance the sodium signals, suggesting diffusion of this species to the surface from the bulk of the material.

Based on some previous studies of bulk concentrations of metals in fibers [6] and extractive methods of removing metals [29], attempts were made to determine the amount of sodium which was present both in the bulk of the fiber and on the fiber surface. The results of this effort are presented in Table 2-3. Acid extraction with a 10% HCl solution for 140 minutes at 20°C and Soxhlet extraction with water for 24 hr were tried. The liquor from each was analyzed by atomic absorption and sodium was found to be present in significant quantities in each. The amount of sodium extracted was converted to concentration based on fiber weight and is listed in the table. Also listed is the fraction of a monolayer the sodium would occupy based on a surface density of 10^{15} molecules/cm^2 and if all of the extracted sodium was from the surface. For comparison, a bulk analysis of these fibers conducted by emission spectroscopy is included.

TABLE 2-3 Sodium Concentration in Graphite Fibers

Acid extraction (10% HCl, 140 min, R.T.)			
			θ
AU	25 ppm		0.23
AS	7		0.12
Pitch	1		0.01
Water extraction (Soxhlet-extracted 24 hr)			
AU	60 ppm		0.74
AS	32		0.39
Pitch	2		0.03
Bulk analysis (emission spectroscopy)			
	Na	K	Ca
AU	2000 ppm	40 ppm	100 ppm
AS	5000	80	100
Pitch	2	5	30

It is evident from the data of Table 2-3 that the extractive techniques removed only a very small amount of sodium from the fiber (<1%). The variation in amount extracted with technique is most probably related to the experimental conditions associated with each technique. For example, the water extraction was conducted at 100°C while the acid extraction was done at 20°C. More sodium might be expected from the higher-temperature technique if the extraction is diffusion controlled. A patented scheme for alkali extraction of metals from graphite fibers [29] indicate that temperature on the order of 700 to 1000°C are optimum. Likewise, the chemical environment of the fiber surface can affect the ability of these techniques to extract sodium. The surface treated AS fiber has significantly more sodium present in the bulk and on the surface, but the metal might be expected to be bound more tightly in the surface layer if it was introduced during the surface oxidizing conditions. Included for comparison is data on an experimental pitch fiber. The values for the HMU fiber were similar to the pitch fiber data.

Comparison of the extracted data with the bulk analysis indicates that a large reservoir of sodium exists in the bulk of these fibers and that only a minute portion has been removed. In terms of surface density, however, the sodium can occupy a significant portion of the surface and may be responsible for the degradation of fiber surface with air exposure detected in the energetic measurements.

The presence of alkali metals in graphite has been attributed to the formation of alkali-graphite compounds with the sodium ions intercalated between graphite layers [26,27]. It has also been demonstrated [28] that sodium has a preference for the irregular spaced areas of graphite where the basal spacing is large. The amount of sodium that intercalates decreases with increasing degree of graphitization and might be expected to be high in a fiber with a low degree of graphitization such as the type A fiber.

4. SUMMARY

The surfaces of Hercules type A PAN-based graphite fibers, both treated (AS) and untreated (AU), were investigated using thermal desorption, krypton adsorption thermodynamics, and ISS and +SIMS techniques. Surface treatment enhanced the surface energetics of these fibers on 30% of the surface over the untreated fiber. In addition, the surface treatment increases the affinity of the fiber surface for adsorbed material by a factor of 8 over the untreated fiber surface. Surface spectroscopy has shown that one surface constituent is sodium which is present in very large concentration on these fibers. Type HMU high-modulus PAN-based fibers possess a surface free of sodium and display adsorption behavior quite similar to an energetically homogeneous graphite surface.

5. CONCLUSIONS

As a result of this study, some conclusions regarding the graphite reinforcing fiber surface can be made.

First, significant quantities of volatile components can desorb from the graphite fiber surface at temperatures up to 200°C. Surface oxidative treatments to enhance adhesion increase the amount of material desorbed by a factor of 8 over the untreated surface. This amount of material, unless adequately removed, has the potential for causing a substantial amount of voids in a composite. In any case, the amount desorbed is indicative of the possible presence of a surface contaminant layer which, unless removed, prevents true interaction between the fiber and matrix from taking place.

Second, surface treatments affect only a portion of the fiber surface, as indicated by an increase in the isosteric heats of adsorption on only 30% of the treated surface over the untreated one.

Third, surface-treated fibers exhibit a marked sensitivity to exposure to the environment. Isosteric heats of adsorption of krypton on surface-treated fibers are lower after environmental exposure. A portion of the high-energy sites generated by the surface treatment remain after air exposure. The untreated fibers, however, are only marginally affected by air exposure.

Fourth, the type A fibers used here possess significant quantities of sodium which is present on the fiber surface and is detectable by surface-sensitive techniques such as ISS and +SIMS. Thermal treatment enhances the sodium signal, indicating a diffusion of this material from the bulk. The presence of these high levels of hygroscopic material suggest that its presence may be responsible for enhanced moisture diffusion along the fiber axis in composites.

Fifth, while all fibers do not exhibit the high surface concentrations of sodium, this cation and other bulk contaminants can migrate to the surface and must be accounted for in any attempts to either explain the mechanism of adhesion or environmental degradation of the fiber-matrix interface.

Sixth, the morphology of graphite fibers appears to exist at the fiber surface. The energetically homogeneous basal plane surface one might expect on a high modulus fiber does indeed appear to exist, as indicated by the krypton adsorption thermodynamic behavior.

ACKNOWLEDGMENT

The author would like to thank Ms. Judith Mescher of the University of Dayton Research Institute for the adsorption isotherms, Mr. W. L. Baun of the Air Force Materials Laboratory for the ISS and +SIMS spectra, and the Air Force Materials Laboratory for supporting this in-house research program.

REFERENCES

1. A. M. Lovelace, AIAA Paper No. 74-242, AIAA Library, New York, 1974.
2. R. J. Diefendorf and E. W. Tokarsky, AFML-TR-72-133, Parts I, II, and III, 1971.
3. R. J. Diefendorf, D. M. Riggs, and I. W. Sorensen, AFML-TR-72-133, Part IV, 1975.
4. D. H. Kaeble, P. J. Dynes, L. W. Crane, and L. Maus, J. Adhesion 7:25-54 (1975).
5. J. V. Larsen, T. G. Smith, and P. W. Erickson, *Carbon Fiber Surface Treatments*, NOLTR 71-165 (1971).
6. J. Hertz, *Investigation into the High Temperature Strength Degradation of Fiber-Reinforced Resin Composite During Ambient Aging*, NASA-CR-124290 (1973).
7. C. S. Brooks, G. S. Golden, and D. A. Scola, Carbon 12:609-617 (1974).

8. T. Kwan, M. P. Freeman, and G. D. Halsey, Jr., J. Phys. Chem. 59:600 (1955).
9. S. Ross and J. P. Olivier, *On Physical Adsorption*, Interscience, New York, 1964.
10. F. A. Putnam and T. Fort, Jr., J. Phys. Chem. 79:459-467 (1975).
11. L. T. Drzal, F. A. Putnam, and T. Fort, Rev. Sci. Instrum. 45: 1331-1334 (1974).
12. L. T. Drzal and J. A. Mescher, *A Volumetric Gas Adsorption System for the Characterization of the Surfaces of Adherends and Reinforcing Fibers*, AFML-TR-76-84, 1976.
13. Hercules Inc., Wilmington, Del. 19899.
14. Uthe Technology International, Sunnyvale, Calif. 94086.
15. E. Steinhagen, S. Abrahamson, and F. W. McLafferty (eds.), *Atlas of Mass Spectral Data*, Wiley, New York, 1969.
16. A. Miller, J. A. Mescher, and L. T. Drzal, *Calculations and Algorithms in the Data Reduction of Adsorption Isotherms* (to be published, AFML-TR-76).
17. D. M. Young and A. D. Crowell, *Physical Adsorption of Gases*, Butterworths, Washington, D.C., 1962.
18. A. V. Kiselev and A. G. Bezers, Russ. J. Phys. Chem. 40:311 (1966).
19. 3M Company, Minneapolis, Minn.
20. W. L. Baun, J. Adhesion 7:261-267 (1976).
21. G. A. Beitel, J. Vac. Sci. Tech. 8:647-657 (1971).
22. S. Brunauer, P. H. Emmett, and E. Teller, J. Amer. Chem. 60: 309 (1938).
23. L. T. Drzal and T. Fort, J. Colloid Polym. Sci. (in press).
24. A. W. Adamson, *Physical Chemistry of Surfaces*, Wiley, New York, 1967.
25. A. V. Kiselev and A. A. Isirikyan, J. Phys. Chem. 65:601 (1961).
26. D. Guerard and A. Herold, Carbon 13:337-345 (1975).
27. F. Beguin and R. Setton, Carbon 13:293-295 (1975).
28. J. G. Hooley, Carbon 13:469-471 (1975).
29. D. W. Gibson, U.S. Patent No. 3,413,094, November 26, 1968, Method of decreasing the metallic impurities of fibrous carbon products.

3

Aluminum Adhesive Bond Permanence

J. Dean Minford

Alcoa Laboratories
Alcoa Center, Pennsylvania

1. INTRODUCTION

The joining of materials through adhesive bonding depends on the phenomenon of adhesion. This property may be further divided into mechanical and specific adhesion.

An example of mechanical adhesion is the anchoring of an organic polymer in a rough surface profile as shown schematically in Fig. 3-1. This so-called hooking effect is of paramount importance in bonding porous materials such as nonglazed ceramics, rigid foams, woods, or other fibrous structural materials. Its effect in bonding nontextured smooth metal surfaces is shown by the polymer adherence achieved to wire-brushed, wire-abraded, or sand- or grit-blasted metal surfaces. Acidic or caustic etching of metals also may produce a rough surface profile on a micro scale. For example, investigations by Bijlmer [1] at the Netherlands Fokker Company have led to their characterization of the most durable bonded aluminum joints using chromic-sulfuric etching. Their scanning electron microscopy (SEM) photographs show that the surface is covered with what they describe as "microetch pits."

The term "specific adhesion" refers to the forces of attraction between atoms which may be further divided into primary bonding forces with bonding energies in the range of 15-170 kcal/mol and secondary or van der Waals forces of 0.02-10 kcal/mol.

While specific adhesion forces are tangible entities and numbers can be measured and assigned to them, there is no satisfactory unified theory of how they quantitatively relate to bond permanence. The predictable bond strength and potential service life of adhesive bonded joints based on adhesion factors alone would be much higher and longer,

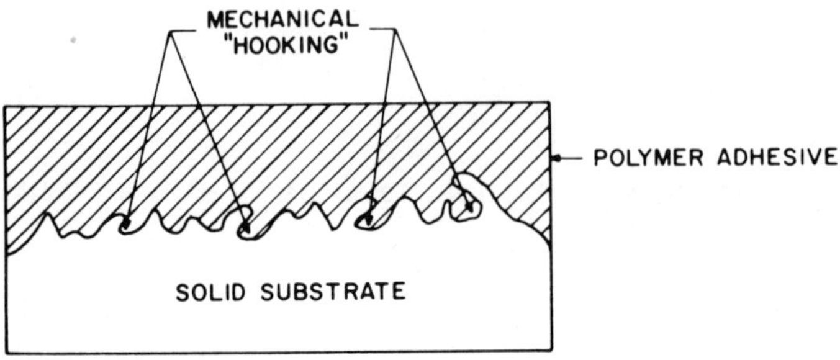

FIG. 3-1 Schematic representation of mechanical "hooking."

respectively, than measured in the real world. This is better understood when it is recognized that the surface of materials to which the organic polymer must adhere is quite different from the idealized surface conditions conceived for theoretical calculations of the interacting forces between individual or collected pools of atoms, dipoles, and ions. No single theory has clearly explained all aspects of the adhesion phenomena. Consequently, the best bonding practice has been to generate the cleanest and most highly reactive surface that can be practically established and accept the degree of bond permanence that results from that practice. It should be possible to achieve some additional benefit from creating a larger surface area with some contribution from mechanical interlocking.

Many factors contribute simultaneously to the overall durability of structural adhesive-bonded aluminum joints. A description of some of these factors has been published by the author [2] and can serve as an introduction to the subject for those less interested in the specific technical details of the subject.

2. EVALUATION OF ADHESIVE BOND PERMANENCE

Alcoa has been engaged in evaluating durable adhesive-bonded aluminum joints, as well as composite aluminum to other metallic and nonmetallic material joints, for many years [3-5]. In particular, Alcoa has been interested in the relative durability of adhesive-bonded aluminum joints as demonstrated by the long-time testing of bonded joints under various accelerated laboratory weathering environments and natural weathering conditions. The general conditions for testing adhesive joints have included environmental factors such as humidity, water soaking, natural weathering, cyclic conditions involving soaking, freezing, and thawing, high-temperature oxidation, continuous and intermittent salt water exposures, and simultaneous application of stress while exposing to various of the above conditions. Instead of the usual 7- to 30-day-exposure test results usually shown in the manufacturer's literature, it has been typical for Alcoa to test as long as 2 years under laboratory-accelerated weathering conditions and up to 8 years or longer in natural weathering environments.

The aircraft industry has considered the deoxidizing of aluminum alloys with the Forest Products Laboratory (FPL) etch, i.e., hot dichromate-sulfuric acid, as a standard for excellence since World War II. Alcoa has used the durability performance of chromic-sulfuric-acid-etched joints as a basis for reference for more than 25 years. Alcoa has used a chromic acid-sulfuric acid etching solution at 82.2°C (180°F), known as Alcoa 3 treatment, rather than the FPL etch. All available evidence indicates that both treatments give the same joint durability results. Over a period of many years, Alcoa has evaluated a large variety of two-part room-temperature- and one-part elevated-temperature-cured paste adhesives and heat-cured tape and film adhesives using room temperature water immersion; 100% humidity at 51.7°C (125°F) soaking; soaking, freezing, and heat cycling exposures; and industrial and seacoast atmospheric exposures [3-5]. Since 1966, the effect of stressing the joints also has been included and the highly significant difference between various adhesive families and between different members of the same chemical polymer family has been demonstrated [2-4].

More recently the work of Bethune and Thrall [6-8] on phosphoric acid anodize pretreatment has caused many to consider this as the optimum for bond durability. Minford [9,10] has recently published Alcoa long-term durability results on phosphoric-acid-anodize-treated surface joints which confirms the excellent performance of such joints.

3. BONDLINE CHARACTERISTICS AFFECTING BOND PERMANENCE

3.1. Modulus of Elasticity

The high modulus of various materials coupled with a difference in the coefficient of thermal expansion can be important factors in determining bond durability. Where large differences exist between the coefficients of expansion of the adherends, it is usually beneficial to use an adhesive of low modulus such as a nitrile rubber-phenolic adhesive as a stress relief interlayer.

3.2. Interfacial Imperfections

DeBruyne and Bascom [11,12] have furnished examples in the literature where an imperfection such as a trapped air bubble can be the site for high localized stresses. These stresses can add to the effect of any desorbing agent from the service environment such as water in significantly reducing overall bond permanence.

Some degree of interfacial imperfection can arise as a result of the normal curing of a structural joint. After wetting the surface of the adherend, the adhesive develops recognizable strength by solvent evaporation, cooling, or chemical polymerization curing processes. Plueddemann [13] has listed these stresses due to adhesive shrinkage along with coefficient of expansion differences as among the principal factors responsible for bond failure.

A different situation can develop if the bond interface is altered due to swelling of the adhesive. For thermoplastic-type adhesives, the absorption of organic solvent can produce a volume change of sufficient magnitude to create significant stress in the interfacial area. Metal adherends offer the worst situation since they neither absorb solvent nor change in volume to help compensate for the volume change in the adhesive. The failure of such adhesive bondlines will be inevitable and usually rapid where swelling of the adhesive by water absorption is accompanied by water desorption of the adsorbed adhesive at the adhesive-metal interface.

3.3. *Heat Curing*

The permanence of bonded joints can be positively affected in several ways by the heat curing of the adhesive. At elevated temperature, the lowered viscosity of the adhesive can improve adhesion through better wettability of the surface. Even with some soiling present, better bond durability may be achieved because the lower viscosity adhesive can absorb, dissolve, disperse, or desorb the soil. Finally, the higher temperature could increase the degree of chemical interaction between the adhesive and a primer or the adherend surface.

In general, the higher the temperature of the cure, the greater the durability of the joint. Among the present commercially available adhesives, the 177°C (350°F) cure materials are the most durable with the 121°C (250°F) systems showing a lesser degree of durability. Where a 121°C (250°F) cure adhesive has been cured at a higher temperature than suggested by the manufacturer, an increase in durability can usually be demonstrated. Whether this increase is from an increase in modulus or a change in the number and/or type of attachments to the substrate is not clear. The fact that within a given family of adhesives the bond durability increases as the creep resistance increases seems to favor the modulus increase argument [14].

3.4. Pressure

The positive effect of pressure in promoting bond permanence can occur in several ways. Applied with heating, pressure may promote better surface wetting and spreading of the adhesive or there could be the physical factor of forcing extra adhesive into the rough profile of an irregular surface or actual pores of a substrate such as wood or ceramic materials. The good durability of joints made with adhesive polymers which can form volatile products under elevated temperature curing conditions such as the vinyl-phenolic adhesives is highly dependent on maintaining high pressure on the bondline. Failure to apply adequate and uniform pressure to a bondline also may result in both nonuniformity of the bondline cross-section and less than optimum adhesive flow.

3.5. Mechanical Energy

The mechanical energy expended in rubbing an adhesive into a metal surface has been correlated by Cassidy and Yager [15] with measured increased bond strength values. It was presumed that the mechanism may be one of displacement of a contaminant film on the surface. If this is the true explanation for this effect, then this factor must be considered a minor one in promoting the good bond durability of uncontaminated acid-etched aluminum joints.

3.6. Cohesive vs. Adhesive Failure Observations

It is generally agreed that the metal-bonded joints with highest durability potential will initially fail cohesively. The degree of durability will be related to the retention of the cohesive failure property after long periods of exposure to weathering service conditions. All observations made at Alcoa in evaluating the bond permanence of aluminum structural joints have shown that the initial presence of a cohesive bond failure prior to exposure is never a reliable indication of good bond durability. Similarly, the initial level of bond strength with apparent cohesive bond failure without exposure is not predictive of any relative durability value [3-5,9,10]. An example of both these situations is cited by DeLollis [16,17] in a 2-year water-sensitivity study of the relative bond permanence of aluminum bonded joints fabricated with nitrile-phenolic or nylon-epoxy adhesives. Whether exposed to continuous high humidity or water immersion, nitrile-phenolic joints showed a high degree of durability, exhibiting cohesive failure under all exposure conditions, while the nylon-epoxy joints with initial cohesive bond failure showed adhesive-type failure within 2 months.

Initial lap shear strength of the nylon epoxy joints was approximately 3 times higher. Thus, regardless of how the bond fails initially or how high the initial bond strength may be, the rate of desorption of the adhesive at the interface by the action of water has a most significant influence on bond permanence. The most conclusive studies relating to the bond desorption by water theory have been investigations of adhesion and bond failure between adhesive resins and glass [18-21].

Prominent adhesive scientists such as Bikerman [22,23] and Sharpe and Schonhorn [24] have stated that there is only the degree of cohesive failure to consider in failed joints since the attractive forces involved at the interface are invariably greater than the cohesive strength of the adhesive and sometimes the cohesive strength of the adherend. Bikerman points out that bond rupture so rarely proceeds exactly between adhesive and adherend that these events (i.e., failure in adhesion) need not be considered in any theory of adhesive joints. It is true that apparent adhesion failures are common, but they take place so near the interface that the adhesive remaining on the adherend after the bond rupture is not observable. Good [25] also reported that his analysis of adhesive joint failure led him to conclude that interfacial separation seemed highly improbable, particularly when true chemical wetting of a surface had taken place. Sharpe [168] has considered the role of weak boundary layers (WBL) in determining the breaking stress of adhesive joints. Acceptance of these ideas offers some complication to acceptance of the bond desorption by water theory described by a variety of investigators. The study by Sterman and Toogood [26] in 1965 showing the increase in bond permanence of glass joints in high-temperature water exposures through the use of silane adhesion promoters furnished further proof that the desorption of the adhesive by water and failure in adhesion at the adhesive-adherend interface was a viable mechanism. How fast the water can permeate the adhesive to the interface as compared to diffusing along the adhesive-adherend interface has been investigated by Laird [27]. Laird showed that diffusion of water along the interface could be as much as 450 times faster than permeation through the adhesive.

Adhesive product data sheets and the literature often cite the durability of bonded joints in various organic solvents. Because adhesives are often polar in nature, they can be sensitive to some of the stronger polar organic solvents. Any absorption of the solvent which results in swelling of the adhesive can cause stress to develop at the interface with metal or ceramic adherends since the metal or ceramic will not swell and the adhesive-adherend interface will become a plane of significant stress concentration.

The bondline thickness must also be recognized as a factor which can affect both initial bond strength and durability of joints. As regards initial bond strength effects, the data can readily be obtained for any adherend/adhesive system by merely controlling the adhesive thickness in the bondline in lap joints over a range of values and stressing to failure in a tensile tester and generating a graph of thickness versus strength values. As regards bond durability, however, other factors such as fatigue and crack propagation behavior of the adhesive or the development of critical bondline flaws as the thickness changes can be very important. An analytical investigation on the influence of bond thickness upon the stress distribution in single-lap adhesive joints has been made by Ojalvo and Eidenoff [28]. They identified two antisymmetrical adherend-bond interface points at which the shear stresses are sufficiently high to cause the growth of joint failures.

Another interfacial factor can be the presence and relative concentration of nonadsorbable or nondesorbable contaminating films at the interface. The most common contaminants are grease films, industrial oils, forming lubricants, and certain plasticizer chemicals. In assembling metal structures by adhesive bonding certain metal-forming operations often precede the assembly of the parts which result in surface contamination by forming lubricants. Best bond durability can only be expected after a separate operation to remove the lubricant prior to adhesive application. Also, metal sheet or extrusions in storage will often have corrosion-resistant films on their surfaces. Industries such as the automotive manufacturers choose adhesives which have the highest tolerance for these surface contaminants. Minford [29] has shown that less success can be expected in achieving high-durability bonds in the presence of these contaminants as compared with the permanence of joints using clean and etched adherend surfaces. The vinyl-plastisol-type adhesives remain the best commercially available adhesives for achieving modest strength and reasonable bond durability at lowest cost in such situations. Adding vinyl plastisol to higher-strength epoxy resins has been a further step toward achieving higher-bond-strength joints with better toleration for surface oil contaminants than with epoxies alone. The significant lowering of the bond durability by the presence of these oils indicates they should be removed prior to bonding, and special surface preparations for promoting better bond durability should be encouraged.

Admittedly, it is difficult to determine—even using visual or microscopic examination—after bond failure whether an apparent adhesive failure occurred at the interface due to improper wetting or at some new interface leaving behind a thin layer of adhesive. In comparing surface features after bond failure with the original adherend surface, the resolution limitation of about 1×10^{-8} m (100 Å) for scanning electron microscopes may not always detect the thin film of adhesive closely reproducing the original surface profile. Optical and staining methods described by Brett [30] to determine the presence of such films are only applicable to fairly thick films, since the optical technique uses interference phenomenon. Films a few angstroms thick are still largely undetected. The use of certain highly specialized analytical tools such as secondary ion mass spectroscopy (SIMS), ion-scattering spectroscopy (ISS), and Auger electron spectroscopy (AES) in recent years by investigators such as Baun et al. [31] has vastly improved the opportunity for deducing surface compositional changes. Even when the surface films are only on the order of atomic dimensions or when the failure occurs near the original interface and includes parts of both the adhesive and adherend, these tools may provide useful information on the locus of failure. Of these procedures the AES has been found to be the least useful primarily because of heating effects of the electron beam which may cause decomposition of adhesive surfaces. The SIMS by itself has been shown to be a powerful tool for elemental surface characterization by Benninghoven [32] and Schubert and Tracy [33]; however, uncertain or rapidly changing secondary ion yields due to changes in chemical bonding make quantitative analysis virtually impossible using SIMS alone [34,35]. SIMS is, however, most helpful when combined with other techniques such as ISS and AES which use a beam of ions of correct energy for combined use with SIMS. The majority of apparent adhesive failure surfaces which have been examined using these tools show cohesive failure in both adhesive and adherend has probably occurred. An elemental analysis method must be used to confirm whether the surface contains both adhesive and adherend. A test for the amount of branched polyethylene adhesive on a copper oxide surface as performed by Blair and coworkers [36] does not in itself prove that the mode of failure was only cohesive in the polymer. Some failures classed as pure adhesive have been examined by ISS/SIMS in which the adherend surface was obviously not wet by the adhesive and no trace of adhesive on the adherend could be found. Further examination has usually disclosed in this situation that a thin layer of contaminating elements caused the adherend surface to be most accurately described as "dirty." Certainly this kind of bond failure at the interface should not be considered adhesive since real bonding between the materials had never really occurred.

4. CHARACTERISTICS OF THE ALUMINUM SURFACE

What is usually referred to as a metal surface is actually a metal oxide surface, as shown schematically in Fig. 3-2. Thus, the adhesive material in the joint is not really involved in any metallic bonding. The oxide on aluminum is remarkably stable in air, extremely thin, tightly adherent, and transparent. Growth of this natural oxide film

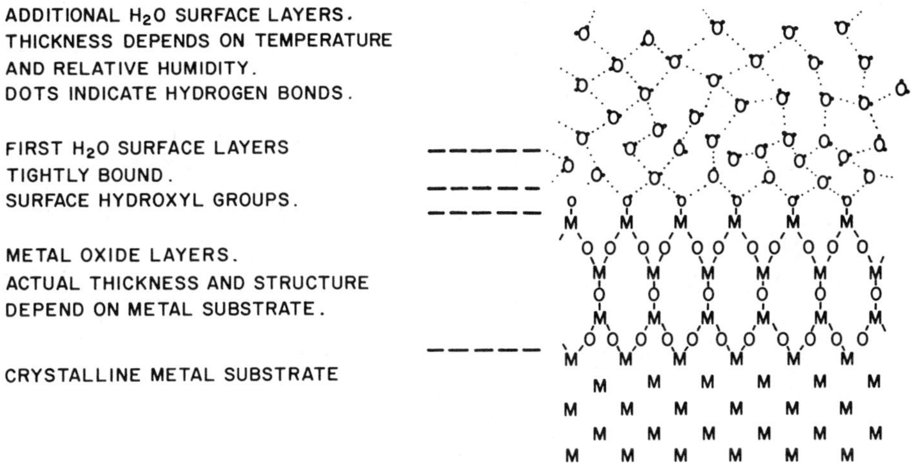

FIG. 3-2 Schematic representation of water and oxide layers on metal surface. M = metal atom, O = Oxygen, O = -OH, O = H_2O.

is self-limiting to some degree depending on the weathering conditions. It may reach 1×10^{-7} m (1000 Å) depending on temperature conditions, alloy type, environmental conditions, and length of exposure time. This covering layer eliminates the attractive forces and interfacial effects ordinarily associated with the underlying aluminum atoms. Only by knowing more about the surface chemistry of aluminum oxide can we deal with interaction between the adhesive polymer and the bondable interface of aluminum adherends.

Since all adhesive polymers show some degree of mobility in the uncured state, the adhesive will flow over the interfacial area in a kinetic process eventually filling all void areas in the substrate surface as shown in Fig. 3-3. If the viscosity of the adhesive is very high, it is possible to trap air and create voids, as shown in Fig. 3-4. If the viscosity is low, such entrapments are less likely. This is one explanation for the superior durability of many heat-cured epoxy joints as compared to room-temperature-cured epoxy joints in highly corrosive environments. The degree of improved chemical wetting of the aluminum surface due to the lower-viscosity melt condition of a one-part epoxy resin during the high-temperature-curing step is shown by the enhanced bond durability in Table 3-1.

In general, the wettability or tendency of an adhesive to absorb on an adherend surface is assumed to be intimately related to the bond joint durability.

Chemical theory predicts that all liquids with surface tensions smaller than the critical surface tension of the adherend will completely

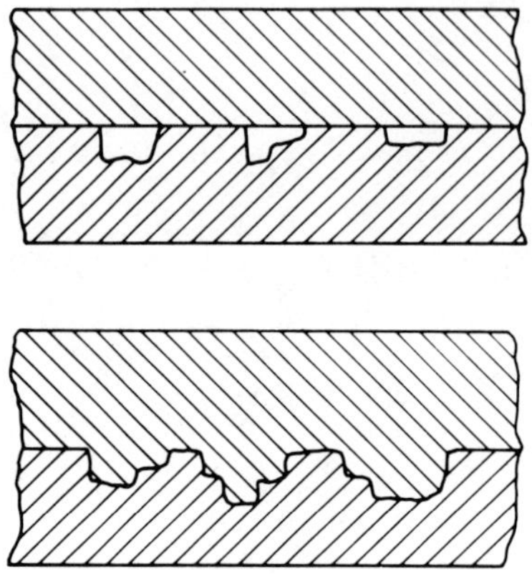

FIG. 3-3 Wetting of a solid substrate by a fluid adhesive: (top) substrate incompletely wetted; (bottom) substrate completely wetted.

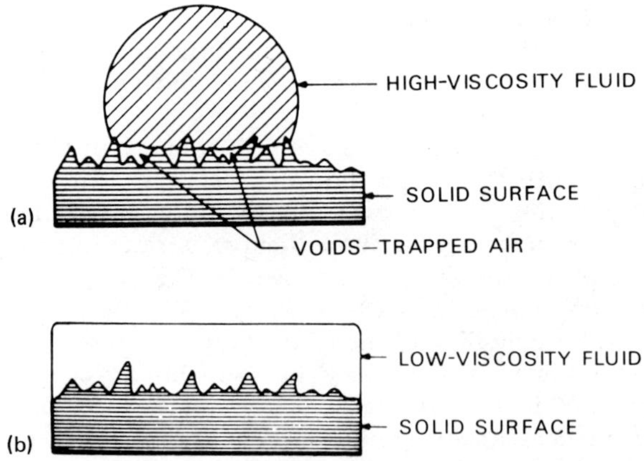

FIG. 3-4 Effect of adhesive viscosity on interfacial contact: (a) low area of interfacial contact resulting from high viscosity of fluid; (b) lack of voids and high area of interfacial contact resulting from low viscosity of fluid.

TABLE 3-1 Effect of Better Surface Wetting on Bond Permanence

Surface treatment	Adhesive	Type cure	Range of bond survival times in seacoast exposure (days)
Vapor-degreased	Two-part epoxy	RT	70-80
Vapor-degreased	One-part epoxy	Heat	71-270
Alodine 1200	Two-part epoxy	RT	1158-1440
Alodine 1200	One-part epoxy	Heat	>2920
Chromic-sulfuric	Two-part epoxy	RT	270-760
Chromic-sulfuric	One-part epoxy	Heat	760-1440

wet the surface of the adherend. In practice, however, the best procedure to achieve highest wetting is through the use of various surface cleaning processes. It also can be concluded that rough surfaces with small dimensions will be most rapidly wetted with a liquid whose surface tension γ_L is the smallest as compared with γ_C, the critical surface tension of the solid surface. This offers one explanation for the long-term durability of aluminum joints where the surface is in a "micro etch pit" condition as from chromic-sulfuric etching. In addition, the adhesive should offer a low surface tension in the melt condition at elevated temperature. One estimate of the wettability of surfaces has involved the measurement of contact angles on those surfaces. Ideally the contact angle of water or most organic solvents for the highly polar aluminum oxide surface should be zero if the surface is clean [11]. In practice the contact angle does not differentiate between the basic adsorption characteristics for many liquids and so presumably may not differentiate for adhesives. It can be assumed that any contact angle deviations from zero are possible indications that surface contamination may be present. Sometimes the degree of contamination may be indicated by the magnitude of the deviation. The value of the contact angle also does not indicate whether any surface contamination present can be desorbed by the adhesive or which contaminants would most likely resist desorption. If a low contact angle measured on an aluminum oxide surface has any predictable meaning, it is most likely that the bond once formed can be ultimately displaced by water in the environment [17]. Even in aircraft bonding where a low-contact-angle wettable aluminum surface condition is realized using the best current technology, a bonded joint may experience localized debonding by water after approximately 20,000 hr of flying time. Thus, the next important goal may be to alter or convert the aluminum oxide surface in such a way as to extend the time to experience disbonding by water to 40,000 hr or more of service.

Various interactions between surface oxide and environmental factors such as water, temperature, and atmospheric chemicals can create bonding complications. Although the oxide surface can be apparently dehydrated at elevated temperature, under normal ambient bonding conditions the outermost surface oxygens will hydrate to form a layer of hydroxyl groups which in turn adsorb and retain molecular layers of bound water. On aluminum, iron, and copper surfaces, for example, the presence of up to 20 molecular layers of water has been shown on so-called dry surfaces [37]. The presence of these various polar constituents on the aluminum surface can be fortunate for promoting bonding to polar organic resins such as epoxies. If this polar surface is not present, as can be shown on a fractured metal surface in a moisture-free and oxygen-free environment, then the metallic surface cannot be wet by polar organic liquids. Admitting a small amount of oxygen to permit surface oxidation on an oxide-free surface permits the polar liquid to spread out and wet the surface. The polarity of the surface-hydrated oxide also can cause any polar organic contaminants to adhere to the aluminum substrate, as shown in Fig. 3-5. Also, nonpolar organic contaminants will tend to be displaced by polar compounds so that nonpolar contaminants constitute still another possible intervening layer involved in determining adhesive bond durability. This nonpolar layer assumes considerable importance in present-day bonding of metal parts in the automotive industries where it is common to bond over lubricant- or oil-contaminated surfaces. At the bottom of all these potentially present layers is the actual metallic surface.

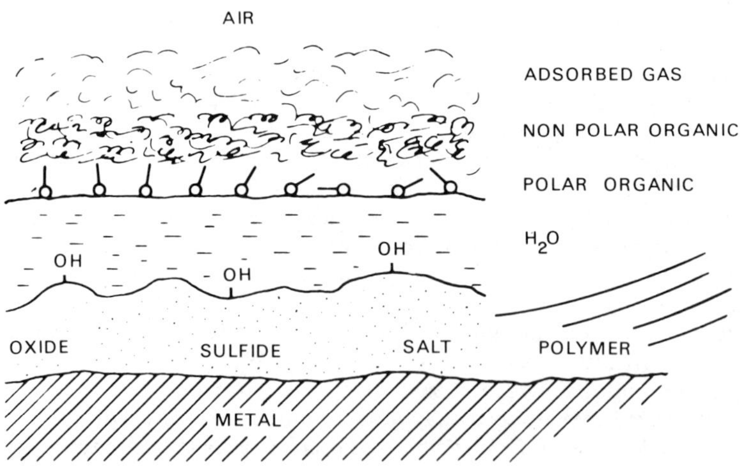

FIG. 3-5 Hierarchy of spontaneously adsorbed layers on a metal surface.

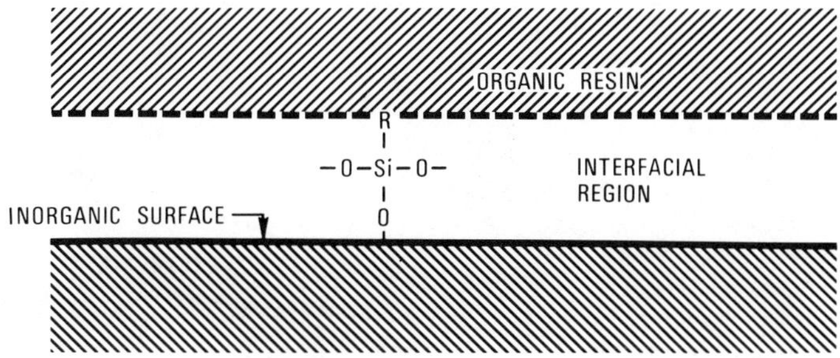

FIG. 3-6 Silane orientation in bondline.

In view of the polar nature of aluminum oxide, it might be appropriate to refer to the orientation of a silane in the bondline of an aluminum and epoxy joint. The silane is a bridging molecule which has preferred orientation chemical linkages on each end of the molecule. One end bridges to the inorganic oxide interface and a reactive organic group on the other end links with the organic adhesive resin, as shown in Fig. 3-6. Considerable data have been gathered in support of the ability of silanes to increase joint durability, especially in water-soaking-type environments.

What happens to the native oxide on an aluminum surface when the surface is mechanically abraded or acid- or alkaline-etched is very important. It is possible to deoxidize an aluminum surface using any of these methods, but new oxide instantaneously regrows so that an oxide-free surface is never present in any practical bonding situation. In Fig. 3-7 it is shown that the rate of oxide regrowth under ambient laboratory exposure conditions, as measured by contact resistance, can take place at different rates on different alloys. We also find that the rate of oxide regrowth will be a function of the environmental conditions. Increasing temperature and humidity and the availability of oxygen can all positively affect oxide regrowth. This means that sufficient oxide to cover the bonding interface is invariably present regardless of how fast the bonding is accomplished after deoxidizing. The effect of longer exposure of the deoxidized surface under varying conditions of humidity, temperature, etc. is primarily thickening of the oxide and increasing the probability of finding additional surface contamination layers.

Characterization of the different aluminum alloy surfaces using analytical tools such as electron microscopy, scanning electron microscopy, and electron diffraction is not new to the aluminum pioneers at Alcoa. However, the use of these tools to study aluminum surfaces before bonding and after breaking bonds has only been undertaken

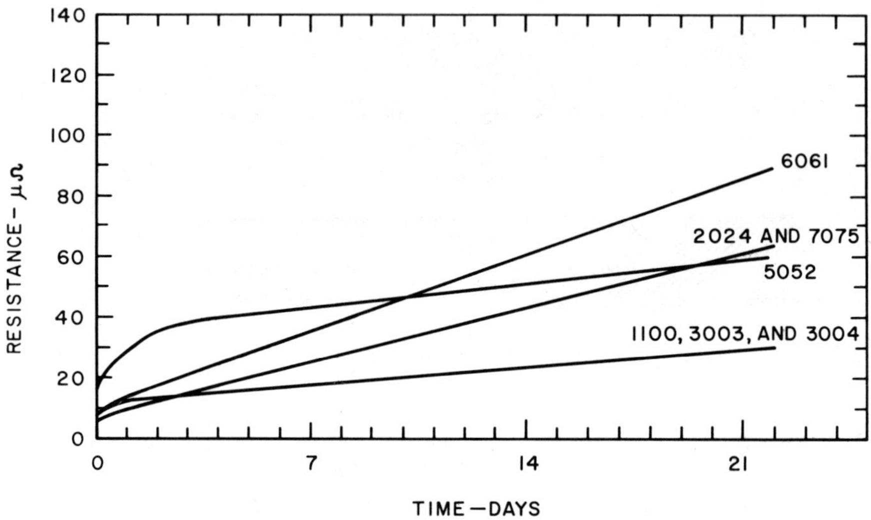

FIG. 3-7 Contact resistance vs. time—surfaces treated in H_2SO_4 - CrO_3 (5 min).

in recent years. Smith [38] and Pattniak and Meakin [39] have pursued such studies as well as Bowen and associates at Northrup [40-44], mostly as support to technical programs of interest to U.S. government agencies. Bowen used the scanning electron microscope (SEM) to determine aluminum surface character, oxide layer thickness, and surface reproducibility. With high-resolution electron diffraction, he sought to determine the crystallographic surface structure, and with Auger spectroscopy the surface chemistry situation. A new technique was used to evaluate the thickness and oxide layer character of aluminum surfaces by bending the prepared surface either 90° or 180° and viewing the bent area with the SEM. The oxide layer on "as-received" or vapor-degreased aluminum was identified as consisting mainly of the bayerite or beta alumina trihydroxide form with a varying oxide thickness of 1×10^{-8} m (100 Å) to 1.5×10^{-7} m (1500 Å). After alkaline cleaning at 76.7°C (170°F), the surface oxide layer remained primarily bayerite but with noticeable slight chemical attack character. Mill contaminants had been effectively removed. When the surface was acid-deoxidized or treated with a spot weld etching solution, the oxide present was still identified as bayerite but the oxide thickness was now reduced to a point below the resolution capability of the SEM, probably 6×10^{-9} m (60 Å) to 1×10^{-8} m (100 Å). It should be pointed out that these three kinds of surface pretreatments are those commonly used in the pretreating of aluminum prior to anodizing. Marceau et al. [45] and McMillan et al. [46] have both demonstrated the good durability of phosphoric-acid-anodized aluminum joints.

The author also has offered evidence on the durability of joints made with either chromic-acid- or phosphoric-acid-anodized surfaces. Both stressed- and nonstressed-type specimens were exposed for long time periods in water soaking, alternate salt water immersion, and natural atmospheric weathering environments [9,10].

The surfaces produced by anodizing has a typical thin barrier oxide layer at the base. The actual surface involved in bonding is a much thicker oxide layer than the barrier layer and consists primarily of boehmite or beta aluminum oxide hydroxide as boehmite is more correctly described [47]. It is not unexpected that this form of bonding oxide should bond durably since an FPL-etched aluminum surface, while showing a much thinner coating than the anodized oxide layer, is also predominantly boehmite in character.

There is the additional possibility that a form of oxide less stable at ambient temperature conditions, such as boehmite, may convert to a more stable ambient temperature form such as bayerite. Bowen [48] has suggested the necessity to generate a controlled thickness film of boehmite on an FPL-etched surface by following the FPL etch with a 90-min seal in boiling water-sodium dichromate solution. Other ways to generate a controlled-thickness boehmite film with good adhesion to both substrate and adhesive would be to use a chromic-acid- or phosphoric-acid-anodizing process. While the technical point might be argued about whether the oxide form identified by Bowen is boehmite or a related form of pseudoboehmite, the fact remains that superior bond performance can be obtained by these pretreatments of the aluminum surface.

The use of anodic coatings on aluminum as surface preparation for bonding in recent years has generated many studies of the oxide layer in detail not hitherto required for other commercial application of anodized aluminum. Since the major commercial application of anodizing has been for protection against weathering of exposed aluminum parts, the use of quite thick oxide layers of 250,000 Å has been common. Most literature has therefore dealt with anodic oxide films in this range. For adhesive bonding, the upper oxide thickness of 5000 Å has been accepted as a thin enough oxide to better transfer the load without fracturing. This is the thickness of oxide generated by Boeing's phosphoric acid anodization process.

Based on the above background, McDevitt and Baun [49] have investigated the chemical characteristics of the composite oxide film found on 2024-T3 aluminum when an anodic barrier layer was formed after certain chemical surface treatments. It was quite apparent from their data that chemical elements which could be present in the surface layer formed by the surface cleaning end up in the final surface layer when anodizing. This should be expected when it is recognized that the thickness of the anodized layer is generated continuously at the metal surface so that the original surface is the top surface after anodizing. This has not been generally recognized, however, because most of the literature and present-day experimentation has been performed on electrochemically polished, high-purity aluminum surfaces which do not have the complex surface chemistry of highly alloyed aluminum such as 2024 or 7075 alloys used for aerospace bonding.

Other surface analyses of 2024 aluminum or 6061 and 7050 alloys have been made by McDevitt and Baun [50] in the first instance and McDevitt et al. [51] in the latter. In all cases the ISS, SIMS, and AES instrumentation was used and the surfaces were pretreated in a variety of solutions. In an earlier work [52], McDevitt et al. had showed the surface layer of 2024 alloy to be magnesium-rich and Smith [53] had showed 6061 alloy similarly had a magnesium-rich surface. By choosing a 300-g H_2SO_4/40-g CrO_3 ratio acid etching solution, McDevitt and Baun found all the surface magnesium was removed to below detectable limits providing at least 4- to 10-minute etching time was used.

McDevitt found distinct differences in the surface chemistry of both 6061 and 7050 alloys depending on the surface pretreatments of degreasing, alkaline etching, nitric-hydrofluoric etching, chromic-sulfuric etching, ammonium bifluoride or alodining. What these differences mean in terms of bond durability must be determined from actual durability testing under various weathering or combined stress and weathering conditions. Correlations need to be developed between these analytical chemistry results and environmental studies.

Considerable work has been performed by Venables et al. [54] on the morphology of the oxides on aluminum surfaces by high-resolution scanning transmission electron microscopy. They have obtained striking visual evidence for fine oxide protrusions which apparently can interlock with the adhesive to enhance bond strength and could account for the exceptional performance of the phosphoric anodize surface-treated joints.

The phosphoric acid anodize surface condition has also been studied by McDevitt et al. [55] using ellipsometry, AES, and ISS analytical techniques. They found the voltage and concentration of acid had an important influence upon the structure, morphology, and properties of the oxide film. Under the conditions of the Boeing PA treatment, a barrier layer of approximately 105 Å will be formed, but these authors found they could increase the depth of the nonporous barrier layer up to 806 Å by increasing acid concentration and voltage.

5. EFFECT OF WATER ON THE PERMANENCE OF ALUMINUM JOINTS

5.1. *Introduction*

Probably the most important environmental factor in the general deterioration of bonded metal joints is the invasion of the boundary layer of the joint by water molecules. Brockmann [56] has studied the change of strength and deformability of the adhesive layer itself under hot and humid climate conditions and concluded that the structural adhesive part of the joint is involved in a nearly reversible process and can reach a predictable equilibrium condition. However, Brockmann states that the deterioration of the actual adhesive forces in long-term aged joints is an irreversible process proceeding in an unpredictable manner toward catastrophic destruction. In the case of aluminum bonds, the water stability of the boundary layer in the joint is dependent both on the chemical structure of the adhesive and the type of surface to be bonded. Two of the most effective ways to increase the stability of the boundary layer to humidity is through the use of special aluminum surface pretreatments plus corrosion resistant primers. However, it must be admitted that all these special surface treatments and primer developments have been carried forward without an exact knowledge of the nature of the adhesion forces existing between the organic adhesives and the aluminum surface. Brockmann [56] has described the typical metal surface as containing metal atoms, reaction products of the metal in the form of oxides, hydroxides, phosphates, chromates, and a strong adsorbed-water layer. Unfortunately, most surface analysis processes change or destroy the real surface layer during the measuring which makes it difficult to exactly define the nature of the surface. The analysis of a surface layer by Electron Spectroscopy for Chemical Analysis (ESCA) or SIMS methods, for example, while developing some important information for describing a surface, still gives no indication of the real reactivity of the surface which is the adhesional property of the surface. Nevertheless, the evidence presented by Lewis and Forrestal [57] and Brockmann [58] seems acceptable that the adhesion between structural adhesives and most metal surfaces is due to chemical bonds between the adherends. After wetting the relatively rough metal surface in its low-molecular and reactive form, the adhesive builds up chemical bonds with the oxides and hydroxides on the surface. In addition to the direct chemical linkages such as phenolates or alcoholic bondings with phenolic resins or epoxide resins, there are a large number of hydrogen bonds that can occur in the boundary layer.

The special question now becomes, in what manner does this adsorbed resin layer react in the environment of water? In tests when polyester or phenolic resins have been adsorbed on aluminum oxide, up to 50 and 25%, respectively, of the total resin has been desorbed when contacting pure water. This would indicate that while the bonds between structural adhesive and metal surfaces may be chemical in nature, they are not absolutely water-stable. Actually, there is no reason to expect that phenolate or alcoholic bonds or hydrogen bonds should not deteriorate in the presence of water. Naturally, the adhesion deterioration rate in a real metal bond with large bonding areas and long diffusion paths for the water will be much slower than observed under the above-described experimental water-soaking conditions. At this stage, it is time to consider in which way it would be possible to replace the chemically stable but water instable bonds with more-water-stable chemical links. Brockmann [58] has been the most recent investigator in this area using 8-hydroxyquinoline or 5,7,2,4-tetrahydroxyflavonene (Morin) as substances which can react with aluminum oxide in alcoholic solutions and form water-stable aluminum complexes. After drying the treated surfaces and rinsing them in methanol, either phenolic resins or nitrile epoxides were cured on the converted surface. A significant increase in bond strength retention following water-soaking exposures was clearly evident.

Kinloch [59] has investigated the interfacial bond relationships under water-soaking conditions and has attempted to develop a mechanism of failure of the bondline under environmental exposure to water conditions. He concluded that the failure usually occurs at or close to the adhesive-adherend interface during which time certain steps can be identified. First, displacement of the adhesive on the oxide governed by the rate of diffusion of water through the adhesive to the interface. Second, loss of strength and failure of the oxide due to subtle changes in its nature. Third, under the special conditions where corrosive water is present there can be gross corrosion of the substrate as a failure mechanism. Obviously, more stable oxides and stronger interfacial forces must be forged from better surface pretreatments and adhesive selections to resist this rupture by water.

While we are primarily discussing the effect of environment water on the bonded joint in this section, it may be appropos to include a discussion of the effect of water on the oxide which is to be bonded. Included as a part of all chemical treatments of the aluminum surface is a final rinsing step designed primarily to remove the etching chemicals, but it has been shown many times that the nature of the rinse water as well can be influencing on bond durability. Wegman [60] considered the fact that while oxide films can be removed by mechanical or chemical means, there is an instantaneous regrowth of a new oxide film. This new oxide film has a different set of properties than the original oxide and can be variable depending on the posttreating procedures. When mechanically removed by silicon carbide abrasion, the bond strength was highest to the immediately bonded surface and decreased with increasing amounts of adhesive-appearing failure with increasing oxide regrowth time in a 20% oxygen and 50% relative humidity (RH) atmosphere. If the humidity was reduced to 4%, then the sharp dropoff in bond strength was not noted as for 50% RH. Wegman and Dickinson [61] had reported earlier that conditioning of sodium-dichromate-sulfuric-acid-etched aluminum in 50% RH for 72 hr still permitted good bonding, so factors of how the oxide was removed and the conditions of oxide regrowth were both influencing. Wegman also found that etched aluminum posttreated with hot tapwater produced high-strength cohesive bond failures while treating in distilled water led to low-strength adhesive-type bond failures. In a later report, Wegman et al. [62] showed that he could individually add multivalent compounds to the distilled water and achieve the high-strength bonds and cohesive-type bond failure found with tapwater posttreatment.

5.2. *Unstressed Aluminum Joints*

Water is such a dominant factor in determining the permanence of aluminum joints and the evidence about the effect of water is so voluminous that it seems appropriate to discuss separately the effect of water exposures on unstressed and stressed joints. Some of the earliest investigations on the unique effect of water on the durability of epoxy-bonded aluminum joints was by Falconer et al. [63] in 1964 with subsequent confirmation by many other investigators [16,64,65]. The experiment of Kerr, MacDonald, and Orman in 1967 comparing the relative effect of dry ethanol or water on the cured epoxy polymer (whether inside or outside the bond joint area) furnished specific evidence for the preferential action of water at the aluminum oxide-polymer interface. While alcohol was more readily absorbed into the polymer matrix, reducing the tensile strength by 30%, there was no bond strength deterioration in the corresponding aluminum-epoxy joint after soaking in alcohol for the same period. By contrast, water in the environment was barely absorbed at all by the cured epoxy, and the epoxy tensile strength was lowered by only 8%, yet the aluminum joint strength was decreased by 50%, as shown in Fig. 3-8. Orman and Kerr reported further studies on the effect of water on the durability of aluminum-epoxy joints in 1969 [66]. Typical bond strength decay curves developing over a 50-day exposure period in 100% RH at 40°C (140°F) or 100°C (212°F) are shown in Fig. 3-9. The variation in the bond strength decay values with increasing temperature after 50 days exposure are shown in Fig. 3-10. These investigators also showed that the adverse effect of water on bond strength was somewhat reversible when the water exposure was terminated and the joints were exposed to heating at 90°C (194°F) in vacuo for up to 24 hr as shown

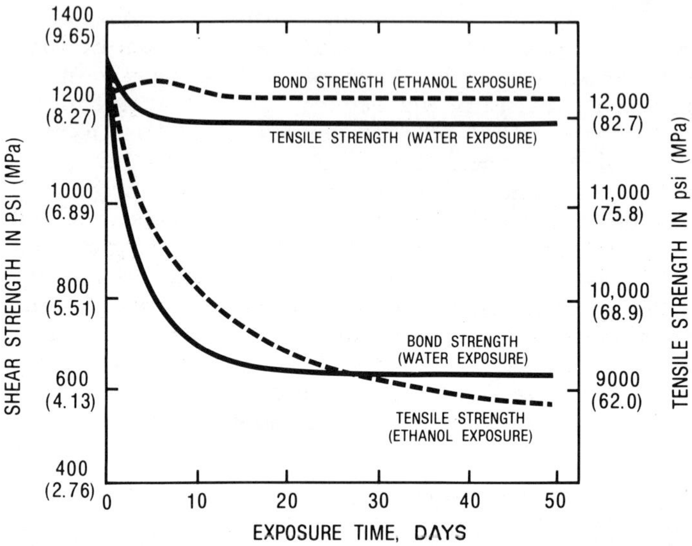

FIG. 3-8 Bond strength of aluminum-epoxy joints and tensile strength of epoxy resin after exposure to water and ethanol at 90°C (194°F).

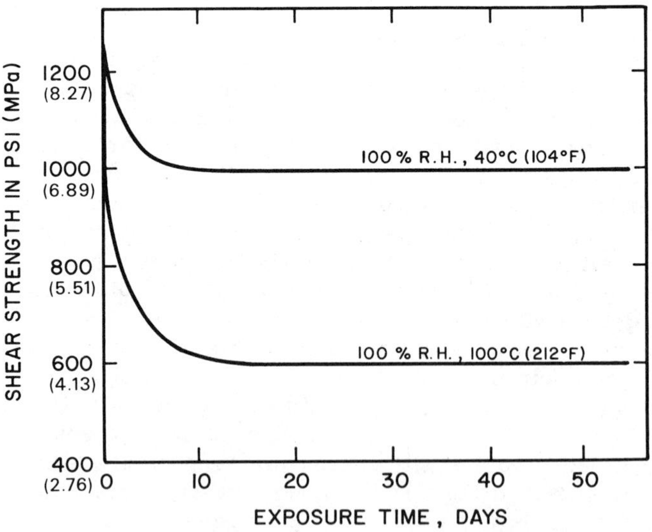

FIG. 3-9 Typical bond strength decay curves of aluminum-epoxy bonds.

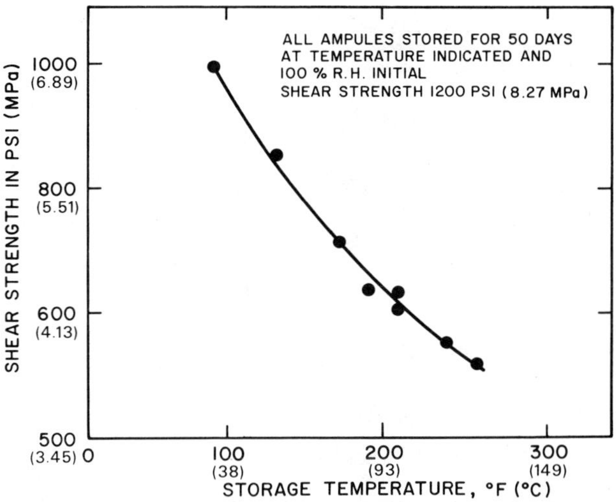

FIG. 3-10 Effect of storage temperature on bond strength.

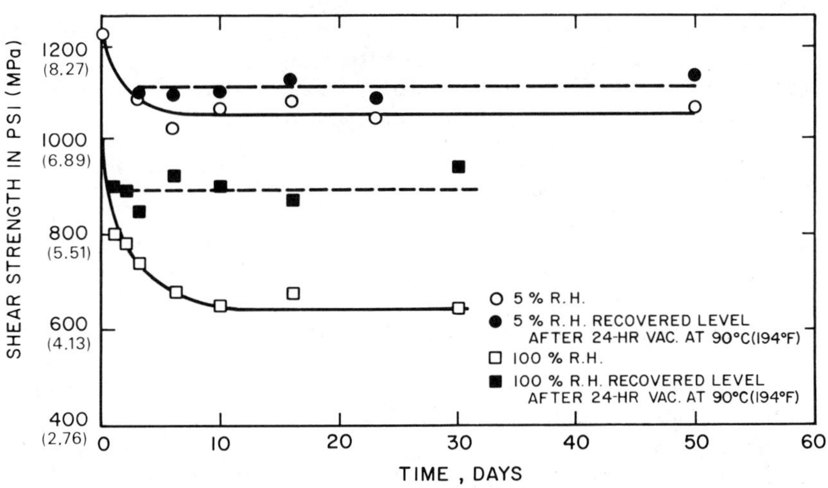

FIG. 3-11 Exposure of aluminum-epoxy bonds at 90°C (194°F).

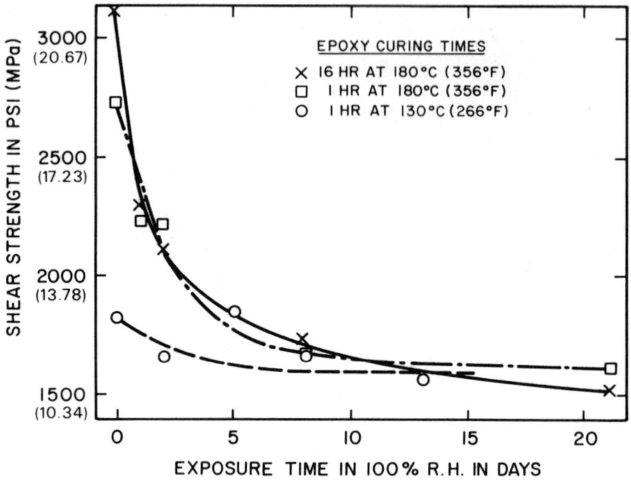

FIG. 3-12 Effect of degree of cure on aluminum-epoxy joint durability.

in Fig. 3-11. Orman and Kerr also investigated the effect of varying epoxy cure on loss of bond strength during high-humidity exposure with the somewhat unexpected results shown in Fig. 3-12. While overcuring raised initial bond strength, it had no effect on the magnitude of bond strength decline with increasing water exposure time. While showing a much lower initial bond strength, the undercured epoxy bondline was relatively unaffected by moisture. This led the investigators to conclude that the possibility of direct chemical bond rupture must be considered more tenable than interference with hydrogen bonding by the water which was another possible mechanism for explaining the dramatic effect of water on joint permanence.

5.3. *Stressed Aluminum Joints*

MacDonald [67] attempted in 1969 to establish the existence of time-temperature relationships for adhesive bonding in a manner similar to that used by Kwei [68] in 1966 for examining the fracture characteristics under dead-loading conditions of aliphatic-amine-cured epoxies. Subsequently, Orman [69] referred to a rapid stress corrosion test developed at the Atomic Weapons Research Establishment, Aldermaston, Berkshire, which eliminated time as one variable. One disadvantage of the normal stress-and-humidity-type testing has been that the stress application apparatus may be occupied, particularly at the lower stress levels, for very long time periods. For example, some Alcoa stressing tests for high-strength heat-cured modified epoxy joints at stress levels of 1.72 MPa (250 psi) have been continued in atmospheric weathering for periods as long as 8 years without failure of the joint. In the Orman procedure, the joints are strained at a constant rate while held in a controlled environment, and the mechanical properties recorded. Comparisons were made as shown in Fig. 3-13 between the failing load performance of nonequilibrated joints tested at some indicated temperature and joints equilibrated to similar temperatures but tested at 20°C (68°F). There appeared to be a reinforcement effect resulting from the simultaneous application of stress and water above 90°C (194°F). At temperatures lower than 90°C (194°F), the fully equilibrated joints failed at lower loads than the nonequilibrated joints, as shown in Fig. 3-13. The surfaces of broken joints showed differences when examined by SEM. In areas which showed apparent adhesion failure, the joints fractured in air showed almost complete adhesion failure as did the joints tested in water at 95°C (203°F). However, at water temperatures lower than 90°C (194°F), the evidence was in favor of boundary layer failure with considerable quantities of polymeric debris remaining on the aluminum surface. Orman and Kerr [66] also showed that at very high or low pH values such as 1 or 13, respectively, the adverse effect of water on the bond strength is greatly diminished when exposed and tested at room temperature.

Most recently, a method has been suggested by Levi et al. [70] whereby hot water aging with mechanical testing at intervals can be

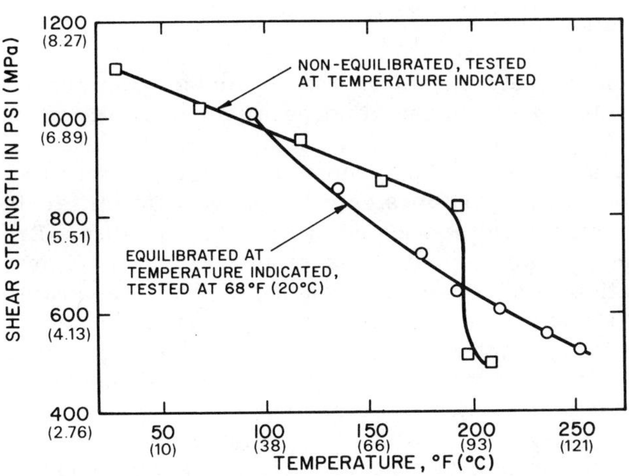

FIG. 3-13 Comparison of equilibrated and DNE (rapid stress corrosion procedure) test data for aluminum-epoxy joints.

used to estimate lifetimes of aluminum-adhesive bonds under constant stress. The method involves semilogarithmic plots of lifetime or aging time against mechanical property and use of a shift factor. An important factor in the test is to determine the tensile strength of the joints which have been soaked in 140°F water at the same 140°F test temperature. The usual procedure would be to test at room temperature and report the loss in tensile-shear strength of the joint due to the exposure under that condition. By using data totally developed at 140°F, the relationship of the log of bond failure time vs. stress/absolute temperature resulted in a straight line in every case for constant stress data. This is a requirement if the application of reaction rate theory to the mechanical behavior of adhesive bonds is to be valid [71]. These investigators showed that only a single data point relating bond lifetime as a function of constant stress need be determined in order to establish the shift in the aging time-strength straight line.

This procedure undoubtedly can minimize the test chamber and equipment tie-up time to produce some kind of answer. The assumption must be made, however, that such a single data point does have sufficient reliability to closely predict the performance that would be obtained from testing a significantly larger number of identically fabricated joints.

6. BOND PERMANENCE AS A FUNCTION OF SURFACE TREATMENT

6.1. General Relationships

The effect of surface pretreatment on the aluminum oxide surface is critical and intimately related to the bond permanence of aluminum joints [1,5,72-75]. It is difficult, however, to distinguish the benefit of different surface treatments in the absence of some environmental exposure since the joints will fail cohesively for many different surface-treating conditions when tensile-shear-tested without exposure. Extended exposure to accelerated or natural weathering conditions provides discriminating evaluation of most surface treatments. But the order of effectiveness of different treatments in promoting maximum bond permanence can vary significantly based on which weathering condition is used for testing. For example, the acid etching pretreatments rate high in promoting bond permanence in water-soaking-type testing. In corrosive weathering environments like salt water or seacoast atmospheres, the chromate conversion coating or anodizing pretreatments produce the more durable bonded joints.

Stressing plus environmental weathering can produce still a different rating among aluminum surface pretreatments. At the highest stress levels the magnitude of the stress tends to be more significant than the degree of weathering, resulting in early bond failures even under relatively mild water-soaking conditions. At low stress levels, and particularly in high humidity/high temperature or corrosive-type environments, the type of weathering condition can be the most significant factor in determining bond durability.

There has been increasing interest in recent years in nonaerospace structural adhesive bonding for many manufacturing applications. For these situations there has been a need to develop bond durability data for a wide variety of surface pretreatments with better cost effectiveness and less sophistication than FPL etching. Alcoa has long recognized that there could be more interest among the nonaerospace manufacturing producers in the simple vapor-degreased surface condition than in the FPL etch. Both kinds of pretreated joints can constitute a standard for comparison with all other kinds of surface pretreatments. Bond durability studies at Alcoa have usually included comparative durability testing with a given alloy and adhesive combination both with vapor-degreased and FPL-type etched adherends. Most recently, interest has been increasing in evaluating bond permanence in the presence of deliberate contamination on the aluminum surface such as the oils and lubricants used for forming parts.

Several joining situations also should be mentioned which have potential application both in aerospace and other manufacturing involving the combination of spot welding, riveting, or metal stitching with adhesive bonding. Once again, investigations have been conducted, including a range from sophisticated surface-cleaning pretreatments to those adding or leaving some contamination on the surface [76,77].

6.2. Effect of Specific Surface Pretreatments

In 1950 some of the earliest bond permanence results were published involving a variety of surface pretreatments by Eickner and Schowalter [78]. The use of a hot sulfuric acid-sodium dichromate solution for cleaning Alclad 2024-T3 aluminum alloy was shown to give bonds of more consistent high quality and salt water spray resistance than the use of abrasive, solvent, or alkaline cleaning methods.

Chromic acid anodizing as a surface pretreatment for aluminum bonding was reported in the literature by Eickner [79] as early as 1955. While not determining any joint durability values involving exposures, these initial bond strength results are still important because subsequent durability testing by Fokker, Bell, and other aerospace manufacturers has prompted the acceptance of chromic acid anodizing as a durable surface treatment for aluminum joints. Excellent bond durability test results also have been obtained at Alcoa with chromic-acid-anodized joints after exposure in room temperature water immersion and natural weathering in both the unstressed and stressed joint condition [9,10]. Eickner also concluded in these early investigations that the concentration of chromic acid (5 or 10%) and the length of the anodizing period (20 or 40 min) did not seem to significantly alter the adhesive bonding properties. Recent Alcoa results show that good bond permanence values based on a variety of exposure conditions were achieved with 5% chromic acid and 40 V held for 30 min to achieve a coating thickness of about 0.12 mil. Eickner also showed that boiling water sealing of the chromic-acid-anodized coating greatly interfered with bonding. In recent Alcoa tests, a tapwater rinse was used followed by a 10- to 15-sec hot deionized water rinse. Any attempt to boiling-water-seal such coating usually produces sealing smudges that can interfere with bonding.

Again in 1960, Eickner [80] reported on the bonding of bare and clad aluminum alloys with varying surface pretreatments as well as bonding stainless steel, chrome-molybdenum steel, and titanium. Superior bonding was secured with the aluminum alloys as compared with any of the other metals. The sulfuric acid-dichromate etch continued to show optimum results. Sulfuric acid anodizing followed by an Alumilite Process No. 205 sealing and chromic acid anodizing followed by a boiling water seal were compared with the acid-etched surface pretreatment. The bonds to the sealed sulfuric-acid-anodized surfaces appeared to be only slightly lower in bond strength than the acid-etched joints and were significantly higher than the sealed chromic-acid-anodized joints. Using present processing procedures for these same anodizing pretreatments, Alcoa results [9,10] show that the bond strengths and durability attainable with the chromic-acid-anodizing pretreatment is generally superior except in the corrosive-salt-water-type exposures. Also, the thicker anodic coating on sulfuric-acid-anodized joints invariably causes lower bond strengths to be observed using any adhesives capable of lap-shear strengths in excess of 27.6 MPa (4000 psi). However, the bond permanence of all anodized joints under corrosive salt-water-weathering conditions is much superior to the best acid- or alkaline-etching pretreatments, including the sulfuric-acid-dichromate-etched joints.

A number of bond permanence studies have been conducted at Alcoa which serve to demonstrate the effect of specific surface pretreatments. Because the aircraft industry has considered the deoxidizing of aluminum alloys with FPL etch, i.e., dichromate-sulfuric acid, as a standard since the time of World War II, it seemed appropriate to evaluate the performance of a wide variety of adhesive chemical classes on etched aluminum as a frame of reference. At Alcoa we have used a chromic acid-sulfuric acid mixture which should offer the same deoxidizing as the commonly employed FPL etch. Over a period of years a large variety of two-part room-temperature- and one-part elevated-temperature-cured pastes and heat-cured tape and film adhesives have been evaluated in three accelerated laboratory weathering and two natural atmospheric weathering environments [4].

For applications outside the aerospace industry, it seemed more apropos to compare the standard chromic-sulfuric-etching procedure with other surface treatments, including vapor degreasing, grit blast, amorphous chromate conversion coating, anodizing, field-applied acid paste etch, and phosphoric acid-alcohol etch. Tests were conducted for as long as 8 years under natural weathering conditions in an industrial setting and a seacoast environment. The laboratory-accelerated test conditions included continuous immersion in water or a wet-freeze-thaw-type cycle test for 2 years.

In Figs. 3-14 and 3-15, the effect of 2-year soaking in room temperature water can be compared using seven different surface pretreatments with a room-temperature-curing two-part epoxy. Among the nonetching-type surface treatments, the grit-blasted and anodizing pretreatment joints showed little change in bond strength at the end of 2 years soaking, while the vapor-degreased and amorphous-chromate-conversion-coated joints were both significantly lower in bond strength and continuing to degrade. Only the chromic-sulfuric-etched joints

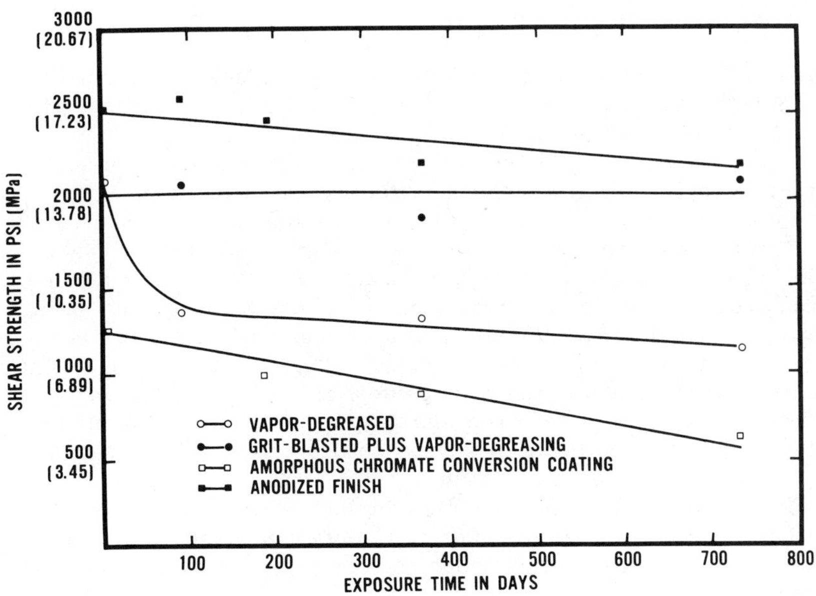

FIG. 3-14 Room temperature water immersion.

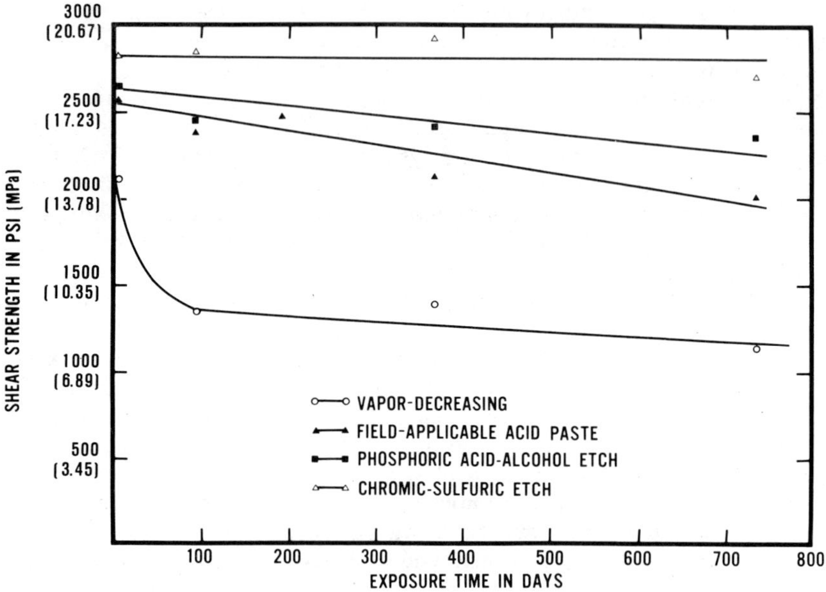

FIG. 3-15 Room temperature water immersion.

showed no degrading among the acid-etching surface pretreatments. The field-applied paste cleaner and the phosphoric acid-alcohol joints slowly lost strength during the 2-year immersion in water.

In Figs. 3-16 and 3-17, the effect of the more aggressive soak-freeze-thaw cycle environment [repeated cycles of 24 hr in 73.9°C (165°F) water, 24 hr in -34.4°C (-30°F) freezer, and 24 hr in 76.7°C (170°F) air oven] are shown. The vapor-degreased joints all failed within 100 days, while it took approximately 2 years for the grit-blasted or amorphous-chromate-conversion-coated joints to reach this same deteriorated condition. The sulfuric-acid-anodized surfaces were obviously much superior in performance. Again, the chromic-sulfuric-etching pretreatment was the best performer among the etchant pretreatments. As anticipated, the phosphoric acid-alcohol and field-applied etching paste treatments were both poorer surface pretreatments based on bond permanence, although the phosphoric-acid-alcohol-etched joints were the more durable of these two. Recent investigations at Alcoa [9,10] show that the anodized surface pretreatments formed the most durable joints of all under this very aggressive set of weathering conditions.

Bond permanence data based on natural industrial atmospheric exposure is shown in Figs. 3-18 and 3-19. Conversion coated and grit

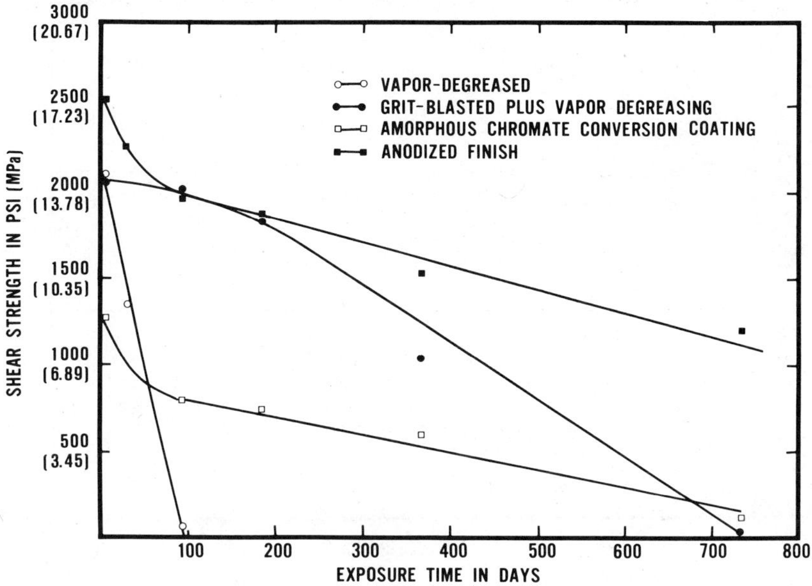

FIG. 3-16 Soak-freeze-thaw exposure.

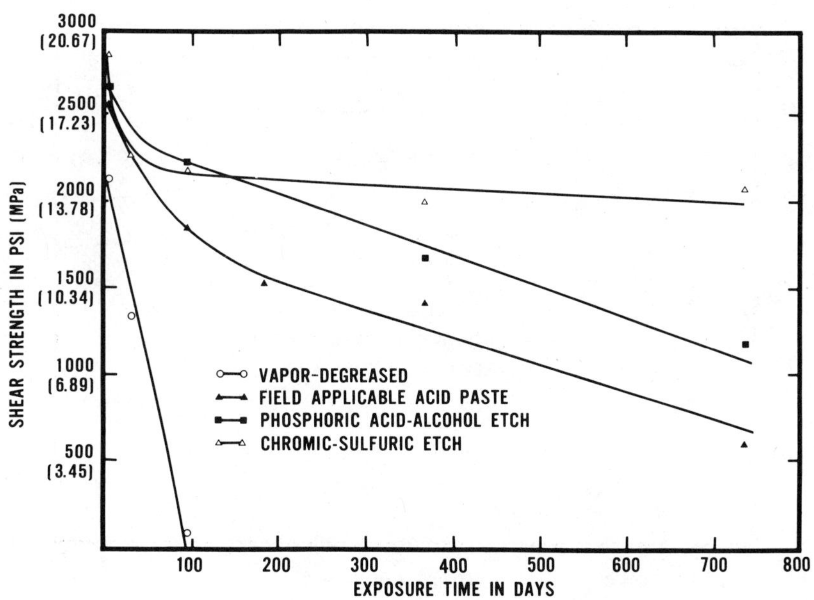

FIG. 3-17 Soak-freeze-thaw environmental exposure.

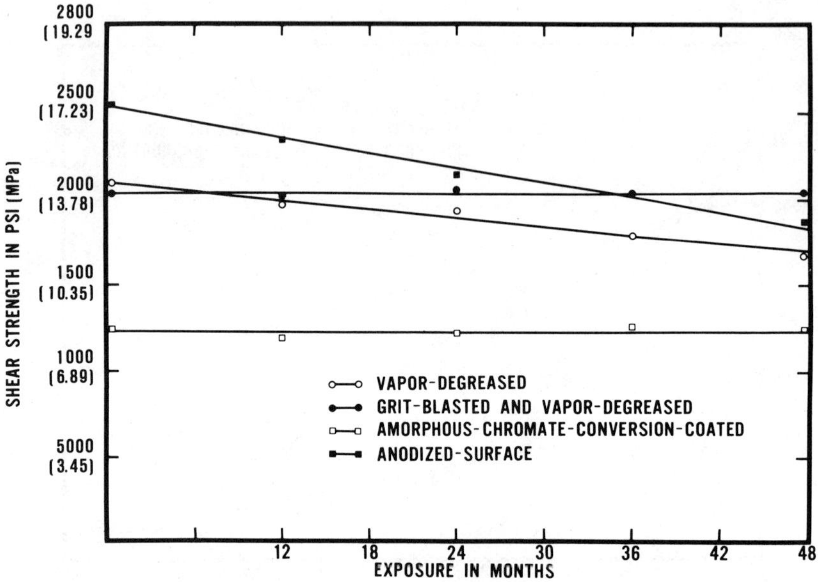

FIG. 3-18 Industrial atmospheric exposure.

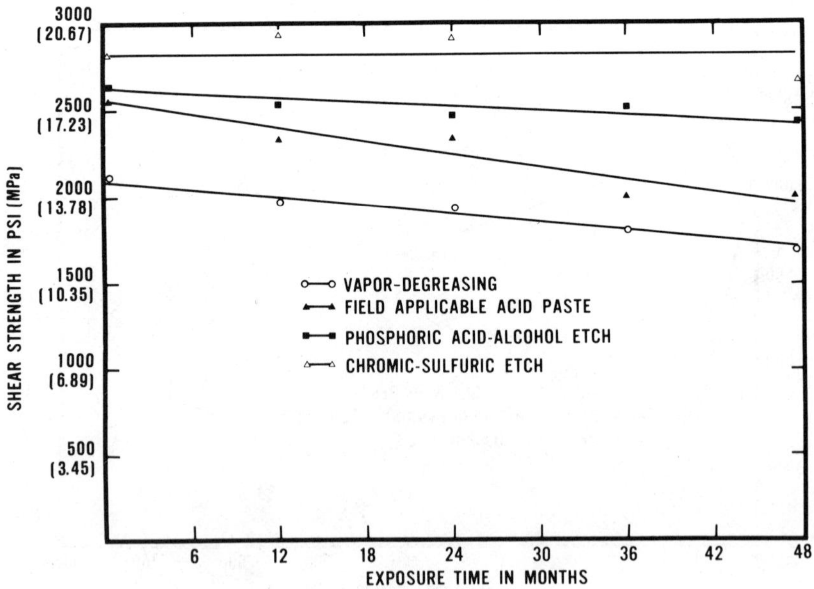

FIG. 3-19 Industrial atmospheric exposure.

blasted joint specimens were both unaffected. The vapor degreased and anodized joints declined only modestly after four years exposure. All the acid etched joints showed only about 10% declines in bond strength after four years exposure.

Seacoast exposure results in Figs. 3-20 and 3-21 show the type of exposure environment most degrading to acid-etched joints. It was anticipated that the performance of the anodized joints would be very good. This is verified by the good bond strength retention after 4 years exposure. The high retention of strength by the grit-blasted joints was rather unexpected since other Alcoa-processed joints exposed to the same environment with a belt-sanding pretreatment had previously performed much more poorly. There may be something unique in surface conditioning affected by the grit blasting which needs to be reevaluated in seacoast-type exposure. Conceivably, a coating of silica could be impinged into the surface, enhancing corrosion resistance. As expected, the vapor-degreased joints failed within a few weeks by undermining corrosion. The decline in strength shown by the amorphous chromate joints also was unexpected. It appeared to be more related to water debonding due to frequent joint wetting than joint weakening due to undermining corrosion. In other testing at Alcoa with joints fabricated with a heat-cured epoxy over chromate conversion-coated aluminum surfaces, there was no decline in bond

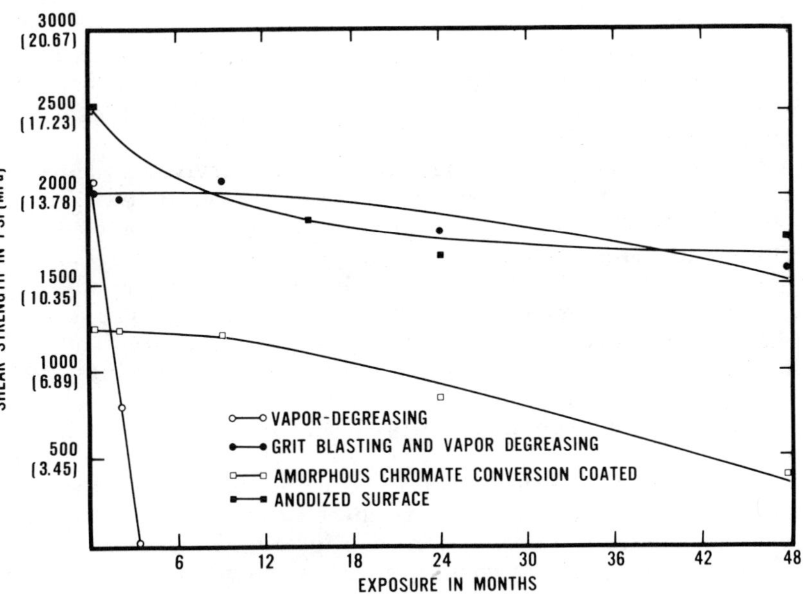

FIG. 3-20 Seacoast exposure.

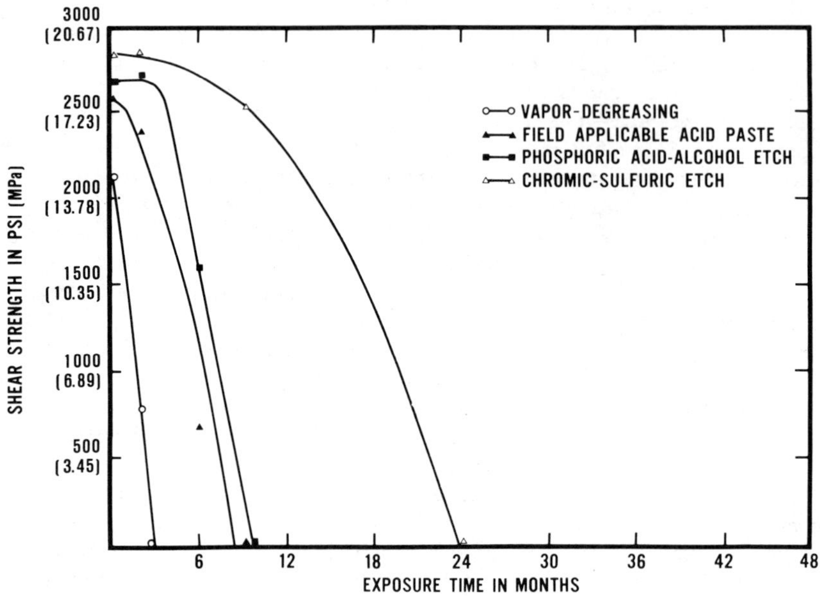

FIG. 3-21 Seacoast exposure.

strength after 8 years exposure in the same seacoast exposure. This demonstrates that superior surface wetting by the adhesive in the high-temperature-curing melt condition can result in significantly higher bond permanence.

A few of the chromic-sulfuric-etched joints survived as long as 22 months. None of the acid-etched joints performed well with all acid-paste-etched joints showing complete undermining corrosion by 10 months and the phosphoric-acid-alcohol-etched joints by 12 months.

Ross et al. [81] have investigated the suitability for bonding for a variety of conversion coatings on aluminum including Alodine 1200, Alodine 600, Iridite 14-2 Clear, and Iridite 14-2 Olive Drab and compared results with the FPL-etched surfaces. Medium thickness of Alodine 1200 coatings were superior to thinner or thicker ones based on peel tests, but tensile-shear tests showed insignificant variation. The joints were exposed to 95% RH at 49°C (120°F), but insufficient data were presented to make definitive conclusions between the coatings as to their long-term durability or comparative durability with other surface pretreatments. Conversion coatings have also been promoted on aluminum honeycomb for bonding durable sandwich panels. Askins and Schwartz [82] have demonstrated that these new technology aluminum honeycomb surfaces with corrosion-inhibiting surface treatments test superior to the "old" technology and bare cores based on environmental aging tests. Minford (see Table 3-1) has shown that Alodine 1200 conversion coating on 6061 alloy aluminum can be a very effective surface pretreatment for bonding, especially in seacoast weathering, provided the adhesive is cured at elevated temperature. Most recently, Minford [174,175] has shown that these good bond durability results for Alodine 1200-treated surface joints can be expected under stress and condensing humidity or stress and corrosive-salt-water-weathering conditions as well.

In recent years, there have been a variety of detailed studies of the effect of etching and rinsing procedures on aluminum which relate specifically to the durability of adhesive bonded joints. Time prohibits review of these many individual efforts in this general review. Therefore, a selection of articles pertaining only to the more fundamental aspects has been included here.

Russell [83] has performed a wide variety of tests involving variations in etching and rinsing procedures which have shown significant differences. In the reference work, he concluded that addition of ferric ion to a new etchant enhanced both strength and durability of 6061-T6 joints. The proof was offered that the quality of the oxide layer was dependent to a significant degree on the rinse water composition and contact time in the rinse-dry cycle following etching. Other approaches suggested for determining the quality of aluminum surfaces for bonding after pretreatment procedures have been the SPD device (Fokker Contamination Tester) and beta backscatter or radiochemical evaporative rate analysis techniques. Bijlmer [84] has concluded that while the surface potential difference (SPD) is not directly related to bondability, it does indicate a certain condition of the oxide layer which is responsible for the bondability of the surface. Jennings [85] has stated that the reproducibility of the surface state of carefully prepared aluminum was indicated by beta backscatter and radiochemical evaporative rate analysis measurements. There was a distinct lowering in backscatter counts from anodizing, and radiochemical retention was equal or greater on anodized surfaces than on chromate-etched surfaces.

Reference already has been made to the work of Bijlmer in Holland in 1973 [1,86,87]. He fully characterized and photographed the surface conditions arising on aluminum when a large number of dichromate-to-sulfuric-acid ratios were compared. It was clear from his pictures that the so-called microetch-type pit surface, most desired for durable adhesive bonding, was only produced by a few ratios. More recently, Russell and Garnis [88] monitored the FPL etching process by measuring the etching rate, the electrochemical potential of the etching reaction, and the electrical contact resistance of the surface after etching, rinsing, and drying cycles were completed. Unclad 6061-T4, 2024-T3, and 7075-T6 alloys were used because the cladding would introduce an uncontrolled variable due to diffusion of alloying elements from the base metal into the cladding during heat treatments. Also, thin spots in the cladding could conceivably etch through, causing a different surface to be exposed to the etchant solution. This present work and previous investigations within the aluminum industry has shown that alloy composition does have a definite effect upon the etching process. Russell and Garnis found that the etching rate, while different for different alloys, was relatively constant after the first minute of etching. It was shown that an initially high etching rate slackened to a constant rate during the first 1/2 minute of the etching cycle. The amount of specimen weight removed during this short time is at least one order of magnitude greater than the oxide ordinarily present on a mill finish surface, so most of the etching in a 10-min FPL etch cycle is involved in removing metal rather than old surface oxide. This latter part of the cycle then is involved with the action carried out to develop the specific type of surface structure that will be important for the bonding process.

In their study of the FPL etching procedure, Russell and Garnis simplified a major problem encountered in the past when new etchant compositions or processes are being evaluated to determine when the etching cycle is completed. This is defined as the point at which the surface structure, characteristic of the etchant-alloy combination, has been formed. An electrochemical cell was used to accomplish this based on the earlier work of Russell and Tanner [89] in which a curve formed by plotting the potential vs. etching time was used to determine the point in time when the etching reaction reached a steady state. The etching time predicted by this so-called Rapid Universal Sensing Cell for the FPL etch was 5.5-6.5 min at which time the same 2024 aluminum alloy surface recommended for bonding by Bijlmer [1] was achieved. This compares with the 10-min etch recommended in the ASTM standard [90].

Further evidence obtained by Bethune [91] using wedge-type crack propagation specimens also must be considered. Bethune demonstrated that freshly prepared standard FPL etchant does not produce optimum surfaces for bonding. Improvement in test performance of the bond was obtained when some aluminum and copper corrosion products were built up in the etchant solution. Both of these would be naturally formed in the continuing etching of 2024 alloy, but a system was developed for preconditioning a new solution prior to production use by deliberate addition of a certain amount of metal scrap. Best performance has been noted with 0.6 pHR aluminum and 0.05 pHR copper [14]. Russell and Garnis [88] developed the corresponding electrochemical curves using both new and preconditioned FPL etchant and showed the curves were significantly different.

The aluminum surface produced by etching is reactive. It is therefore expected that the time and environmental factors acting on this surface after etching can be critical ones as regards adhesion. This would naturally include the rinsing and drying operation which must follow each etching procedure. A common way to follow surface change on aluminum as regards film regeneration on the surface has been to measure the surface contact resistance changes. Russell and Garnis found that the longer the rinse-dry cycle, the greater the contact resistance. This clearly indicates that a reaction was occurring between the rinse water and the surface of the aluminum. The reaction was found to become more rapid and greater in degree as the purity of the rinse water increased. Comparing different alloys, these same authors found the reactivity in terms of increasing surface resistance varied significantly with the alloy reactivity in order 5052 > 6061 > 2024 > 7075. Since the rinse water was composed of either distilled or deionized water, the coating formed on the aluminum surface could only be forms of aluminum oxides. These kind of experiments clearly show that rinse water quality and rinsing time are important factors controlling the structure, composition, and thickness of surface oxide. In this regard, the work of Wegman et al. [92] showing the effect of water quality, rinse time, and temperature on bond strength is pertinent.

One of the other main approaches to improving the durability of aluminum bonded joints through alteration of the metal adhesive interface has been through the use of corrosion-inhibiting primers employing chromates. Sell [14] has pointed out the need for rather close control of the primer thickness as part of choosing the right combination of adhesion and mechanical characteristics necessary for applying corrosion-inhibiting primers. In general, the durability of primed joints increases with primer thickness up to the point where the primer begins to act as a separate layer. At that point, the properties begin to level off since the primer now is no longer as tough as the adhesive layer. Also, the primer thickness needs to be balanced with other physical properties such as low temperature peel, noting that the peel properties tend to increase as the primer thickness decreases. Reinhart [93] has recently reviewed the improvements in the area of primers as employed in the bonding of military airplanes and discusses novel concepts for priming metallic adherends for structural adhesive bonding. The relative effectiveness of many inhibitors is presented as they perform in Air Force screening tests.

Although not directly related to surface treatment of the aluminum, the durability of aluminum bonded joints can also be affected by the organic-fabric carrier where supported film adhesives are used. The carrier materials constitute yet another possible weak link in the interfacial bonding bridge area. Generally, nylon or polyester fabrics are used. Neither lend themselves readily to adhesive bonding. Accordingly, this material will usually be the weakest link in the bonding bridge for most bonding situations.

A carrier material usually will be used to improve handling, control bondline thickness, and reduce void population in the bondline in manufacturing bonded structures.

The influence of surface preparation and surface exposure time (SET) on the strength and durability of bonded 5052-H34 joints has been studied by Ross et al. [94] using both peel and shear test procedures. As anticipated, the peel test was the more sensitive to changes in the quality of the surface preparation over a 30-day exposure period. The influence of SET varies considerably with various combinations of adhesives and surface preparation, and the publication should be consulted for details. Similar studies were conducted earlier by the same authors for 2024-T3 [95] and 6061-T4 [96] alloys and again the original papers should be reviewed to obtain specific interacting relationships.

Mention should be made of an investigation by Smith [97] which demonstrates good bond durability potential based on the wedge testing procedure for a simple two-step processing consisting of a special degreasing and a proper tapwater or carbonate soak step. A dark film is formed on the surface of the aluminum, and the bondability is said to be equal to or better than the FPL etch or phosphoric acid anodize pretreatments. The treatment has been given the name STAB. It has been mentioned that one of the best surface pretreatments developed in recent years for aluminum is phosphoric acid anodizing. Ordinarily it would appear that this pretreatment would be difficult to employ anywhere but in a manufacturing situation. Locke et al. [98] have described a nontank phosphoric acid anodizing surface preparation procedure using a gelled phosphoric acid coupled with a stainless steel wire screen as the cathode-to-anodized-aluminum surface by applying a potential of 4 V for 10 min. This has been demonstrated and proved for field repair of aircraft and has ended in the publication of a *Standardized Repair Handbook for Aerospace Structures* by the Air Force Materials Laboratory.

Mention should also be made of the specific effect of fluoride contamination on the bond durability of aluminum joints investigated by Chen et al. [99]. It was found that fluorine contamination in water and atmosphere can accumulate readily on an etched aluminum surface, and only 0.2 monolayer of chemisorbed fluorine on the surface can lead to gross bond failures.

Finally, while again not directly related to aluminum surface treatment, the point should be made that intrinsically different bond durabilities exist in the different generic classes of adhesives. Assuming an adequate surface pretreatment for bonding, Sell [14] has ranked the generic classes of adhesives for metal bonding in a specific order. This list shows nitrile phenolics > high-temperature epoxies > 121°C (250°F) curing epoxies > 121°C (250°F) curing rubber epoxies > vinyl epoxies > two-part room-temperature-curing pastes with amine cures > two-part room-temperature-curing pastes with anhydride cures > two-part polyurethanes.

7. BOND PERMANENCE AS A FUNCTION OF ALLOY

Aluminum alloys have a wide variety of alloying compositions, strengths, and processing conditions that can affect bond permanence. First, there is a definite relation between bond permanence and the general corrosion resistance of an aluminum alloy. This has been repeatedly demonstrated in Alcoa investigations. This can be most readily observed where bond joints have been exposed to a corrosive test environment like continuous or intermittent salt spray. For example, when one of the higher-strength but less corrosion-resistant alloys such as 2024 or 2036 is used to make joints, their survival time in the seacoast must be only 25% as long as 6061-T6 alloy joints mentioned above. A recent investigation at Alcoa has shown that substituting 2036 for 6061 alloy lowered the joint survival time average in intermittent salt spray from 172 to 87 days due to the differences in the inherent corrosion resistance of the two alloys. Although not as significantly different as the salt spray results, the effect of alloy can also be observed in water-soaking-type tests as shown in Table 3-2. The relative corrosion resistance of these alloys is X5085 > 6151 > 2036. We might expect to find the bond permanence performances to fall into the same relative order and the results shown confirm this. Only the 2036 bonded joints delaminated in the salt spray environment and only the X5085 joints maintained good bond strength in the same exposure without special surface pretreatment. We also can see from this same data that the use of a chromate conversion surface treatment can improve the performance of all the alloys because it restricts the possibility for as rapid undermining corrosion of the bondline.

The above relationships are concerned most directly with the variation in alloy compositions and their effect on general corrosion resistance. The different strengths and processing conditions pertinent to different alloys also must be considered. The main importance of alloy strength, as it relates to bonding, can be observed by using different strength alloys to fabricate equal thickness adherend lap joints and testing them in a tensile machine. It becomes apparent that bond strength is a composite of several factors including thickness of the aluminum. With the same adhesive, it might be necessary to use a high-purity aluminum alloy in 0.625-cm (0.250-in.) thickness, a medium strength 6061-T6 alloy in 0.317-cm (0.125-in.) thickness, and 2024-T3 high strength alloy in 0.163-cm (0.064-in.) thickness to achieve approximately equivalent lap shear strength joints [100]. If the adhesive is brittle, then the use of lower-strength alloys at equal

TABLE 3-2 Effect of Surface Treatment[a] on Bonded Joints Using One-Part Epoxy Heat-Cured Adhesives

	Alloy 2036-T4			Alloy 6151-T4			Alloy X5085-H111		
Exposure condition	A	B	C	A	B	C	A	B	C
Int bond strength (psi)	1850	1930	2130	2550	2690	2530	2250	2270	2110
(MPa)	(12.75)	(13.30)	(14.68)	(17.57)	(18.53)	(17.43)	(15.50)	(15.64)	(14.54)
85% RH @ 23.9°C (75°F)[b]	380	313	2200	1350	1550	2410	1860	1890	1910
	(2.62)	(2.16)	(15.16)	(9.30)	(10.68)	(16.60)	(12.82)	(13.02)	(13.16)
100% RH @ 51.7°C (125°F)[b]	20	0	690	920	1050	1100	1260	1520	1230
	(0.14)		(4.75)	(6.34)	(7.23)	(7.58)	(8.68)	(10.47)	(8.48)
5% salt spray @ 35°C (95°F)[c]	0	0	1110	80	150	2090	690	530	1210

[a]A—mill finish; B—vapor-degreased; C—chromate-phosphate conversion coating.

[b]Three months exposure.

[c]Three weeks exposure.

adherend thicknesses will cause higher deflections in the joint at equivalent stress levels, resulting in apparent lower bond permanence values. It has been demonstrated at Alcoa, for example, that the apparent durability in water exposure testing of 0.163-cm (0.064-in.) and 0.317-cm (0.125-in.) chromate-conversion-coated 3003-H14 epoxy-bonded joints was significantly different even in the unstressed exposure condition [101]. Actually, the real difference in measured shear strength was due to the fact that the higher degree of flexing experienced in the testing of a thinner adherend lap-shear joint caused a lower apparent strength bond failure in the somewhat brittle converted oxide surface.

The effect of processing conditions on bond durability is related to the variation in oxide buildup on the adherend surface which can result from special heat treatments. All the aircraft alloys fall into this category. The thorough deoxidizing of the surface which is part of the standard process for adhesive bonding these alloys automatically removes the oxide buildup. Where relatively high-strength alloys may still be desired for general manufacturing purposes, this deoxidizing of the surface may not be affordable and a lower bond durability based on comparative weathering testing must be accepted.

8. BOND PERMANENCE AS A FUNCTION OF JOINT DESIGN

Significant differences exist in the ability of bonded joints to survive. Much depends on the kind of stresses to which the joint is exposed and the design of the joint to resist these stresses. Some of the general principles affecting joint design are shown in Fig. 3-22 where the relative ability of bonded joints to survive high tensile or shear loads is in sharp contrast to their poor resistance to high peel or cleavage loads. The diagram of the stresses imposed in a simple lap joint configuration shows clearly that the combination of decreasing stiffness in the adherend and increasing brittleness in the adhesive is conducive to decreasing bond durability when the bond is under stress. Figure 3-23 shows the effect of tapering the edge areas of the joint where the stress is highest. The resultant stress can be significantly lowered. While zero taper thickness is ideal, it is never realized in practice but the increase in bond permanence as the condition is approached is recognized.

Another special case of joint design is the normal vs. the idealized stress distribution in a butt joint. As can be seen in Fig. 3-24, some mechanical restraint around a butt joint is necessary to resist the cleavage stresses. A typical adhesive-bonded joint between two cylindrical pipes, i.e., a tube-in-sleeve joint, is the practical application of the above principle to use mechanical restraint in a butt joint.

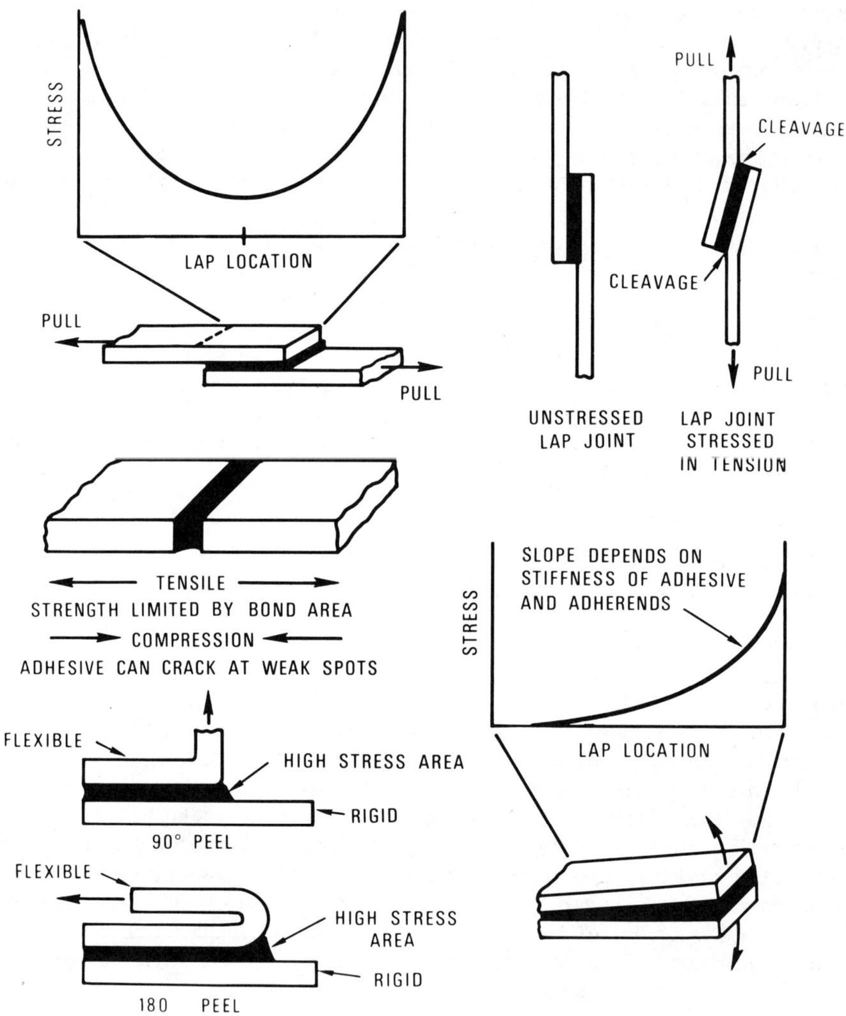

FIG. 3-22 Common stresses met in adhesive joints.

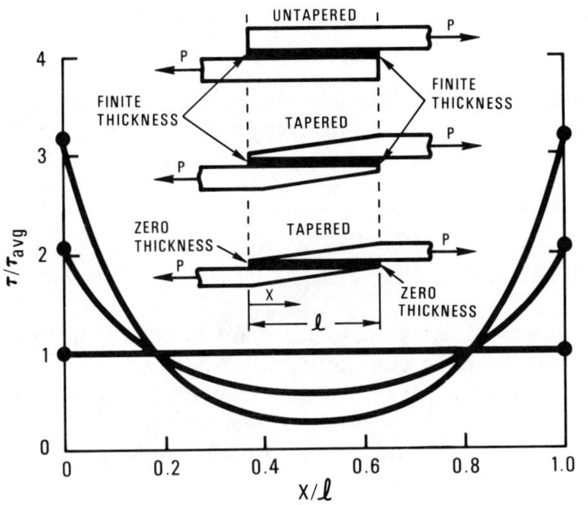

FIG. 3-23 Shear stress distribution in linearly tapered bonded joints.

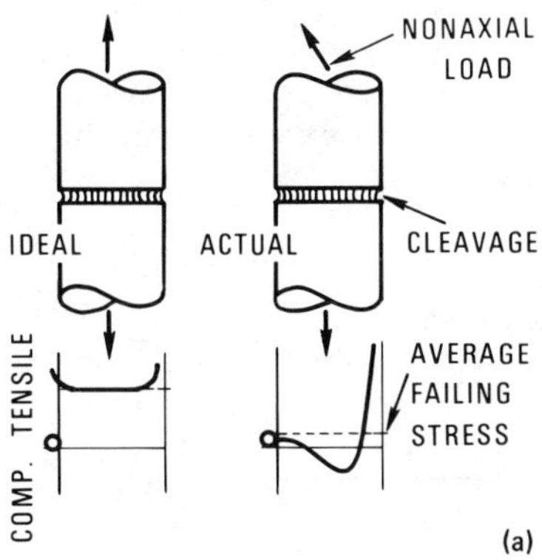

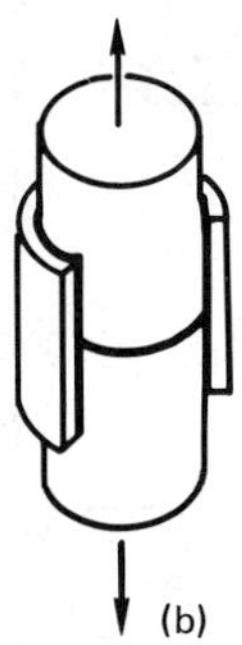

FIG. 3-24 Tensile stress in butt joints: (a) stress distribution under ideal and normal loading; (b) use of mechanical restraint to avoid cleavage stresses.

Joint design considerations for increasing the permanence of bonded joints have been considered in detail in many publications, some of which are offered for the reader's consideration [102-104].

9. BOND PERMANENCE AS A FUNCTION OF ADHESIVE SELECTION AND ADHESIVE CURING VARIATION

Choosing the right adhesive is vital to securing good bond durability. Certain generic adhesive families are conducive to creating higher bond durability than other families [4,14]. Also, there are significant performance differences within a single family of similar chemical description. These differences are readily identifiable by long-term testing.

In this regard, Figs. 3-25, 3-26, and 3-27 show the relative durabilities in aggressive soak-freeze-thaw exposure tests of different commercial epoxy adhesives. Three room-temperature-curing two-part epoxies in Fig. 3-25 show three different reactions to test environment. The joints made with one epoxy failed within 2 years. Another retained more than 50% of initial strength. The third epoxy product joints kept gaining in bond strength throughout the two-year exposure period. Among a group of five one-part heat-cured epoxies evaluated in Fig.

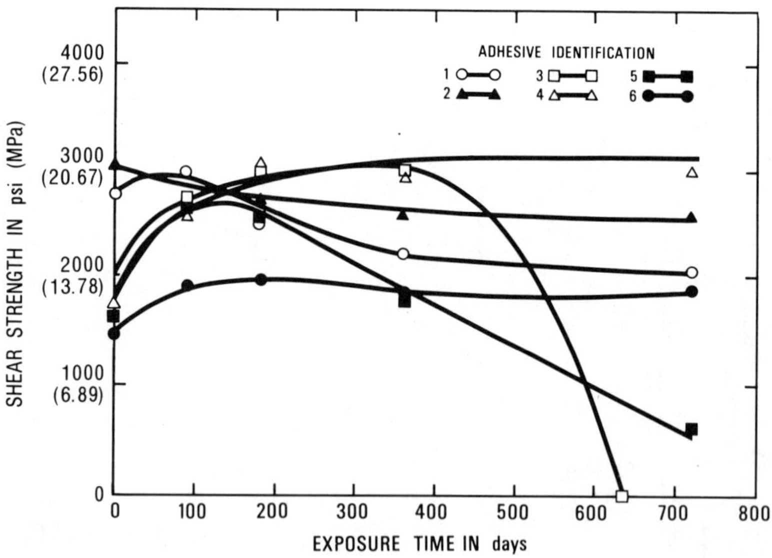

FIG. 3-25 Two-part epoxies—water immersion cycle.

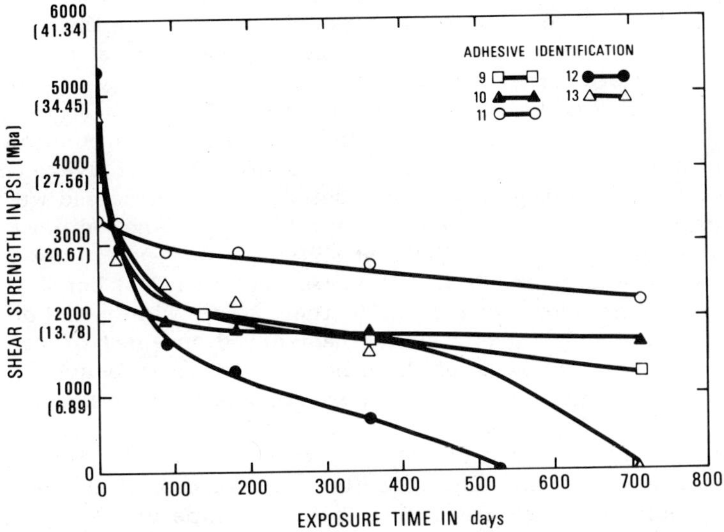

FIG. 3-26 One-part epoxies—water immersion cycle.

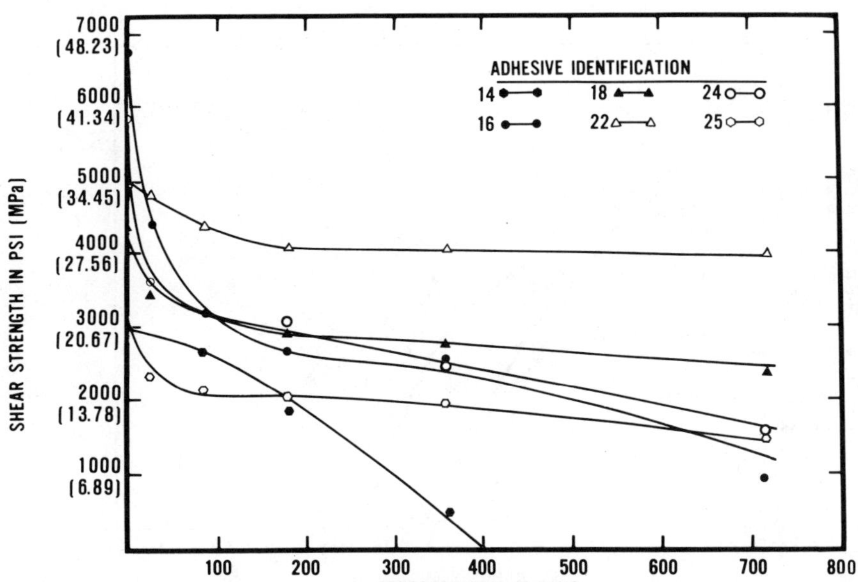

FIG. 3-27 Tape-and-film structural adhesives—water immersion cycle.

3-26, the joints from one product failed within 520 days, another failed within 2 years, while three others all maintained bond strengths in the vicinity of 13.78 MPa (2000 psi) after 2 years aggressive accelerated weathering. The more durable bond performance of a variety of proprietary tape and film adhesives is shown in Fig. 3-27.

Similarly, results shown in Fig. 3-28 are based on Alcoa bond durability tests at a very aggressive seacoast exposure site. Of five room-temperature-curing two-part epoxies evaluated, joints fabricated with four of the products had either failed within 2 years or showed very little shear strength retention. The best epoxy joints showed only modest decline in bond strength after 2 years, but then lost bond strength rapidly. As shown in Fig. 3-29, there was a wide variation in bond durability among a group of five heat-cured one-part proprietary epoxy adhesives. While all of these heat-cured epoxy bonds generally survived longer than the two-part room-temperature-cured epoxy joints shown in Fig. 3-28, only one out of five of the adhesives provided joints which survived the 4-year exposure period. Finally, in Fig. 3-30 we note that the joint durability performance of all the tape and film heat-cured adhesives were distinctly superior as a group, with all but one adhesive type joint surviving 4 years or longer.

The variation between different chemical families in some Alcoa test exposures in the seacoast atmosphere are shown in Fig. 3-31.

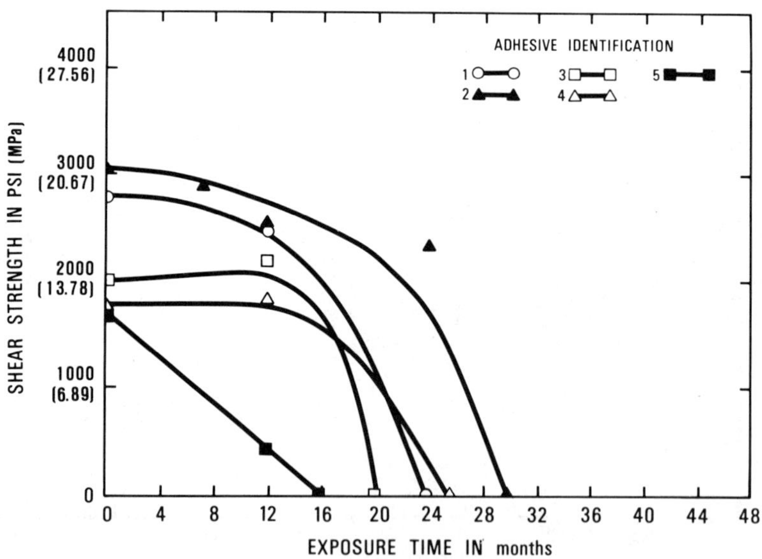

FIG. 3-28 Two-part epoxies—marine atmosphere.

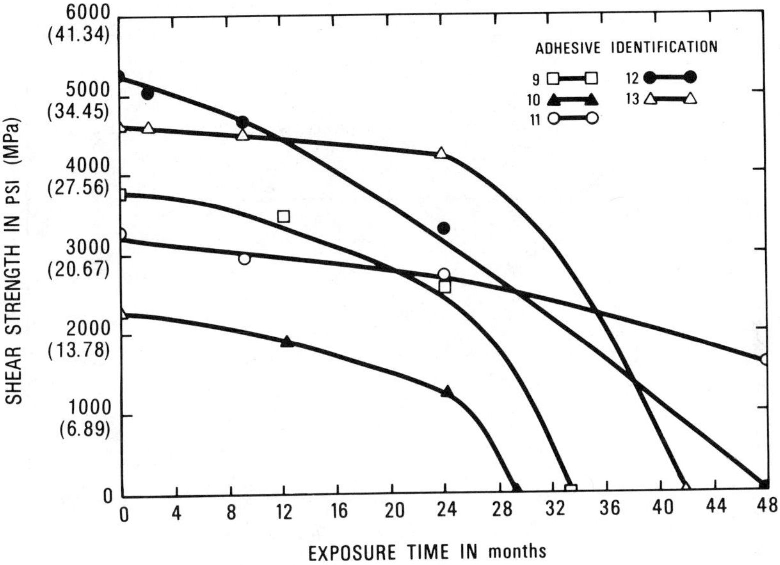

FIG. 3-29 One-part epoxies—marine environment.

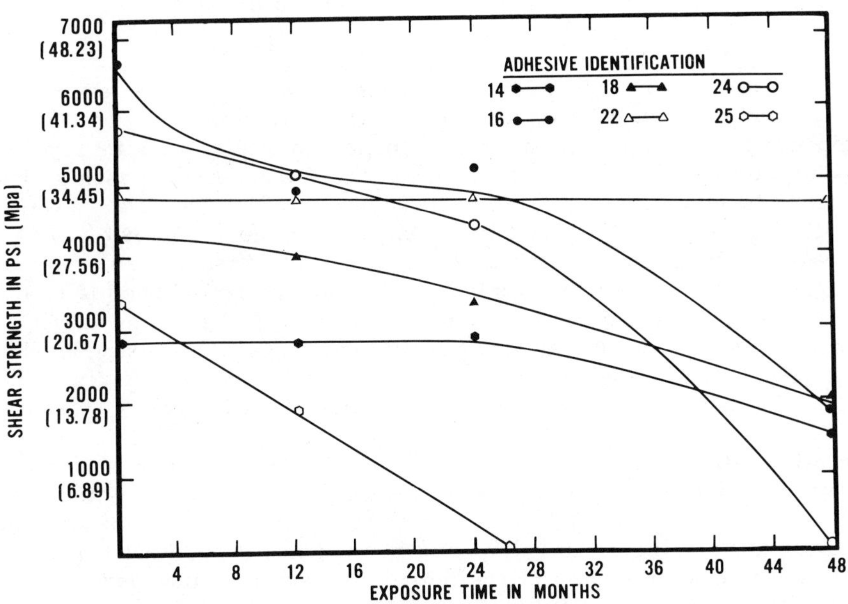

FIG. 3-30 Tape-and-film structural adhesives—marine environment.

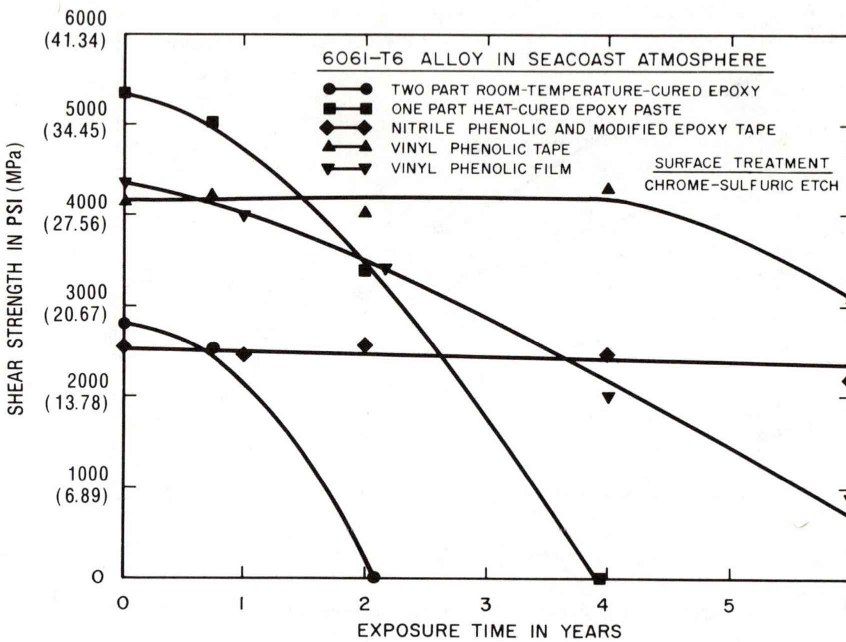

FIG. 3-31 Effect of adhesive variation on bond durability.

Bond durabilities ranged from 2 years with a two-part room-temperature-cured epoxy to virtually 100% bond strength retention after 6 years exposure for a combination nitrile-phenolic and modified epoxy tape. Seventy-five percent bond strength retention was obtained with a vinyl-phenolic tape adhesive.

Heat curing of a two-part room-temperature-curing epoxy may increase initial bond strength and overall water resistance due to increasing the polymer crosslink density. But there is little information in the literature about the possible deleterious effect on bond permanence of overheating an adhesive during the curing. An example of this was recently observed in an Alcoa study of adhesive-bonding automotive body sheet alloys using a one-part modified epoxy adhesive recommended by the manufacturer for this application. In this manufacturing situation, heat-curing conditions for the adhesive were not selected from the manufacturer's data sheet but rather by the temperatures and times required for the primer and paint bake on the vehicle. In Table 3-3, we note the lower bond permanence encountered with this adhesive when curing at the higher temperature and approximate time accumulated in the multiple primer and paint baking ovens. Preference would be for the second one-part epoxy shown in Table 3-3 with its broader curing spectrum. This adhesive produced joints with equal durability performance regardless of which curing schedule was used.

It might be appropriate to cite a recent study of the comparative durability between four potentially durable adhesives considered for bonding military hardware. Wegman [105] evaluated the durability as a function of stress and environment under both static and dynamic loading conditions. The use of the hot water soak-residual strength test to prescreen and predict durability rankings was deemed highly successful. Wangsness [106] has compiled durability data on a variety of different adhesive systems utilizing a sustained load-stress-testing technique which offers ranking with respect to the general chemical composition of the adhesives. Very detailed comparative evaluation of different structural adhesives under equivalent bonding and testing conditions has been conducted by Askins [107,108] as part of an Air Force contract. It was shown that the adhesives reacted differently on the optimized FPL or phosphoric-acid-anodized surface, but the primer under the adhesive had to be considered in explaining some of the performance differences.

TABLE 3-3 Variation in Bond Durability of Epoxy Adhesives with Different Heat Cures

2036-T4 Alloy

			Retained bond strength, %	
Epoxy	Cure conditions	Initial shear strength, psi (MPa)	100% RH @ 51.7°C (125°F), 4 weeks	5% salt spray @ 35°C (95°F), 3 weeks
A	1 hr @ 149°C (300°F)[a]	1670 (11.50)	100	0[b]
A	2 hr @ 177°C (350°F)[c]	1900 (13.09)	25	0[b]
B	1 hr @ 177°C (350°F)[a]	1930 (13.30)	64	55
B	2 hr @ 177°C (350°F)[c]	1950 (13.44)	61	58

[a]Manufacturer's recommended cure.

[b]All bonds failed within 7 days.

[c]Simulation of cure cycles for priming and painting.

10. CLAD vs. BARE ALLOYS AND BOND PERMANENCE

Clad aluminum alloys have been used in aircraft since the Second World War. It was only during the Vietnam War that a controversy arose regarding the durability of clad vs. bare aircraft alloy joints. The location of bonded joints on exposed surface areas was responsible for this problem. In addition, the concentration of aircraft in the Vietnam conflict along the seacoast areas and on aircraft carriers accelerated development of the problem. Cladding aluminum alloys was specifically developed to improve their corrosion resistance. Riel [109,110], Greer [111], and Rogers [112,113] have reported in the literature that clad alloys are more susceptible than bare alloys to bondline corrosion under aggressive salt water exposure conditions. This has resulted in publication of information of the type shown in Fig. 3-32. McFarlen [114] also has reported the results of an investigation of the same question using a variety of primer and adhesive combinations. He found no evidence in 5% salt spray exposure for preferential attack along the cladding interface.

Close study of Rogers' and McFarlen's papers coupled with recent Alcoa investigations has indicated the possibility that both investigators

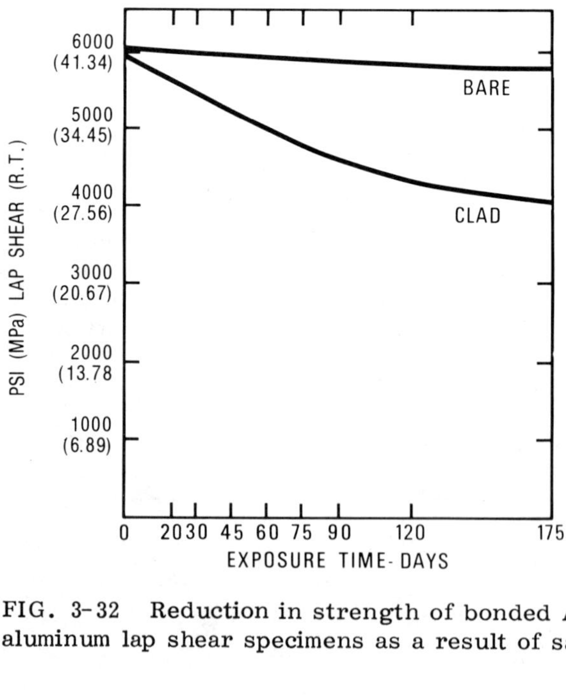

FIG. 3-32 Reduction in strength of bonded Alclad and bare 2024-T3 aluminum lap shear specimens as a result of salt spray exposure.

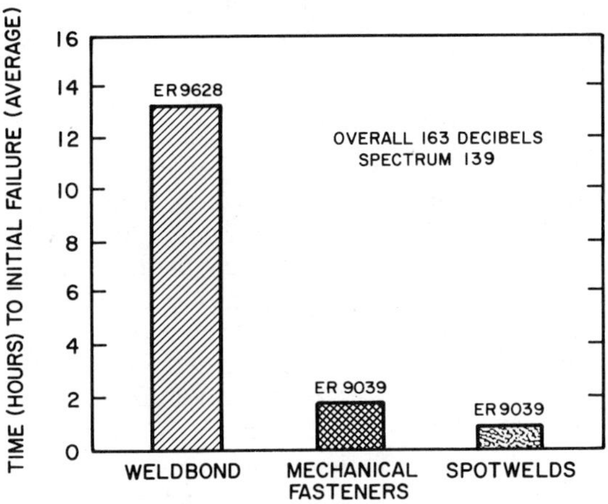

FIG. 3-33 "In a sonic fatigue endurance test, weldbond came through with flying colors when compared with mechanical fasteners and spot welds." (Source: Ref. 116.)

correctly interpreted their data. The kind of adhesive used and the degree of chemical wetting of the substrate surface secured can both be very important factors in determining whether the preferential corrosion occurs along the cladding interface at the bondline or elsewhere in the cladding layer.

11. BOND PERMANENCE OF COMBINATION ADHESIVE AND RESISTANCE SPOT-WELDED JOINTS (WELDBOND)

A joint produced by spot welding through an uncured adhesive should have some of the better properties of each joining method. Although a number of technical reports have been issued in recent years from various U.S. Air Force-sponsored programs, very little of this information can form the basis for direct comparisons of bond permanence with only adhesive bonded joints.

About 1967, attention was directed toward weldbonding in this country by the issuance of a 147-page translation by the federal government of a Russian document, *Glue-Weld Joints in Machine Building.* Government contract work at Lockheed-Georgia Company referred to the joint as a weldbond. In 1969, an article [115] focused on some of the initial conclusions of Lockheed-Georgia, including comparative sonic fatigue endurance data as shown in Fig. 3-33. Some weldbond durability performance attributed to Lockheed-Georgia was summarized in *Materials Engineering* in 1971 [116] stating that weldbond joints retained much of their strength in high humidity. For example, 70-day, 24-hr-cycle tests consisting of 16 hr at 100°F and 95% RH followed by 8 hours in ambient air reduced the strength of overlap joints to roughly 2700 psi, a 10% reduction. Weldbond joints also were successfully tested under low cycle flexure in liquid hydrogen without brittle fracture.

In 1971, Lockheed-Georgia issued an interim technical report [117] describing the evaluation of a variety of surface pretreatments, primers, adhesives, and weldbond processing variables. Relatively short-term weathering cycle tests and wedge bond testing after high humidity soaking were used to discriminate among the surface treatments and adhesives. In another technical report [118], Lockheed-Georgia showed significant differences in weldbond joint durability in salt spray depending on the type of surface pretreatment. While the spot weld etch was preferable for optimizing spot welds, joint permanence was significantly lower than for FPL-etched weldbond joints.

Mahon et al. [119] at Grumman Aerospace were working separately on methods for achieving durable weldbonds for 350°F service. They concluded that immersion chemical pretreatment of aluminum with HNO_3/Na_2SO_4 solution at 79.4°C (175°F) to 87.8°C (190°F) was preferable to immersion in $H_2SO_4/Na_2Cr_2O_7$ solution at 83.2°C (180°F) for development of class A spot welds and tough adhesive joints. Stress durability testing was conducted in spring-loaded fixtures with conditioning in 100% RH at 60°C (140°F) in accordance with MIL-A-83377.

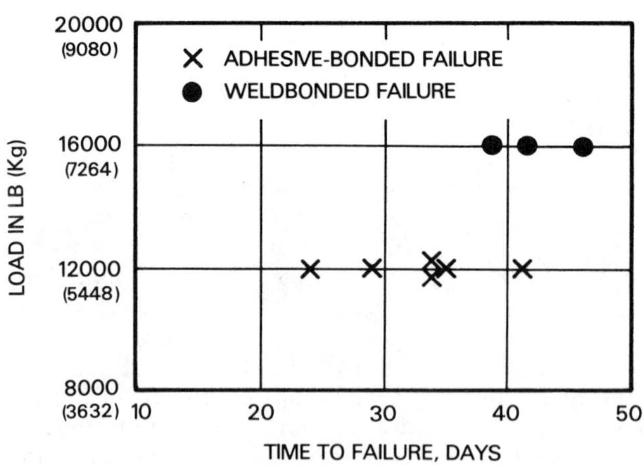

FIG. 3-34 Durability tests.

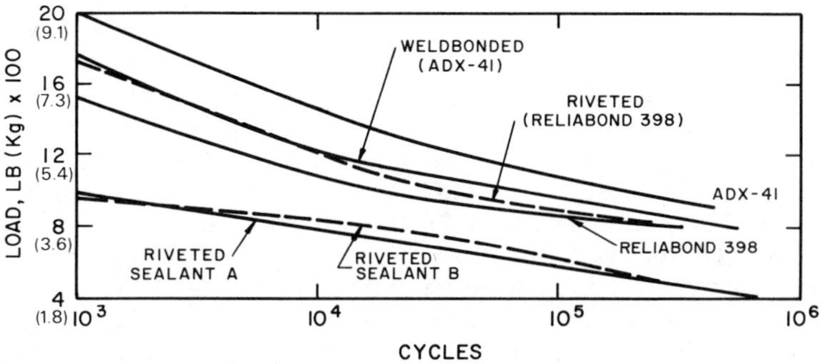

FIG. 3-35 Comparative fatigue strengths—overlap shear joints.

Joint survival times were compared with corresponding adhesive-bonded joints, with the results shown in Fig. 3-34. The average time to joint failure values were slightly longer for the weldbond joints at 80% of the respective ultimate bond strengths. The same investigators also compared the relative permanence of weldbonded vs. adhesive bonded joints fatigued per the general requirements of MMM-A-132, with the results shown in Fig. 3-35. The adhesive joints fabricated with one-part epoxy ADX-41 actually showed somewhat higher fatigue strength than the same adhesive in a weldbond configuration. A different adhesive, Reliabond 398, used with and without riveting showed the rivetbond to have slightly superior fatigue resistance.

Very recent technical information relating to aluminum weldbonding for aerospace applications is that of Wu and Bowen [120]. These investigators sought to establish a production method for weldbond including optimization of surface preparation, adhesive application, and spot welding variables to assure environmental durability and consistent spot weld quality. Earlier investigations by the same authors [121] involved the development of several improved surface treatments and selection of best adhesives including elucidation of the mechanism of weld expulsion with spot weld etched as compared to FPL-etched adherends. At first, Bowen and Wu had advocated the use of an FPL etch plus a 60-min sodium dichromate seal as the best surface pretreatment for aerospace weldbonding. Later, an improved treatment was developed consisting of degreasing, hot alkaline cleaning, deoxidizing in nitric acid, and anodizing in phosphoric acid-sodium dichromate solution at room temperature [121]. Based on durability in salt water exposure and minimum crack growth in wedge test specimens in 100% RH at 48.9°C (120°F), Bowen and Wu suggested B. F. Goodrich 0500PE130 adhesive with phosphoric-acid-sodium-dichromate-anodizing at 1.5 V for 20 min as the surface pretreatment.

Interest in weldbonding has spread in recent years to truck manufacturers [122]. Weldbond and adhesive bond joint durability comparisons were made by Hall [123] in 1974 referring to the current interest by the truck manufacturers but still using the aerospace FPL etch as surface pretreatment for the weldbonding and adhesive bonding. No difference in bond durability was noted between the adhesive bonded and weldbonded joints in 100% RH at 71.1°C (160°F), and both declined steadily in strength over a 6-month exposure. Some significantly higher bond strength retention by the weldbond-type joint was shown after 10 months as compared with the adhesive-bonded joint when cycled daily between 60% RH at 26.7°C (80°F) and 100% RH at 71.1°C (160°F).

Bond durability information seemed needed involving no special surface pretreatment of aluminum for those manufacturing operations such as truck and cars which cannot afford the cost of aerospace techniques. Accordingly, an investigation was undertaken at Alcoa. A technical paper in 1975 compared relative permanence of weldbond and adhesive-bonded joints [76]. This work was a continuation of earlier Alcoa investigations dealing with adhesive bonding of auto body sheet [29]. Some of this information has been summarized in Tables 3 4, 3-5, and 3-6. In Table 3-4, the positive effect on initial bond strength and on joint durability in high humidity exposure is shown incorporating a vinyl plastisol adhesive in a weldbond-type joint where no special surface pretreatment had been used. A more rigid and potentially higher-adhesive-strength polymer like a one-part heat-cured epoxy does not produce a similar positive effect in a weldbond, as shown in Table 3-5. In Table 3-6, the relative durability of weldbond joints in a continuous salt spray environment has been summarized for a vinyl plastisol, two-part polysulfide-modified epoxy and a one-part heat-cured epoxy. Both the vinyl plastisol and polysulfide-modified epoxy show the increase in comparative initial strength mentioned previously. Only the vinyl plastisol weldbond joints have superior salt spray resistance. None of these weldbond joints was made with special surface pretreatment of the aluminum. Because vinyl plastisols are not strong enough adhesives for aerospace applications, it is understandable why this positive effect of a weldbond combination on durability has apparently not been noted in the aerospace reports on weldbonding.

Another argument for weldbonding relates to fatigue resistance of joints. The improved fatigue life of adhesive bonded joints as compared with assembling aluminum aircraft with rivets has been long recognized. Manufacturers of trailers, vans, and trucks have, however, primarily used riveted construction because they have not been able to afford more sophisticated adhesive bonding techniques available to the aerospace industry. In the sequence of Figs. 3-36, 3-37, 3-38, and 3-39, the current fatigue strength data developed by Alcoa on comparisons between spot-welded, adhesive-bonded, and weldbond joints is shown.

TABLE 3-4 Effect of Weldbond on Vinyl Plastisol Heat-Cured Joint Durability in High Humidity Exposures[a]

Type joint	Average initial bond strength, lb (kg)		Overall strength, retention of metal joints, %			
			2036-T4		×5085-H111 alloy	
	2036-T-4	×5085-H111	85% RH at 24°C (75°F)	100% RH at 51.7°C (125°F)	85% RH at 24°C (75°F)	100% RH at 51.7°C (125°F)
Adhesive	910 (409.50)	860 (387.00)	76	70	78	70
Weldbond	1450 (652.50)	1510 (679.50)	99	97	89	93

[a]Exposure period was 3 months.

TABLE 3-5 Effect of Weldbond on Durability of One-Part Epoxy Heat-Cured Joint in High Humidity Exposures[a]

Type joint	Average initial bond strength, lb (kg)		Overall strength, retention of metal joints, %			
			2036-T4		×5085-H111 alloy	
	2036-T4	×5085-H111	85% RH at 24°C (75°F)	100% RH at 51.7°C (125°F)	85% RH at 24°C (75°F)	100% RH at 51.7°C (125°F)
Adhesive	1610 (724.50)	1220 (549.00)	100	59	94	76
Weldbond	1270 (571.50)	1210 (544.50)	100	64	95	74

[a]Exposure period was 3 months.

TABLE 3-6 Effect of Weldbond on Joint Durability in Salt Spray Exposure[a]

Type joint	Average initial bond strength, lb (kg)		Overall strength retention of metal joints, %			
			2036-T4 alloy		×5085-H111 alloy	
	2036-T4	×5085-H111	1 week	3 weeks	1 week	3 weeks
Vinyl plastisol	910 (409.50)	860 (387.00)	85	35	99	94
Weldbond (vinyl plastisol)	1450 (652.50)	1510 (679.50)	96	75	95	93
One-part epoxy	1610 (724.50)	1220 (549.00)	96	78	84	86
Weldbond (one-part epoxy)	1270 (571.50)	1210 (544.50)	100	77	84	86
Polysulfide epoxy	700 (315.00)	—	—	69	—	—
Weldbond (polysulfide epoxy)	1175 (528.75)	—	—	47	—	—

[a]5% salt continuous spray at 35°C (95°F).

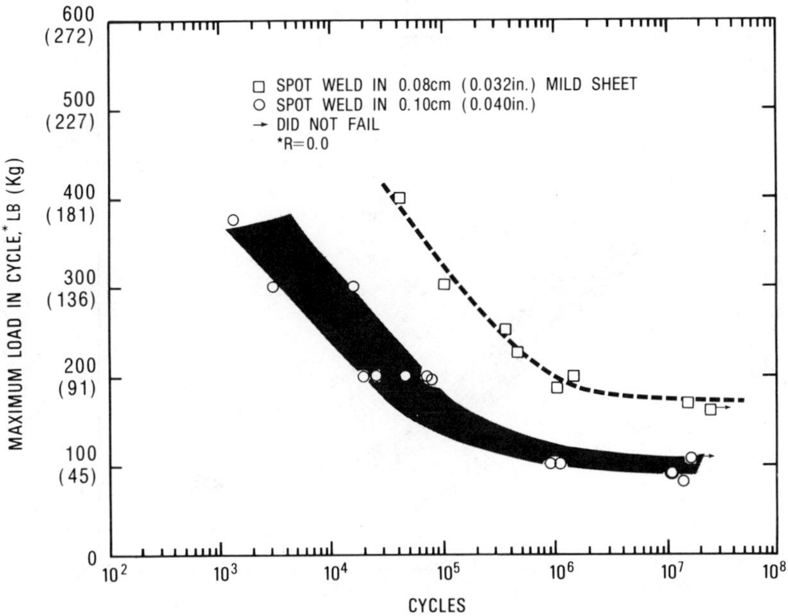

FIG. 3-36 Fatigue strength of lap joints in 0.10-cm (0.040-in.) 2036-T4 sheet spot weld.

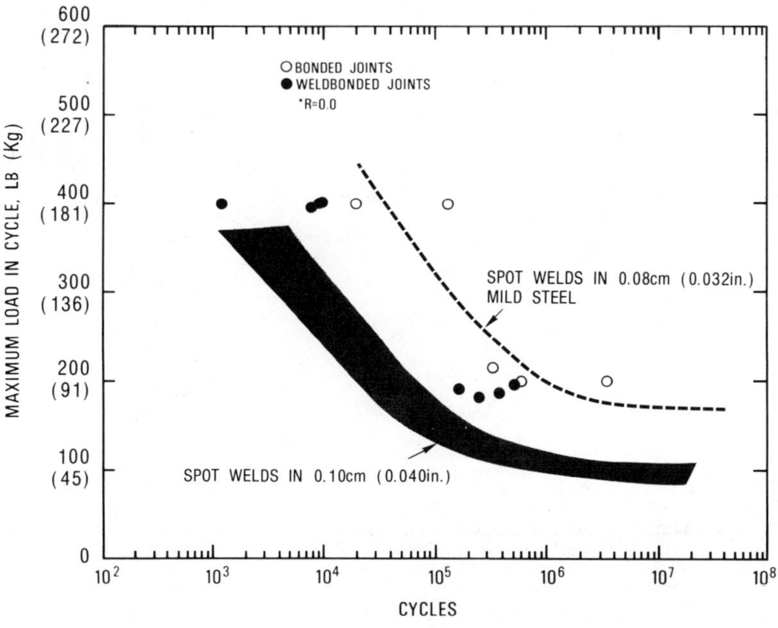

FIG. 3-37 Fatigue strength of lap joints in 0.10-cm (0.040-in.) 2036-T4 sheet polyamide-modified epoxy.

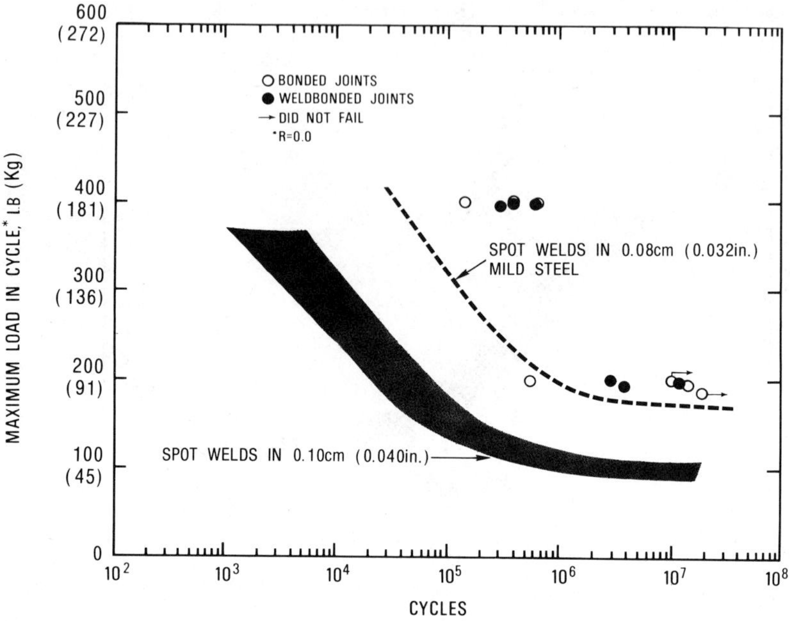

FIG. 3-38 Fatigue strength of lap joints in 0.10-cm (0.040-in.) 2036-T4 sheet high-shear/high-peel epoxy.

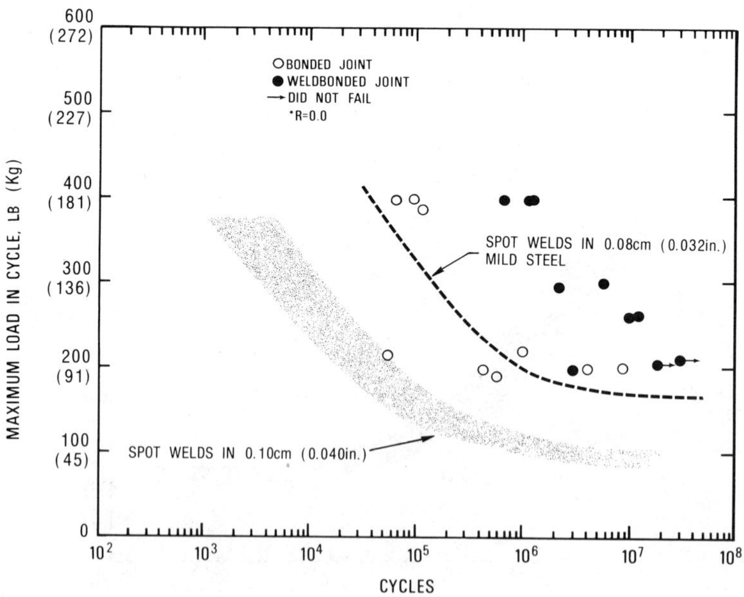

FIG. 3-39 Fatigue strength of lap joints in 0.10-cm (0.040-in.) 2036-T4 sheet room-temperature-curing polysulfide-modified epoxy.

In Fig. 3-37, the effect of weldbond with a two-part polyamide-cured epoxy, widely sold as a relatively flexible commercial epoxy, is shown. All the adhesive-bonded joints and the majority of the weldbonded joints had superior fatigue resistance as compared with the spot-resistance-welded joints. At the same time, a few selected bonded joints actually showed an order of magnitude of greater fatigue resistance than the corresponding weldbonded joints. Corresponding data for the highest peel strength two-part room-temperature-curing epoxy commercially available (39 pli T-peel) is shown in Fig. 3-38. These weldbonded joints had slightly better fatigue resistance at higher load levels but bonded joints had comparable fatigue resistance at the lower loads. Finally, in Fig. 3-39, data are shown for a two-part polysulfide-modified epoxy adhesive used prominently in the trailer and van industries for stiffening roof sheet by bonds to the extruded roof ribs. The most significant improvement in joint permanence, as based on higher fatigue resistance, was shown for the weldbonded joints fabricated with this adhesive. Thus, the properties of the adhesive seem to have an important effect in determining whether a weldbonded joint will be superior to, equal to, or inferior to an adhesive bonded joint in fatigue resistance.

12. EFFECT OF ENVIRONMENTAL TEMPERATURE ON BOND PERMANENCE

Because temperature is a common factor in all service environments, its effect on bond permanence should be considered. At the time of the Second World War, service temperature range of primary interest was about -55°C (-67°F) to 93.3°C (200°F). With the more recent development of aerospace applications, the range of interest has expanded from -252.8°C (-423°F) to 537.8°C (1000°F). Problems reflecting on bond permanence with varying temperatures are the result of stress concentrations and gradients developed within the joint. Some of the causes of stress concentration development are differences in the thermal coefficients between adhesive and adherends; shrinkage of adhesive in curing; trapped gases or volatiles evolved during bonding; differences in modulus of elasticity and shear strengths of adhesive and adherends; differences in thermal conductivity of adhesive and adherends; and residual stresses in the adherends as a result of release in bonding pressure. Operating at cryogenic temperatures, for example, can intensify stress concentrations generated as a result of the magnified differences in the physical and mechanical properties of adhesives and adherends at these very low temperatures. For example, a low-modulus adhesive may be able to stress-relieve itself at room temperature by deformation, but at cryogenic temperatures the modulus of elasticity may increase to the degree that this is no longer possible.

Some of the earliest comparisons made between adhesive joints at -252.8°C (-423°F) were published by Frost [124] in 1959. Of the 10 proprietary adhesives selected for test, the epoxy-phenolic adhesive showed the most uniform bond strength throughout the range from -252.8°C (-423°F) to 25.6°C (78°F), while a filled room-temperature-curing epoxy also showed a nearly flat strength curve slightly rising in bond strength as the test temperature was raised. A variety of vinyl-phenolic and rubber-modified phenolic adhesives all showed twice the bond strength at 25.6°C (78°F) as compared with -252.8°C (-423°F) and hence would not be recommended for service under cryogenic exposure conditions. Hertz [125] in 1961 reported additional testing under cryogenic conditions confirming good performance of epoxy-phenolics and ambient cured epoxies and lesser performance of nitrile-phenolics at -252.8°C (-423°F). The superior performance of room-temperature-cured polyurethane and heat-cured epoxy-nylon adhesives were also demonstrated at -252.8°C (-423°F).

The probable causes and mechanisms for the deterioration of metal adhesive bonds at temperatures up to 260°C (500°F) was investigated by Black and Blomquist [126] in 1958. These authors showed the significant effect of the metal surface conditions on the permanence of joints made with phenolic-epoxy-type adhesive. For example, stainless steel joints deteriorated rapidly at 260°C (500°F), while aluminum joints using the same adhesive were very stable. This bond durability of the aluminum joint was lost rapidly if iron or copper was deposited on the aluminum surface as a pretreatment. In earlier investigations [127-129], the same authors had been attempting to develop more-heat-resistant adhesives using phenolic and epoxy resins with significant results. It was found that bond permanence and strength at 315.6°C (600°F) was significantly affected by varying the ratio of the resins, the acid accelerators present, and surface stabilizers, i.e., chelating agents used to pretreat the aluminum surface.

The permanence of bonded metal joints at possible missile reentry temperatures was investigated by Olson and Lulling [130] in 1962. Testing at temperatures up to 537.8°C (1000°F), these investigators again confirmed a lesser bond permanence for stainless steel joints as compared to aluminum joints when tested out to 192 hr at 260°C (500°F) with selected adhesives. Only one adhesive had sufficient retained strength at 537.8°C (1000°F) to be evaluated. A significant negative effect of rapid heating also was noted where special means were used to achieve the peak test temperatures within 10 min as compared with the slower heating of the test joints in an air oven held at test temperature.

The state-of-the-art concerning adhesives for high and low temperatures was covered by Kausen in a review article in 1964. His summary comparing the best anticipated bond permanence based on the heat resistance and thermal aging properties of the most promising adhesive

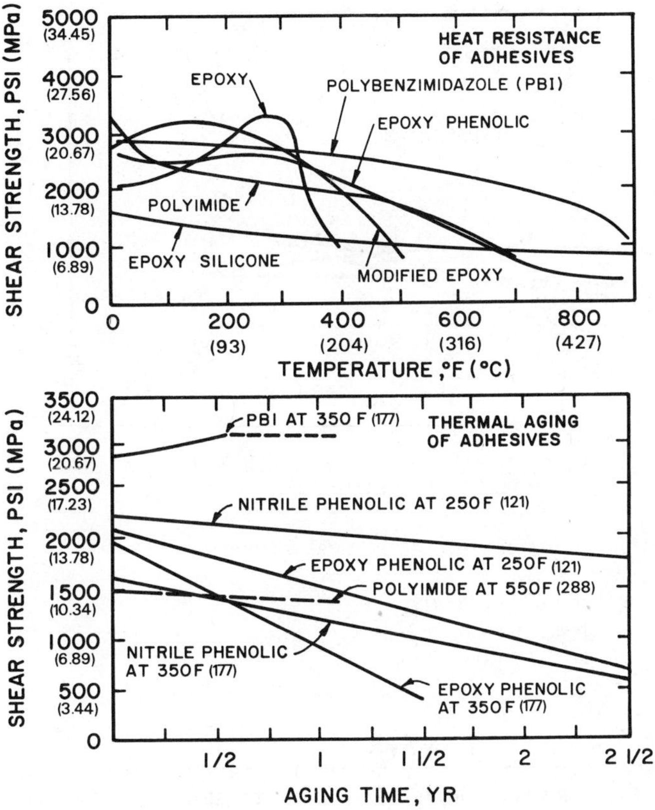

FIG. 3-40 Comparison of high-temperature structural adhesives shows that the polyaromatics (PBI and PI) have highest shear strength at temperatures above 427°C (800°F) (top) and best retention of strength on aging (bottom).

resin families is shown in Fig. 3-40. For best permanence under cryogenic service condition, the various adhesive resin families compared as shown in Fig. 3-41.

By 1965, remarkable progress had been made in developing heat-resistant resins and polymers as reported by Krieger and Politi [131]. Epoxy-phenolic adhesives were shown to be well suited for short time exposure even to 537.8°C (1000°F), with approximately 6.89 MPa (1000 psi) retained shear strength. While the short-term performance of this class of adhesives is very respectable even at 537.8°C (1000°F), these adhesives are still degraded sufficiently at 260°C (500°F) to be unsuitable for long-term applications. Krieger and Politi found that

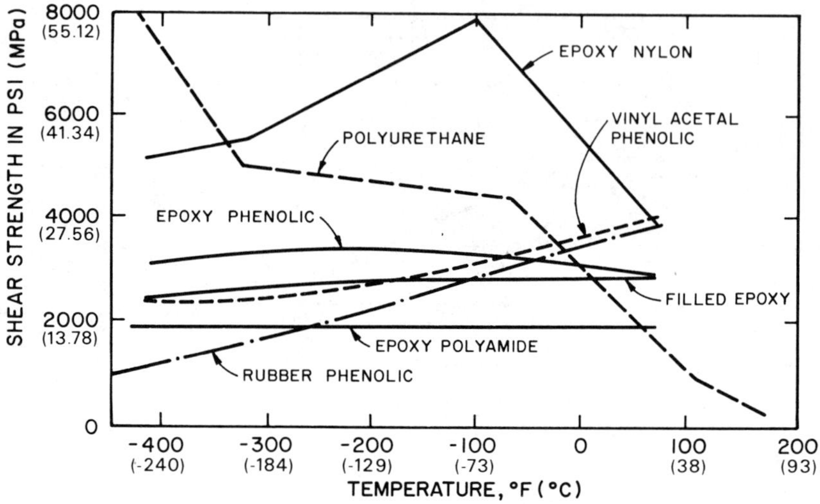

FIG. 3-41 Properties of cryogenic structural adhesive systems.

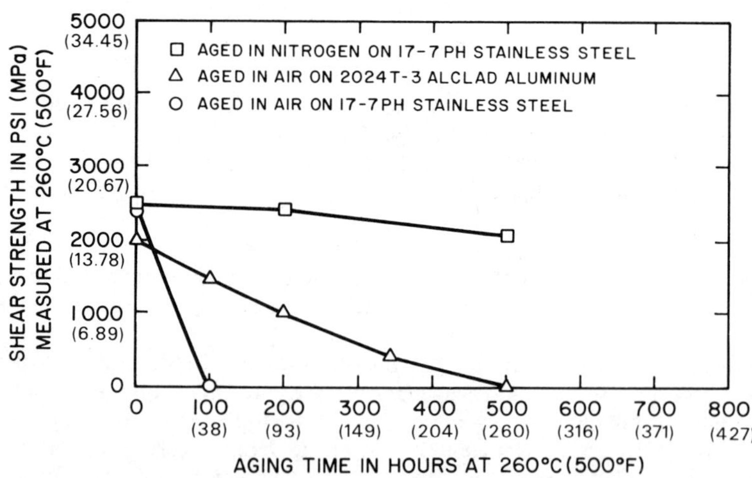

FIG. 3-42 Effect of aging at 260°C (500°F) on HT-424 (epoxy phenolic).

aluminum joints bonded with these adhesives exhibited significantly better bond permanence properties than found for bonded steel or titanium joints, as shown in Fig. 3-42. Other polymer modified epoxies, such as nitrile- or nylon-modified epoxies, though cured at 176.7°C (350°F), do not have good strength retention above 176.7°C (350°F). Better bond permanence for temperatures up to 537.8°C (1000°F) than the epoxy-phenolics is provided by both polybenzimidazole and polyimide adhesives, with the polyimide joints showing the most permanence after long time exposure at temperatures from 204.4°C (400°F) to 315.6°C (600°F).

13. BOND PERMANENCE IN NATURAL WEATHERING EXPOSURES

Beginning in 1955 and continuing through 1960, Eickner [132-134] began to report on the bond permanence of aluminum bonded joints exposed to exterior weathering conditions as compared with the laboratory type of accelerated weathering testing being conducted in various laboratories to qualify adhesives under military specifications MIL-A-5090B or MIL-A-8431. Two sites chosen for their aggressive environment were a site 50 yd from the seashore in Miami, Florida and another a quarter-mile from salt water in the Panama Canal Zone. General good performance of unstressed aluminum joints was found after several years exposure where the aluminum surface was prepared by etching in hot sulfuric acid-dichromate solution. The next logical step was taken to introduce some stress in the joint and reported by Eickner in 1960 [135]. At the time, in the absence of stressing fixtures capable of applying precisely known load levels, the joints were stressed by deflecting the specimen a measured distance in an arc over a center ridge support bar (Fig. 3-49). Some correlation also was sought in these investigations between laboratory-accelerated weathering testing and atmospheric exposures by exposing duplicate stressed and nonstressed bonded joints in a 30-day salt water spray test; 30, 60, or 90 days in 48.9°C (120°F) and 97% RH; or 15 days at 73.9°C (165°F) and 97% RH. Stressed joints tended to show lower retained bond strength. Some failed during exposure as compared to unstressed joints. The more corrosive accelerated salt water spray and the Miami seacoast exposures were noted as most deteriorating. The short exposure times to 165°C to 97% RH did not generally cause failure of any of the joints, possibly indicating that the stress loads on the joints must not have been very high.

Additional investigations dealing with the permanence of aluminum bonded joints in atmospheric weathering were reported in 1967 by Wegman et al. [136]. Only the use of a Panama exposure site was in common with the earlier testing by Eickner at the Forest Products Laboratory. A temperate climate was represented by exposure at the Picatinny Arsenal location in Dover, Delaware; a dry climate by exposure in Yuma, Arizona; and the Panama exposure was described as a "moderate jungle site." Fourteen commercial structural adhesives were used. Curing temperatures ranged from room temperature to 176.7°C (350°F). One room-temperature-curing polyamide epoxy and one 121.1°C (250°F) curing-modified epoxy adhesive were included in the temperate exposure using mechanical- and acid-etching surface pretreatments. After 1 year, it was concluded that the hot, dry Yuma exposure did not deteriorate and might elevate joint strength levels as compared with the control bond strength values after a year exposure under ambient laboratory conditions. Exposure at Picatinny generally resulted in only slight or modest joint deterioration. The tropical Panama exposure deteriorated all the joints to some degree and a few drastically. Duplicate joints also were exposed to MIL-STD-304 conditions (cycles from low air temperature to elevated temperature and 95% RH conditions) to determine possible predictive correlation between this lab-accelerated weathering test and any of the natural atmospheric exposures used in this investigation. The authors concluded that MIL-STD-304 exposure results tended to predict the trends noticed for joints subjected to high humidity

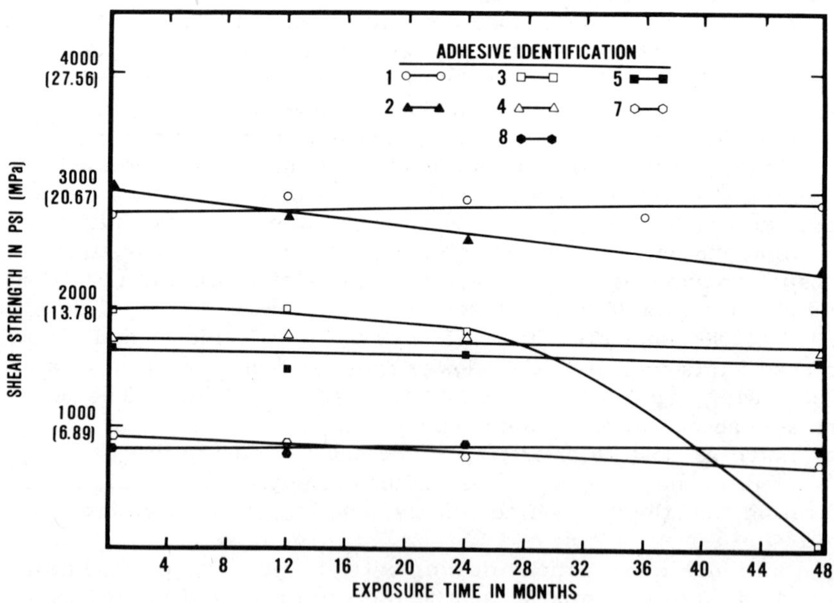

FIG. 3-43 Two part epoxies—industrial atmosphere.

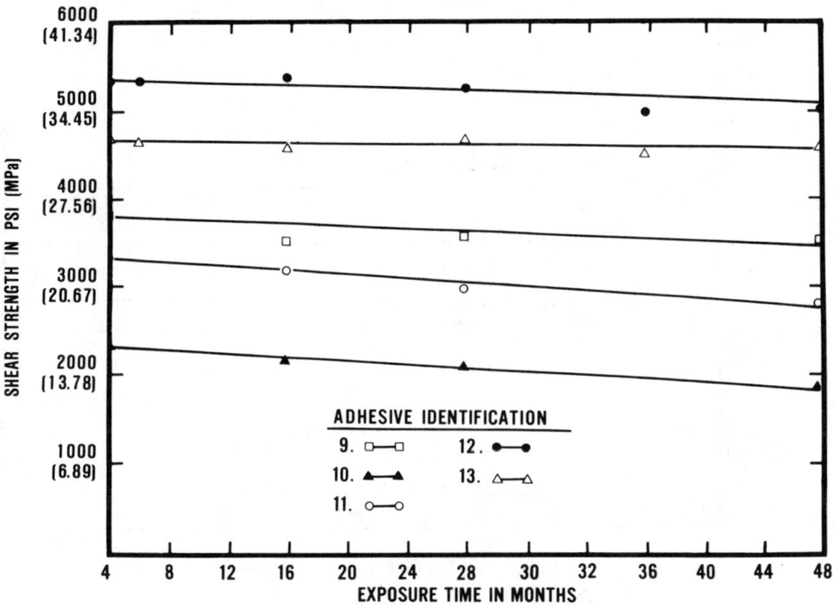

FIG. 3-44 One-part epoxies—industrial atmosphere.

exposure. Comparison of individual values in the report show wide variance. Similar joints in the accelerated lab tests were as much as 450% higher in retained bond strength or 35% lower than the corresponding adhesive joints exposed in Panama. Where the surface treatment was varied, the sulfuric acid-dichromate etching gave the highest initial strength and highest retention of bond strength joints following exposure in the temperate climate weathering environment at the arsenal.

A survey study of the effect of 1-, 2-, or 3-year natural weathering on aluminum joints in southern Florida was conducted by Hause et al. [137] in 1965. Room-temperature-curing two-part and heat-cured one-part epoxy pastes were evaluated as well as heat-cured nitrile phenolics, nylon epoxies, modified epoxies, and vinyl phenolics in film form. All joints were fabricated using acid-etched surfaces. Overlap shear, honeycomb peel, and honeycomb beam flexure testing were conducted. The general conclusion was that these bond strength properties were generally unaffected after 3 years exposure.

Many of the longest time evaluations of aluminum bond permanence conducted under natural atmospheric weathering conditions have been reported by Alcoa beginning in 1972 [3-5,9,10]. Figures 3-43, 3-44,

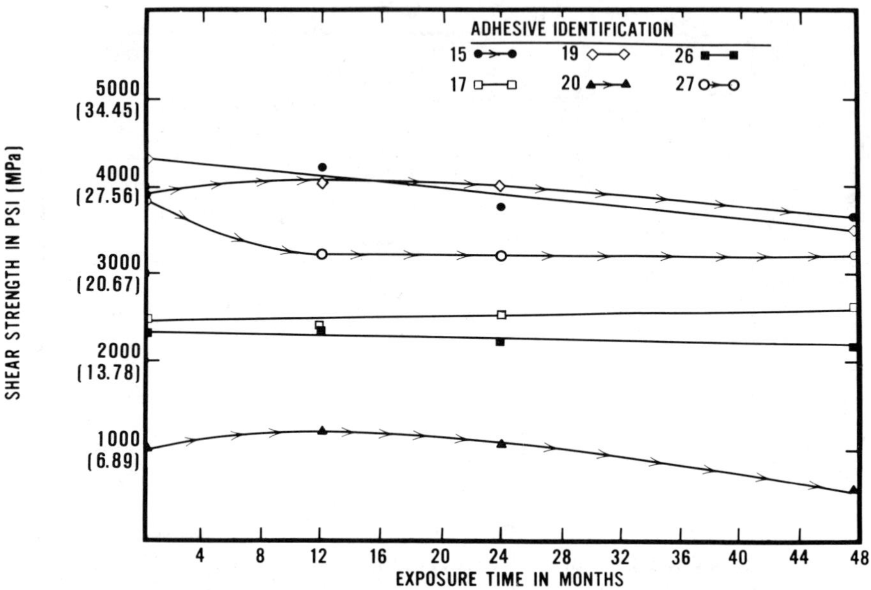

FIG. 3-45 Tape-and-film structural adhesives—industrial atmosphere.

and 3-45 are typical of the 4-year adhesive bond permanence values obtained with a variety of properietary two-part room-temperature- and one-part elevated-temperature-curing epoxies and various tape and film structural aircraft-type adhesives exposed to an industrial-type weathering. Additional Alcoa long-term weathering results in the industrial and marine atmosphere for periods up to 8 years are cited in Tables 3-7 and 3-8, where the relative performances of a room-temperature- or elevated-temperature-curing epoxy adhesive can also be related to variation in surface pretreatment.

14. APPLICATION OF FRACTURE MECHANICS TO STUDY OF ADHESIVE JOINT PERMANENCE

Fracture mechanics as a science has made it possible to define the toughness of materials in the presence of flaws. While originally applied to study of the toughness of solid materials, it has been adapted to study adhesive-b0nded sections which can fail by the propagation of a brittle crack through the bondline. It is probable that the adhesive joints in real structures are always blemished by flaws of a variety of

TABLE 3-7 Relative Effectiveness of Different Surface Treatments on 6061-T6 Aluminum Bonded with a Two-Part Room-Temperature-Cured Epoxy

	Exposure conditions							
	Industrial atmosphere lap shear strength in psi (MPa)				Seacoast atmosphere lap shear strength in psi (MPa)			
	Initial	2 yr	4 yr	8 yr	Initial	2 yr	4 yr	8 yr
Vapor degreasing	1,970 (13.57)	1,530 (10.54)	1,264 (8.72)	907 (6.25)	1,970 (13.57)	0	0	0
Grit blasting & vapor degreasing	2,030 (13.98)	2,060 (14.19)	2,030 (13.99)	1.930 (13.30)	2,030 (13.99)	1,800 (12.40)	1,580 (10.89)	1,350 (9.30)
Chromate conversion coating	1,260 (8.68)	1,230 (8.47)	1.198 (8.25)	1,130 (7.79)	1,260	835 (8.68)	430 (2.96)	0
Anodized[a]	2,530 (17.43)	2,130 (14.68)	1,868 (12.87)	b	2,530 (17.43)	1,660 (11.44)	1,690 (11.64)	b
Chromic-sulfuric acid	2,830 (19.50)	2,900 (19.98)	2,731 (18.82)	2,370 (16.33)	2,830 (19.50)	0	0	0
Phosphoric acid-alcohol etch	2,660 (18.33)	2,470 (17.02)	2,450 (16.88)	2,130 (14.68)	2,660 (18.33)	0	0	0
Acid paste	2,500 (17.23)	2,300 (15.85)	2,099 (14.46)	2,030 (13.99)	2,500 (17.23)	0	0	0

[a]Anodized in 15 weight % sulfuric acid at 21.1°C (70°F) and 12 A/ft^2 and sealed for 20 minutes in boiling deionized water (0.85 mil coating).

[b]Joints still under test after 7 years of exposure.

TABLE 3-8 Relative Effectiveness of Different Surface Treatments on 6061-T6 Aluminum Bonded with a One-Part Heat-Cured Epoxy

	Exposure conditions: Room temperature water lap shear strength in psi (MPa)			Cyclic wet-freeze-thaw shear strength in psi (MPa)		
	Initial	1 yr	2 yr	Initial	1 yr	2 yr
Vapor degreasing	4,330 (29.83)	2,000 (13.78)	1,170 (8.06)	4,330 (29.83)	820 (5.65)	0
Belt sanding & vapor degreasing	4,870 (33.55)	3,100 (21.36)	2,170 (14.95)	4,870 (33.55)	1,050 (7.23)	0
Chromic-sulfuric etching	5,330 (36.72)	4,270 (29.42)	3,530 (24.32)	5,330 (36.72)	1,330 (9.16)	0

	Industrial atmosphere lap shear strength in psi (MPa)				Seacoast atmosphere lap shear strength in psi (MPa)			
	Initial	2 yr	4 yr	8 yr	Initial	2 yr	4 yr	8 yr
Vapor degreasing	4,330 (29.83)	3,100 (21.36)	2,090 (14.40)	450 (3.10)	4,330 (29.83)	0	0	0
Belt sanding & vapor degreasing	4,870 (33.55)	4,130 (28.46)	3,670 (25.29)	3,400 (23.43)	4,870 (33.55)	0	0	0
Chromic-sulfuric etching	5,330 (36.72)	5,300 (36.52)	5,010 (34.52)	4,730 (32.59)	5,330 (36.72)	3,370 (23.22)	0	0
Chromate conversion coating	2,670 (18.40)	2,600 (17.91)	2,730 (18.81)	2,770 (19.09)	2,670 (18.40)	2,665 (18.36)	2,730 (18.81)	2,580 (17.78)

types and sizes, i.e., bubbles, dust particles, or unbonded areas. Furthermore, it is the progressive separation radiating from these flaws that basically controls the strength of the structure. Thus, it is somewhat unrealistic for investigators to evaluate structural bond permanence by using as near perfect joints as they know how to make for measuring the stress needed to fail the joints.

The fracture mechanics approach to evaluating joint permanence basically relates to the early work by Griffith [138] in 1920 with brittle materials like glass. Irwin in 1960 used the concept to measure a material's resistance to fast crack extension in the presence of a flaw, and reference is made to his work for more details [139].

The above background was applied to adhesive joints by Ripling et al. in 1963 [140] with the overall conclusion that the overriding factor in establishing the toughness of any bonded joint is the speed with which the crack moves along the bondline. Low toughness will be characterized by abrupt jumps of the crack, while slow-moving cracks predict a high degree of bond permanence, i.e., toughness.

In Fig. 3-46, the most sophisticated type of test specimen arising out of the Mostovoy and Ripling work is shown [141-151]. Known as tapered double cantiliver beam (TDCB), this bonded specimen can be used for quantitative evaluation of fracture toughness and for evaluations of bond permanence under various environmental weathering conditions. Unlike less sophisticated and less expensive uniform double cantilever beam (DCB) shown in Fig. 3-47, a constant stress on the propagating bond failure can be maintained. In the DCB specimen and the inexpensive specimen used in the "wedge test" (see Fig. 3-48), the stress load drops off as the crack propagates. To make direct comparisons between different sets of bonding parameters involving alloy, surface treatment, and adhesive, these more inexpensive crack-

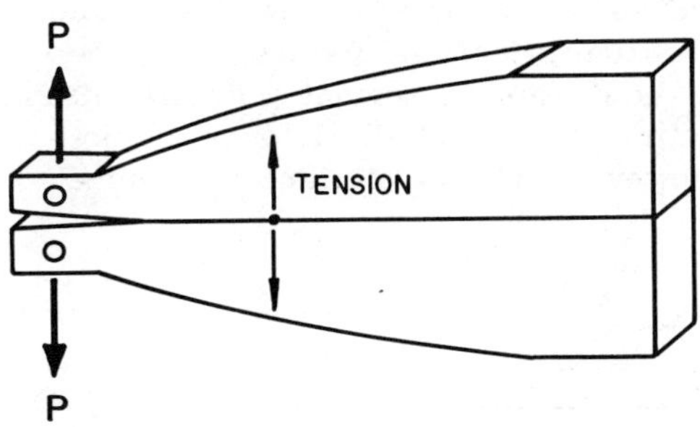

FIG. 3-46 Bonded tapered double cantilever beam (TDCB). Used for quantitative evaluations of fracture toughness and environmental durability effects on bonded systems.

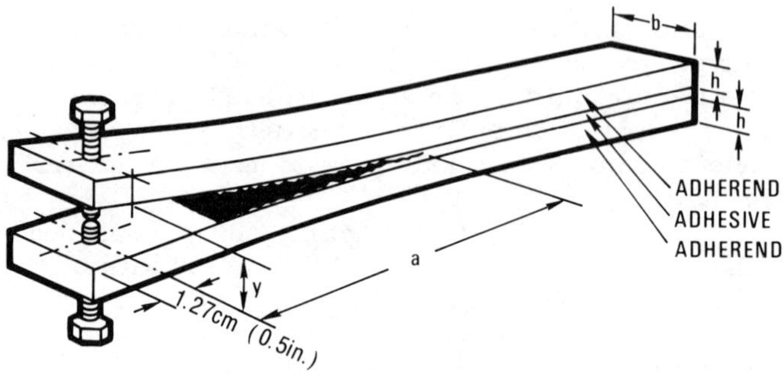

FIG. 3-47 Uniform double cantilever beam (DCB) specimen.

propagation-type specimens are satisfactory and widely used to indicate potential bond permanence.

The general theory states that bonds, like homogenous solids, fail because they contain flaws that cause load variations of the stress field to induce crack growth. Once this preexistence of cracklike flaws is accepted, the problem of bond failure reduces to determining the condition or conditions under which a flaw will enlarge. If the flaw is at the loading edge of the bond, then the stress is intensified at these locations. In studies, a crack is deliberately initiated and allowed to arrest. The whole test device is placed in some environment, and the rate of crack growth with time is reported. The location of the bond failure also can be noted. Bethune at Boeing showed that a 176.7°C (350°F) curing nitrile phenolic adhesive that had performed satisfactorily in aircraft for better than 25,000 hr of operation failed consistently within the adhesive in a crack propagation test. At the same time, a 121.1°C (250°F) modified epoxy which showed sporadic disbonding in aircraft showed eventual interfacial failure using the same procedure. It is rare that the durability service potential of an adhesive is known before the testing is conducted. It does promote confidence in the test method, however, when a distinct difference

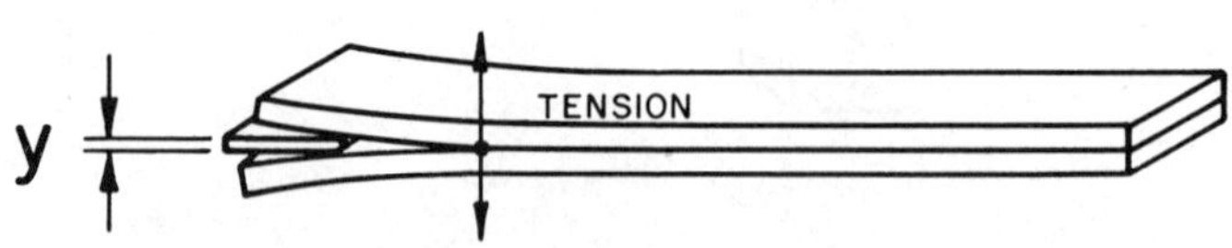

FIG. 3-48 Thin adherend uniform DCB ("wedge test"). Not for quantitative evaluations because of possible plastic deformation of adherends.

in bond failure is reproduced which is like the disbond situation in service.

In recent years, much attention has been focused on the so-called wedge test [152-154]. It can be used to distinguish between adherends, surface treatments, and adhesives by measuring the relative rate of crack growth in the exposure environment. Two experiments tried at Boeing are exemplary. Sections of actual aircraft were removed where disbond was present and other sections where no disbond was evident. When wedges were inserted in the bondline and the structures were exposed to 100% RH at 60°C (140°F), complete fracture of the bondlines occurred in less than an hour for those sections where disbond had been noticed earlier. The sections free from disbond exposed to the same conditions remained intact for many hours.

In a second experiment, the surfaces to be bonded were treated with various chemical conditions present in the sulfuric-acid-sodium-dichromate-etching tanks. The crack propagation rate was studied as the conditions of the bath were varied. The necessity to maintain certain dissolved levels of aluminum and copper were established for treating aluminum aircraft alloys in order to obtain optimum bond permanence values. This procedure has been equally applicable to studying effects of time and temperature in etching and even the effect of rinse delay of the acid-etched adherends. Boeing has used the criterion of only accepting bonding conditions which lead to a crack propagation rate of less than 1.9 cm/hr (0.75 in./hr) in a 100% RH at 48.9°C (120°F) exposure condition.

The fracture mechanics parameters for a variety of different adhesives were studied by Mostovoy and Ripling in 1975 using the tapered-double-cantilever-beam-type specimen [155]. Included in the evaluation were six properietary heat-curing structural adhesives of the following categories:

Adhesive type	Curing temperature, °C	(°F)	Curing time, min.	Pressure during cure, MPa	(psi)	Type film
Modified nitrile epoxy	121.1	(250)	60	0.27	(40)	Supported
Nylon epoxy	143.3	(290)	60	0.27	(40)	Unsupported
Nitrile phenolic	176.7	(350)	50	0.70	(100)	Unsupported
Modified epoxy	137.8	(280)	60	0.14	(20)	Unsupported
Modified epoxy	137.8	(280)	60	0.14	(20)	Supported
Modified epoxy	121.1	(250)	60	0.17	(25)	Unsupported

The overall behavior of the adhesives as a group was similar and all displayed a high level of resistance to monotonically increasing loads (G_{Ic}) over the test temperature range from -17.8°C (0°F) to 93.3°C (200°F). At room temperature and below, none of the adhesives seemed to be sensitive to water under sustained loads; but above 37.8°C (100°F), their resistance to water under static loading decreased rapidly. At 65.6°C (150°F), only the nitrile-phenolic adhesive cured at 121.1°C (250°F) showed good flaw tolerance, i.e., G_{Iscc} less than 1 in.-lb/in.^{2}. The decrease in flaw growth resistance was particularly pronounced for the nylon-epoxy system. Loads above this G_{Iscc} value result in cracking rates that are excessive, and catastrophic failures result in relatively short times. In metals, the fatigue growth rates even above a threshold value G_{th} are sufficiently slow that metal structures can still be designed for finite service life. Adhesives, however, generally crack sufficiently rapidly above the G_{th} value that finite service life design does not appear to be feasible. It was apparent from this work that the critical material property for adhesives is their resistance to fatigue crack growth, particularly if the structure is expected to experience an alternating load.

Bascom and coworkers [156] have used a special combined-mode adhesive joint specimen which is a scarf-type joint to study the development of microcracks and their propagation and interaction with main crack failures. The locus of bond failure in such joints was in the adhesive resin but less than 1000 Å from the interface. This near-interfacial failure was judged to be the result of the stresses directing the failure into the interfacial area. If this is true, then apparent interfacial failures of this type should be expected for many joint designs when the loading results in both tensile and shear stresses acting in the bond.

Much evidence exists that apparent interfacial failures are nonetheless cohesive in that a thin film of the adhesive is left on the substrate. Jennings [157], Mulville [158], and Wilcox and Jemian [159] all achieved such failures using butt joints loaded in tension, butt joints loaded in shear, and thick-bent lap-shear specimens, respectively. Still there have been reported cases of interfacial peel that appear to be true adhesion failure, that is, molecular separation along the adhesive-adherend boundary. Systems reported by Huntsberger [160,161] seem to fall into this category.

Since the development of a crack and its extension is so critical in bond durability evaluations, it follows that some effort needs to be made to establish the feasibility of automatic detection and classification of bondline defects in adhesively bonded structures. Loew et al. [162]

have been working in this area with multiple-layer bonded structures. Detection of a bond flaw was expanded to include its classification as an unbond, void, or porosity condition. The evaluation was a scan with a 0.5-in. transducer in the pulse-echo mode. Flaws could be found with accuracies of 91% as to their presence and 92% as to identifying the type. A so-called false-alarm rate of 19% was found where no defects existed, but in only 2% of the cases was a defect called a nondefect. The procedure worked regardless of the layer depth of the flaw. Clark [163] has worked with specimens with intentionally induced adhesive bondline flaws of the type designed and built in the PABST program representing the highest present state of bonding technology. His tests showed that bondline flaws loaded primarily in shear did not grow, while those loaded by peel or tension forces did exhibit flaw growth. Flaw growth rate was higher at the slow rate of 2 cycles/hr in fatigue testing as compared with a faster cycle rate of 30 Hz.

Jones and Kaeble [164,165] have conducted investigations for several years directed toward the determination of fundamental material parameters that are descriptive of fracture behavior of adhesive joints. Methods were developed to measure physical-mechanical and fracture behavior of adhesive joints under carefully controlled environmental conditions at several temperatures. The high-temperature-resistant adhesives did not display transitions in mechanical properties at room temperature, and thus did not display crack propagation in fatigue experiments conducted at room temperature. However, the crack growth rates could be altered by exposing the adhesive bonds to various chemicals which apparently could favorably compete with the adhesive by adsorption for the various interfaces in the joint. Other chemicals could diffuse into the adhesive and plasticize the adhesive at the crack tip. Stable crack growth was found in fatigue testing in each test media, and at low-cycle rates the test media effect decreases. It was believed that the corrosion effects of water at the adherend interface could increase the fatigue crack propagation rates in long-term fatigue.

Bascom et al. [166] have been part of a Navy program for more than 10 years directed toward developing an adhesive fracture test methodology and data base. Their work has demonstrated that fracture testing of structural adhesive joints is fundamentally more complex than the fracture testing of monolithic structures, as shown especially by the very strong bond thickness effect and the fact that the testing must be done under combined stress conditions. The authors conclude that major work remains to be done on developing the analytical relationships for the effect of joint geometry on adhesive fracture behavior.

15. BOND PERMANENCE AS A FUNCTION OF SIMULTANEOUS EXPOSURE OF THE JOINT TO STRESS AND ENVIRONMENTAL CONDITIONS

15.1. General Considerations

It is difficult to explain why so few bond-permanence-testing investigations involving sustained stress were described in the literature prior to Sharpe's technical disclosures in 1965 and 1966 [167,168]. The effect of simultaneous stressing and weathering on bond permanence was quite dramatic and needed to be studied in more detail. It has certainly always been recognized that periodic or sustained stress was present for every bonded joint in an assembly.

Eickner and coworkers at the Forest Products Laboratory made some of the earliest disclosures about the permanence of bonded joints under various humidity conditions with the joint under stress [135, 169,170]. In these tests, a steady stress was maintained in a rather qualitative though accurately described manner, as shown in Fig. 3-49. The dramatic bond failures shown in the later work of Sharpe was not evident in these early tests. Hence, the significance of high stress with environmental exposure was missed. The quantitative method of imposing and holding a sustained stress using calibrated spring jigs, as employed by Sharpe, is costly and somewhat more artificially contrived than simply bending a lap joint as used by Eickner. The use of the spring-loaded jig does guarantee that a known stress is being imposed and that the stress will not be relieved by accommodations in the joint itself, as can occur in the bent lap joint technique. Some early investigators also used a cleavage lap shear test specimen as shown in Fig. 3-50. Because a lap joint would never be used in pure cleavage by design, these valves would not have much significance.

Another early investigator in simultaneous stress and weathering was Carter [171], who first placed stressed aluminum joints in a Florida natural weathering environment in June 1964. However, the results

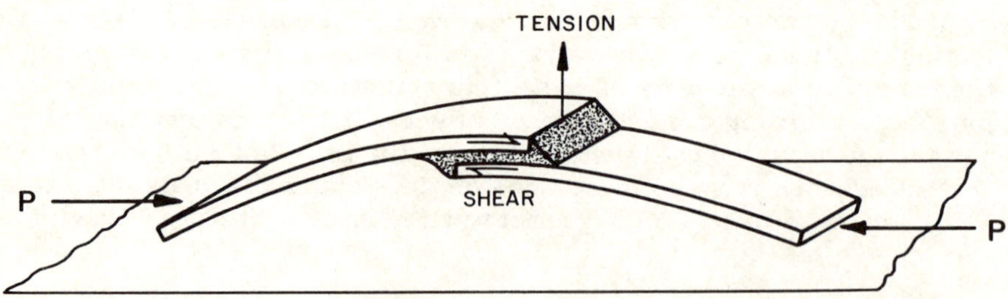

FIG. 3-49 Standard lap shear in bending. Specimen [2.54 cm (1 in.) wide] or entire assembly is bent, and time to failure recorded.

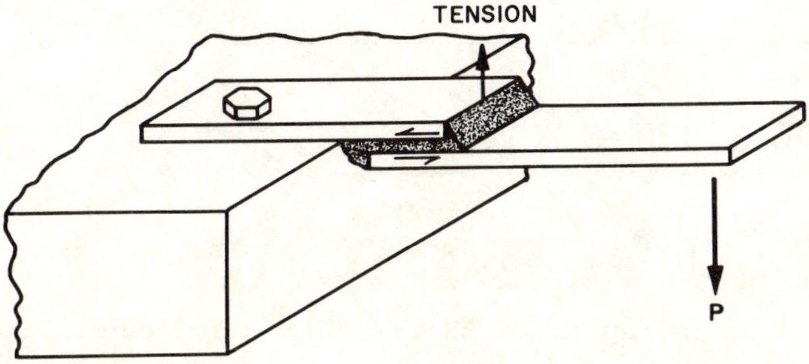

FIG. 3-50 Cleavage lap shear time to failure recorded.

of these tests were not reported in the literature until 1967. It is clear from Carter's 1967 paper that he was aware of the dramatic bond failures that could occur even in a natural weathering environment with epoxy-bonded aluminum joints if the stress on the bond was high. Also in 1966, Carter published a paper describing the rapid failure of adhesive peel joints under stressing in water conditions [172]. Carter further reported that air having a relative humidity as low as 50% could reduce the durability of many structural adhesives under stress. Carter did not employ either the bent lap joint of Eickner or the spring-loaded Sharpe jig in his Florida weathering tests. He bolted groups of six lap joints together and suspended them from a frame with weighed concrete blocks attached to the bottom joint.

More recent evaluations of bond durability under stress and weathering conditions have been conducted by Wegman [173] using a modified Carter T-peel specimen or a stressed weathering frame with multiple joints under spring tension. Minford [3,4,10,174,175] has used the Alcoa stressing ring procedure (see Fig. 3-53), while a variety of Boeing investigators [6,91,152-154,176-179] have worked with various methods of propagating the cracks initiated in the interfacial area (see Fig. 3-46, 3-47, 3-48). Krieger [180] and Coleman [181,182] have also conducted studies involving the rate of crack propagation in bondlines.

Krieger has also been active in developing techniques to make stress analysis of metal-to-metal joints. Yurek [183] has focused on the testing of blister-detection-type lap joints (see Figs. 3-51 and 3-52) strung together in a stressing frame with an attached spring replacing the concrete block dead load stressing employed in earlier investigations. While there is no clear understanding of all the mechanisms involved in adhesive bond stress corrosion-type failures involving aluminum, the work of Bascom [184] nicely summarizes the current state of understanding with an appropriate list of references which the reader may find helpful.

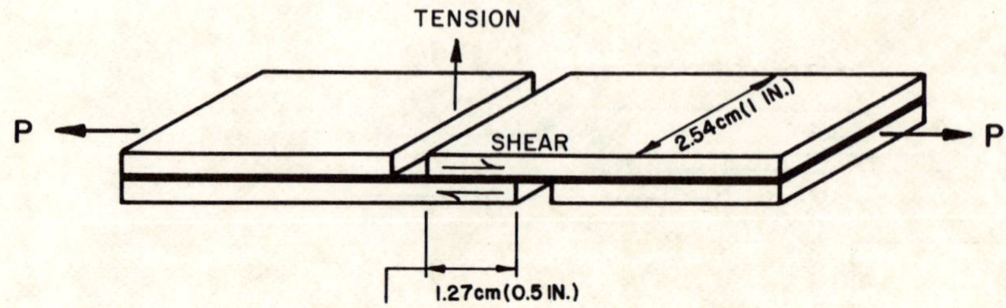

FIG. 3-51 Blister detection lap shear (MMM-A-132) time to failure recorded.

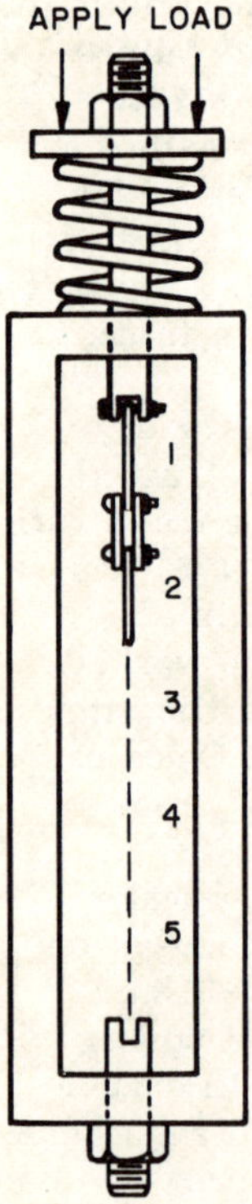

FIG. 3-52 3M's spring-loaded fixture.

All the above comments have referred to the use of a steady stressing situation which can produce durability definition between different conditions of adherend, adhesive, and surface treatment. However, there are many structural situations where a steady load application does not represent the service condition. Certainly this is the case with aircraft and most vehicle structures in which loads are applied and relaxed at intervals. There has therefore been a trend in durability testing of aerospace bonds toward testing procedures in which the load is repetitiously applied and relaxed. This cyclic stressing, in turn, should be applied under environmental conditions consistent with weathering factors in the service environment. Some of the most comprehensive studies in this regard, undertaken by Frazier and Lajoie at Bell Helicopter, have been summarized in an Air Force Materials Laboratory Technical Report [185]. A question arises as to the selection of a load-unload period because adhesives vary as to the rates at which they creep under load and relax when the load drops off.

The principle used at Bell can be demonstrated by comparing a relatively low-temperature-curing adhesive like FM-123-7 with 176.7°C (350°F) curing FM-1000. The FM-123-7 adhesive polymer shows a yield point in its stress-strain curve but no work-hardening properties. Creep under stress measurements showed the almost instantaneous deflection was 0.0069 cm (0.0027 in.), creep in the next 15 min was 0.0008 cm (0.0003 in.), and for the next 45 min was 0.00038 cm (0.00015 in.). When the load was relaxed, all the displacement relaxation was completed within 2 min. For this adhesive a load time of 30 min and an unload time of 5 min would seem to be a desirable cycle. For an extremely tough, high-elongation adhesive like FM-1000, the creep and relaxation properties are different and another cycle is indicated. At the end of an hour, the FM-1000 polymer was still creeping, although at a slower rate. Initial deflection was similar to the FM-123-7, i.e., 0.0061 cm (0.0024 in.), but within 30 min additional creep was 0.0071 cm (0.0028 in.), and an additional 0.0020 cm (0.0008 in.) of creep after 60 min. The same sequence occurred in reverse when the load was relaxed with the relaxation still measurable after 15 min. Based on this information, it was decided that a maximum bond damage cycle for the FM-1000 joint would likely be 2 hr of loading followed by an hour of relaxation. Although these differences can be shown between adhesives, it was necessary for economy of evaluation of a large number of adhesives to set a compromise load cycle for all materials of 27.56 MPa (4000 psi) for 1 hr and unload for 15 min. The selection of a specific load level also creates an additional problem in that a curve of durability vs. stress level cannot be determined.

These procedures for evaluating bonded joints are related to the so-called spectrum stress testing of aluminum alloys for aircraft. In testing metallic components, three different conditions of periodic stressing and vibration can be recognized depending on whether the aircraft is landing, in steady flight, or under severe maneuvering conditions.

15.2. Stress Endurance Evaluations

The cost of maintaining a sustained stress on individual bonded joints in various weathering situations is a limiting factor in the study of the many combinations of adherends, adhesives, and surface pretreatment conditions which have already been reported in the literature under unstressed conditions. Various arrangements have been suggested, including linking a series of specimens in a fixture made from metal angles and stressing the whole series by loading with a tensile machine and then mechanically holding the specimens in the constrained position with coil springs. Alcoa testing has been conducted using a separate aluminum stressing ring for each lap joint specimen. Each ring, individually calibrated, can be accurately loaded without the use of a tensile machine (see Fig. 3-53). In addition, the individual stressing fixture permits separate exposure to the widest possible variety of exposure conditions, including cycle tests. Also, the stress can be checked anywhere the exposure is taking place without removal of the specimens from the testing area.

The typical bond durability plot used with simultaneous stress and environmental testing is shown in Fig. 3-54. Theoretically, the plot is an attempt to determine what level of stress will permit long joint life. The stress imposed on the bond in practice should always

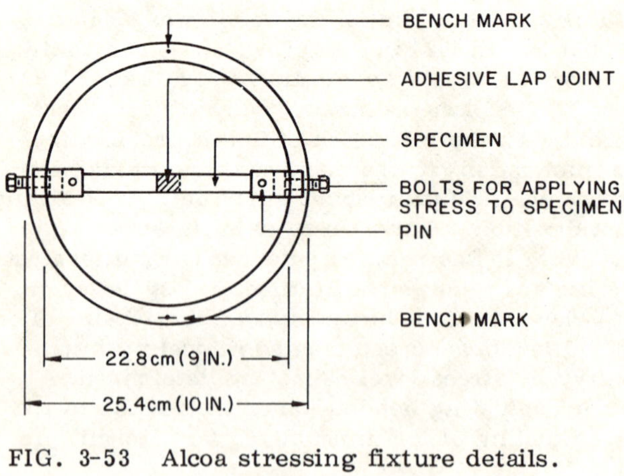

FIG. 3-53 Alcoa stressing fixture details.

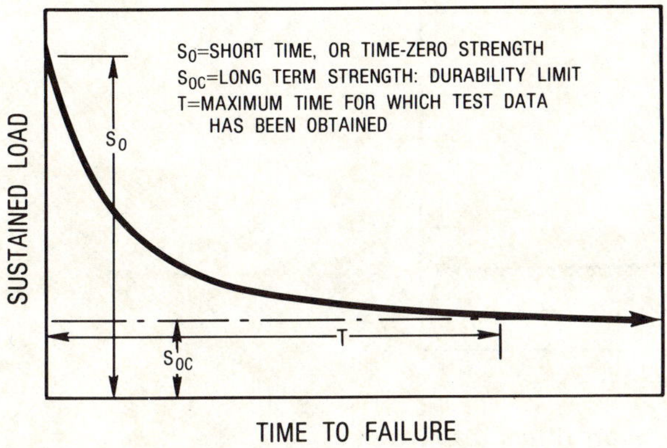

FIG. 3-54 Durability limit curve in shear mode.

be below the value indicated as the so-called stress endurance limit. It is impossible to really prove that the bond will never fail at this stress level unless the test has been conducted for an infinitely long time.

Sometimes deviations from the ideal plot occur, as shown in Figs. 3-55 and 3-56. While there may be a real stress endurance limit, in the case of Fig. 3-55, it is so low in value that it would be dangerous in practice to build any structure using the particular parameters of adherend, adhesive, surface condition, and service environment being investigated. The kind of deviation as shown in Fig. 3-56 also is encountered occasionally but not really understood. It does necessitate long-term testing, however, to determine if it is a possibility. How long a test should be conducted to determine if this kind of deviation might occur is largely unknown.

Reference should be made to those investigators who have attempted to take the constant stress data obtained using the Sharpe jigs and combine it with reaction rate theory as concerns polymer mechanical behavior [186]. McAbee and Levi [187] have shown that such treatment is applicable to some adhesive data under constant rate of loading conditions. Different reaction rate equations were developed for three decreasing relative humidity situations. Under 90-95% RH exposure conditions, the equations were relatively successful in predicting failure times for AF-126 film adhesive joints. Under 50% RH conditions, the data scatter was considerably greater. At 20% RH, the derived equations were not useful. It was reasoned that the lower humidities do not promote significant weakening of the bonds, and hence random flaws in the bondlines probably play a more prominent role in inducing bond failure.

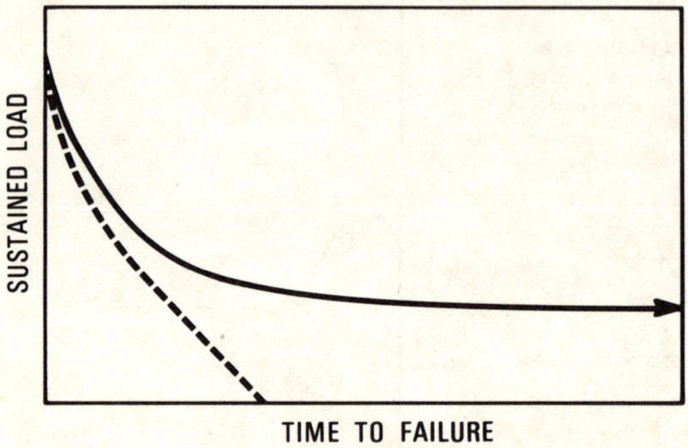

FIG. 3-55 Deviation of bond durability, plot I.

As examples of the effect of simultaneous stress and condensing humidity on the permanence of bonded joints, we can show the relative durability plots that were obtained at Alcoa for a two-part room-temperature-curing polyamide epoxy and a vinyl phenolic paste adhesive. In Fig. 3-57, we note that in a 2-year continuous exposure to 100% RH at 51.7°C (125°F) in the unstressed condition, there was very little change in bond tensile-shear strength when using these adhesives on chromic-sulfuric-acid-etched 6061-T6 aluminum alloy adherends. By contrast, in Fig. 3-58 the effect of imposed stress on the polyamide-cured epoxy caused it to have the appearance of the deviated bond durability plot shown previously in Fig. 3-55. The durability curve with imposed stress for the vinyl phenolic adhesive showed the more typical stress endurance plot with clear indication that the stress endurance level appears to be a little below the 3.45-MPa (500-psi) stress level. With many commercial room-temperature-curing adhesives evalu-

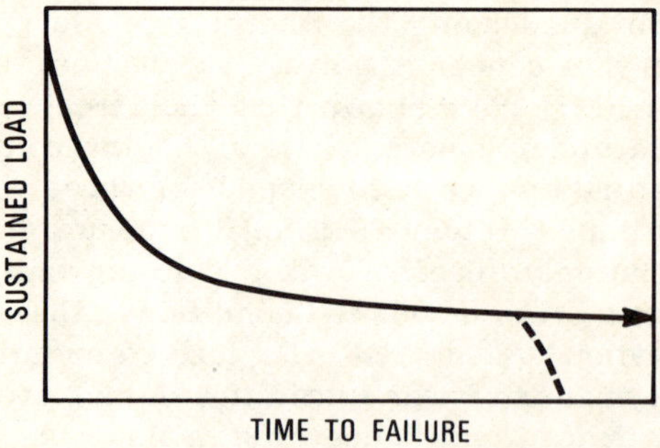

FIG. 3-56 Deviation of bond durability, plot II.

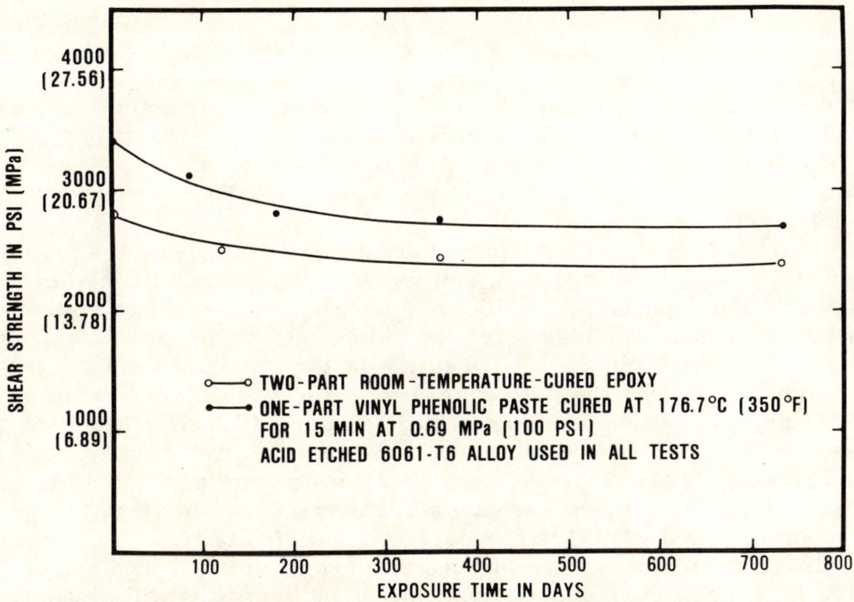

FIG. 3-57 Unstressed specimens in 100% relative humidity at 51.7°C (125°F).

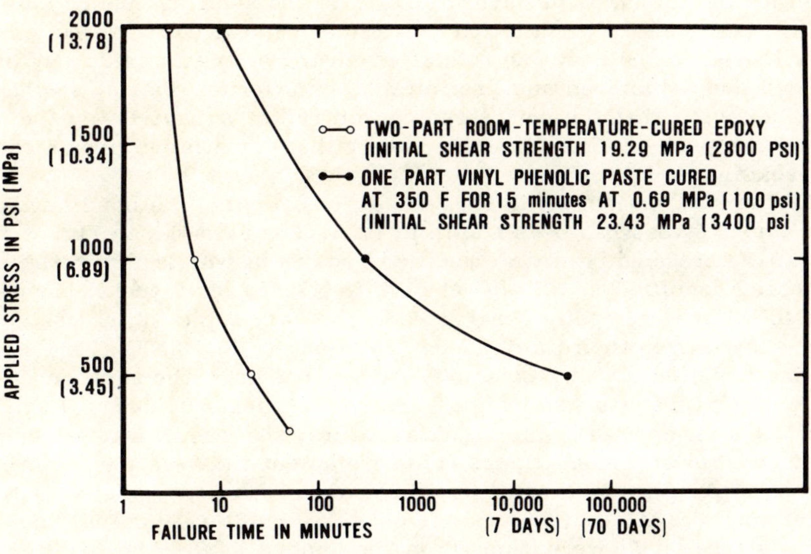

FIG. 3-58 Stressed specimens: Exposure in 100% relative humidity at 51.7°C (125°F)—acid-etched 6061-T6 aluminum.

ated at Alcoa in recent years involving a variety of different chemical classes, it has been necessary to reduce the stress to the vicinity of 0.69 MPa (100 psi) to approach the apparent stress endurance limit. This can be 5% or less of the measured initial lap-shear strength of the bonded joint.

Patrick [188] has conducted a number of studies using the aluminum DCB-type stress test specimen and amine-cured epoxide resin adhesive joints in water exposure conditions to develop a possible explanation of the mechanism of stress corrosion failure of such bonds as mentioned above. Patrick proposes that when the test specimen is immersed in water under stress, the water interacts at the highest stress concentration point and causes cracking to initiate by a solvolysis of the bond between the resin and the aluminum substrate. As the crack propagates, water follows into the crack void, interacting with the aluminum oxide. By some complex mechanism of solvating the aluminum oxide, the aluminate ion (AlO_2^-) is formed. This ion diffuses into the aqueous media. The aluminate ion continues to concentrate until the solution saturation point is reached. Then, aluminum hydroxide will precipitate out and more aluminate ions will diffuse. Conceivably, a steady-state condition could be reached. Erickson [189] had earlier showed the rapid diffusion of polyamines like triethylene tetramine (TETA) and tetraethylene pentamine (TEPA) through epoxide resins. Patrick suggests that such amines as contaminants on aluminum surfaces could accelerate the formation of bayerite oxide on the aluminum under water-soaking conditions. The fact that polyamines can diffuse in these joints simply increases the probable rate at which the aluminum oxide may be hydrated and solubilized and precipitated as bayerite. Patrick emphasizes, however, that the very large concentration of bayerite on the surface is a post-cracking function rather than a causative phenomenon for cracking.

As the environment diminishes in aggressiveness, the bond permanence under stress has been shown to rise. For example, in Fig. 3-59, a known water-sensitive nylon-modified epoxy film was tested in the unstressed condition in 100% RH at 51.7°C (125°F) for 2 years. Sensitivity to water was readily observable in the very rapid initial loss in lap-shear strength from 41.3 MPa (6000 psi) to 13.78 MPa (2000 psi) over the first 100 days of exposure. After that time, the additional drop in bond strength was insignificant for the next 600 days. In the milder room temperature water soak exposure, the loss in bond strength was steady but at a much slower rate throughout the 2-year exposure. The overall drop in bond strength in this case, however, was actually greater than 41.34 MPa (6000 psi) to about 5.51 MPa (800 psi). When various stress levels were imposed on the bonded joints, the results shown in Fig. 3-60 were only partly predictable. Because of the faster penetration of water effect of the condensing humidity at 51.7°C (125°F), the durability plot indicated no stress endurance limit existed

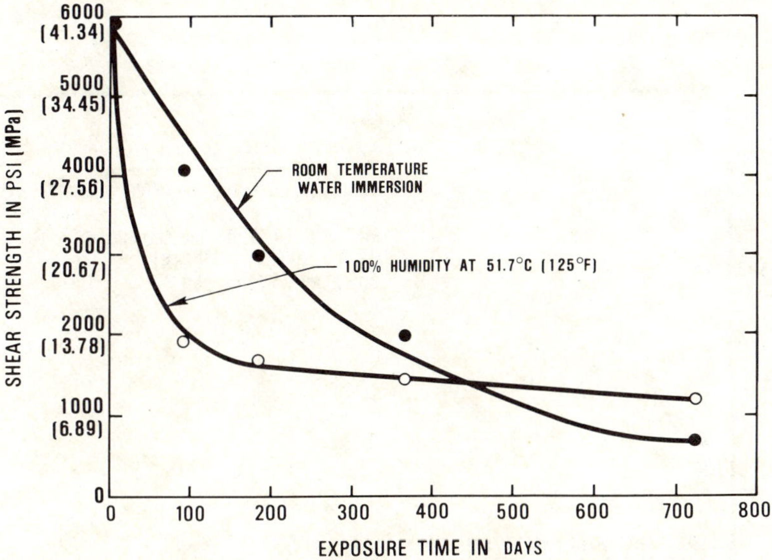

FIG. 3-59 Unstressed specimens: Adhesive no. 21 (nylon-modified epoxy film) cured 1 hr at 176.7°C (350°F) on acid-etched 6061-T6 alloy.

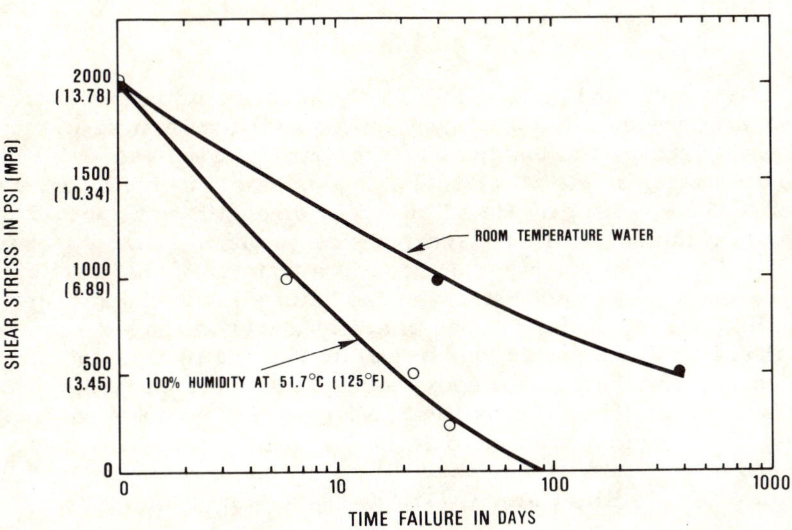

FIG. 3-60 Stressed specimens: Adhesive no. 21 (nylon-modified epoxy film) cured 1 hr at 176.7°C (350°F) on acid-etched 6061-T6.

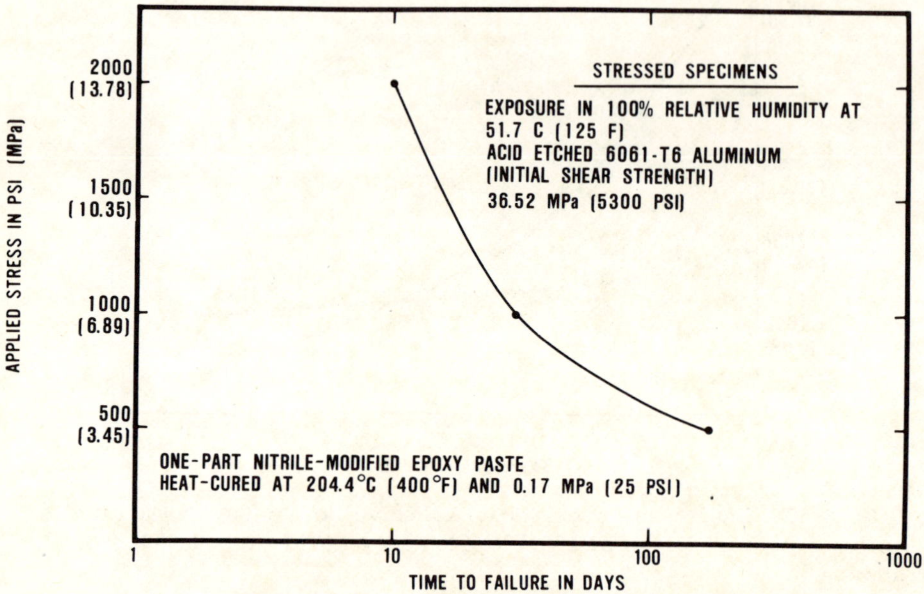

FIG. 3.61 Stressed specimens; Exposure in 100% relative humidity at 51.7°C (125°F), acid-etched 6061-T6 aluminum (initial shear strength) 36.52 MPa—(5300 psi).

even at 1.72 MPa (250 psi), which is only 4% of the initial bond strength. The bond permanence was obviously longer at the same stressing level when simply soaked in room temperature water as compared with condensing humidity at 51.7°C (125°F). In this case, the bond survival time at 3.45 MPa (500 psi) (8% of the initial strength) was approximately 300 days or 15 times as long a survival time as shown at the same stress in 100% RH at 51.7°C (125°F). Other Alcoa tests show that the bond permanence of joints under stress in the natural industrial atmospheric weathering condition would be significantly longer at the same stress level than in a water-soaking environment. In this respect, nitrile-modified one-part heat-cured epoxy joints have survived for more than 8 years under a stress of 6.89 MPa (1000 psi) in Alcoa tests conducted in the natural weathering conditions at the Alcoa Technical Center while showing bond failure in about 30 days when stressed at 6.89 MPa (1000 psi) in 100% RH at 51.7°C (125°F), as shown in Fig. 3-61.

REFERENCES

1. P. F. A. Bijlmer, Influence of chemical pretreatments on surface morphology and bondability of aluminum, J. Adhesion 5:319 (1973).
2. J. D. Minford, Durability of structural adhesive bonded aluminum joints, Adhesives Age, pp. 17-23 (March 1978).
3. J. Minford, Evaluating adhesives for joining aluminum, Metals Eng. Quart. (November 1972).
4. J. D. Minford, Durability of adhesive bonded aluminum joints, in *Treatise of Adhesion and Adhesives*, Vol. 3 (R. L. Patrick, ed.), Marcel Dekker, New York, 1973.
5. J. D. Minford, Effect of surface preparations on adhesive bonding of aluminum, Adhesives Age, p. 24 (July 1974).
6. A. W. Bethune, Durability of bonded aluminum structure, SAMPE J. (July-September 1975).
7. E. W. Thrall, Pabst program test results, Adhesives Age, pp. 22-33 (October 1979).
8. E. W. Thrall, The future of metal bonding, ASTM Symp. Adhesives and Bonding Processes for the Automobile and Aircraft Industries, October 1979.
9. J. D. Minford, Comparison of aluminum adhesive joint durability as influenced by etching and anodizing pretreatments of bonded surfaces, J. Appl. Polym. Symp. 32:91-103 (1977).
10. J. D. Minford, Etching and anodizing pretreatments and aluminum joint durability, SAMPE Quart., pp. 18-27 (July 1978).
11. N. A. DeBruyne, The extent of contact between glue and adherend, Bulletin 168, Technical Service Department, Aero Research Ltd., Duxford, Cambridge, England, December 1956.
12. W. D. Bascom, The origin and removal of microvoids in filament wound composites, NRL Report 6268, May 24, 1965.
13. E. P. Plueddemann, Adhesion through silane coupling agents, 25th SPI Reinforced Plastics Composites Div. Conf., Washington, D.C., February 1970.
14. W. D. Sell, Some analytical techniques for durability testing of structural adhesives, 19th Nat. SAMPE Symp., April 1974.
15. P. E. Cassidy and B. J. Yager, A review of coupling agents as adhesion promoters, Tracor, Inc., NASA Contract NASA-24073, September 1969.
16. N. J. DeLollis, Mode of failure in structural adhesive bonds, Appl. Polym. Sci. (June 1967).
17. N. J. DeLollis, Theory of adhesion mechanism of bond failure and mechanism for bond improvement, Adhesives Age (December 1968-January 1969).
18. J. A. Laird and F. W. Nelson, The effect of glass surface chemistry on glass epoxy systems, SPE Trans. (April 1964).

19. P. W. Erickson, A. Volpe, and E. R. Cooper, Effects of glass surfaces on laminating resins, Modern Plastics (August 1969).
20. J. Outwater and D. Kellogg, A simple experiment to show the origin of water debonding of resin against glass, Rept. for Contract V-3219 (01) (X), September 15, 1961.
21. W. A. Zisman, Surface chemistry of glass reinforced plastics, Symp. Glass-Resin Interface of the Soc. Plastics Industry, 19th An. Exhib. and Conf., Chicago, February 6, 1964.
22. J. J. Bikerman, *The Science of Adhesive Joints* (2nd ed.), Academic, New York, 1968.
23. J. Bikerman, *Recent Advances in Adhesion* (L.-H. Lee, ed.), Gordon and Breach, New York, 1973.
24. L. H. Sharpe and H. Schonhorn, Surface energetics, adhesion, and adhesive joints, Advan. Chem., Ser. 43, p. 189 (1964).
25. R. J. Good, *Recent Advances in Adhesion* (L.-H. Lee, ed.), Gordon and Breach, New York, 1973.
26. S. Sterman and J. B. Toogood, How to promote adhesion with silicones and silanes, Adhesives Age (July 1965).
27. J. A. Laird, Glass surface chemistry for gl5ss reinforced plastics, Final Report, Navy Contract W-0679-C (FBM), June 1963.
28. I. U. Ojalvo and H. L. Eidinoff, Bond thickness effects upon stresses in single lap adhesive joints, SAE Paper No. 770090, February 28-March 4, 1977.
29. J. D. Minford and E. M. Vader, Adhesive bonding of aluminum automotive body sheet, SAE Int. Auto Eng. Cong. Paper No. 740078, Detroit, February 1974.
30. C. L. Brett, J. Appl. Sci. 18:315 (1974).
31. W. L. Baun, N. T. McDevitt, and J. S. Solomon, Chemistry of metal and alloy adherends by secondary ion mass spectroscopy, ion scattering spectroscopy, and Auger electron spectroscopy, Surface Analysis Techniques for Metallurgical Applications, ASTM STP 596, March 1975.
32. A. Benninghoven, Surf. Sci. 28:541 (1971).
33. R. Schubert and J. C. Tracy, Rev. Sci. Inst. 44:487 (1973).
34. R. Schubert, J. Vac. Sci. Tech. 12, No. 1:505 (January 1975).
35. R. Werner, Dev. Appl. Spectroscopy 7A:297 (1969).
36. H. E. Blair, S. Matsuoka, R. G. Vadimsky, and T. T. Wang, J. Adhesion 3:89 (1971).
37. J. C. Bolger, The chemical composition of metal and oxide surfaces and how these interact with polymeric materials, 30th An. Tech. Conf. SPE, Chicago, 1972.
38. T. Smith, The mechanisms of adhesive failure between polymers and metallic substrates, AFML TR-74-73, 1974.
39. A. Pattniak and J. D. Meakin, Characterization of aluminum adherend surfaces, Technical Report No. 4699, Franklin Institute, July 1974.

40. Development of corrosion resistant surface treatments for aluminum alloys for spot-weld bonding, AFML Contract No. F33615-74-C-5027, 1974.
41. B. B. Bowen, et al., Development of corrosion resistant surface treatments for aluminum alloys for spot-weld bonding, Interim Technical Report, NOR 74-93, May 1974
42. B. B. Bowen, et al., Development of corrosion resistant surface treatments for aluminum alloys for spot-weld bonding, Interim Technical Report No. 2, NOR 74-231, August 1974.
43. B. B. Bowen, Preparation of aluminum alloy surfaces for spot-weld bonding, SAMPE Symp., Buena Park, Calif., April 1974.
44. Characterization of surfaces prior to adhesive bonding, AFML-TR-76-118, Northrop Corporation, July 1976.
45. J. A. Marceau, Y. Moji, and J. C. McMillan, A wedge test for adhesive bonding, 21st SAMPE Symp., April 1976.
46. J. C. McMillan, J. T. Quinlan, and R. A. Davis, Phosphoric acid anodizing of aluminum for structural adhesive bonding, SAMPE Quart. 7, No. 3:13 (April 1976).
47. K. Wefers and G. M. Bell, Oxides and hydroxides of aluminum, Technical Paper No. 19, Alcoa Research Laboratories, 1972.
48. B. B. Bowen, Preparation of aluminum alloy surfaces for spot weld bonding, SAMPE Mater. Rev. 7:374 (1975).
49. N. T. McDevitt and W. L. Baun, Some observations of the relation between chemical surface treatments and the growth of anodic barrier layer films, Air Force Materials Lab, AFML-TR-76-74, June 1976.
50. N. T. McDevitt and W. L. Baun, Surface analysis of 2024 aluminum after treatment with sulfuric-chromic acid solutions, J. Electrochem. Soc., p. 1058 (July 1976).
51. N. T. McDevitt, et al., Surface analysis of 6061 and 7050 aluminum alloys after conditioning by chemical treatment, Air Force Materials Lab., AFML-TR-76-13, March 1976.
52. N. T. McDevitt, W. L. Baun, and J. S. Solomon, AFML-TR-75-122, Wright-Patterson AFB, Dayton, Ohio, 1975.
53. J. A. Smith and W. E. Martinsen, Amer. Ceram. Soc. Bull. 52:855 (1973).
54. J. D. Venables, D. K. McNamara, J. M. Chen, and T. S. Sun, Oxide morphologies on aluminum prepared for adhesive bonding, Appl. Surf. Sci. 3, No. 1 (July 1979).
55. N. T. McDevitt, W. L. Baun, G. Fugate, and J. S. Solomon, Surface studies of anodic aluminum oxide layers formed in phosphoric acid solutions, Air Force Materials Lab., AFML-TR-77-55, May 1977.
56. W. Brockmann, The environmental resistance of metal bonds in new industries and applications for advanced materials technology, 19th Nat. SAMPE Symp., Vol. 19, Azuza, Calif., 1974.

57. A. F. Lewis and L. J. Forrestal, Chemical nature of polymer to metal adhesion, Adhesion, Spec. Tech. Publ. ASTM, Philadelphia, 1964.
58. W. Brockmann, Interface reactions and their influence on the long term properties of metal bonds, Bicent. Mater. Progr.—SAMPE, p. 383 (1976).
59. A. J. Kinloch, Interfacial fracture mechanical aspects of adhesive bonded joints, AGARD Lecture Series 102, Wright-Patterson Air Force Base, Dayton, Ohio, October 1979.
60. R. F. Wegman, How oxides of aluminum affect bond strength, Adhesives Age, pp. 20-27 (January 1967).
61. R. F. Wegman and R. Dickinson, Picatinny Arsenal Technical Report 3300, February 1966.
62. R. F. Wegman, W. M. Bodnar, M. J. Bodnar, and M. J. Barbarisi, Effects of deionized water immersion of prepared aluminum surfaces on adhesive bondability, Picatinny Arsenal Technical Report 3495, May 1967.
63. D. J. Falconer, N. C. MacDonald, and P. Walker, Chem. Ind., p. 1230 (1964).
64. C. Kerr, N. C. MacDonald, and S. Orman, J. Appl. Chem. 17: 62 (1967).
65. J. A. Scott, Durability of bonded joints, ASTM STP401, p. 16 (1966).
66. S. Orman and C. Kerr, The effect of water on aluminum epoxide bonds, in *Aspects of Adhesion*, No. 6 (D. J. Alner, ed.), University of London Press, London, 1971, p. 64.
67. N. C. MacDonald, in *Aspects of Adhesion*, No. 5, University of London Press, London, 1969.
68. T. K. Kwei, J. Appl. Polym. Sci. 10:1647 (1966).
69. S. Orman, Atomic Weapons Research Establishment, unpublished data.
70. D. W. Levi, R. F. Wegman, M. C. Ross, and E. A. Garnis, Use of hot water aging for estimating lifetimes of adhesive bonds to aluminum, SAMPE Quart. 7, 3:1 (April 1976).
71. E. McAbee and D. W. Levi, Picatinny Arsenal Technical Report 4105, 1970.
72. R. F. Wegman, Effect of surface preparation and elevated temperature storage on adhesive bonds to aluminum, Picatinny Arsenal Technical Memorandum 1255, May 1964.
73. N. Chessen and V. Curran, Preparation of aluminum surfaces for bonding, Appl. Polym. Symp., No. 3:319-325 (1966).
74. R. F. Wegman, Effects of surface preparation of aluminum on bonds obtained with an epoxidized novolac adhesive, Picatinny Arsenal Technical Notes FRL-TN-69, September 1961.
75. N. Evans, Pretreatment for bonding, Eng. Mater. Des. 13, No. 2: 190-192 (February 1970).

76. J. D. Minford, F. R. Hoch, and E. M. Vader, Weldbond and its performance in aluminum body sheet, SAE Int. Auto. Eng. Cong. Paper 750462, Detroit, February 1975.
77. J. D. Minford, unpublished reports, Alcoa Laboratories, Alcoa Center, Pa.
78. H. W. Eickner and W. E. Schowalter, A study of methods for preparing clad 24S-T3 aluminum alloy sheet surfaces for adhesive bonding, Forest Products Laboratory Reports No.s 1813 and 1813A, 1950.
79. H. W. Eickner, Adhesive bonding properties of various metals as affected by chemical and anodizing treatments of the surfaces, Forest Products Laboratory Report No. 1842-A, February 1955.
80. H. W. Eickner, Adhesive bonding properties of various metals as affected by chemical and anodizing treatments of the surfaces, Forest Products Laboratory Report No. 1842, August 1960.
81. M. C. Ross, R. F. Wegman, M. J. Bodnar, and W. C. Tanner, Strength of joints to selected corrosion resistant finishes, SAMPE J. 12, No. 5:4-6 (September-October 1976).
82. D. R. Askins and H. S. Schwartz, Durability of adhesive bonded honeycomb sandwich in accelerated adverse environments, in *Materials and Processes in Service Performance*, SAMPE, Azuza, Calif., 1977, pp. 329-350.
83. W. J. Russell, A chromate-free process for preparing aluminum substrates for adhesive bonding:A preliminary study, Picatinny Arsenal Technical Report 4861, July 1976.
84. P. F. A. Bijlmer, Surface potential difference measurements on aluminum alloys and their relation to bondability, in *Adhesion 2* (K. W. Allen, ed.), Applied Science Publishers, London, 1978, pp. 45-76.
85. C. W. Jennings, Two techniques for characterizing surfaces for adhesive bonding, Sandia Laboratories, SAND76-0137 Report, May 1976.
86. P. Bijlmer, Pickling aluminum alloys for adhesive bonding, Metal Finishing (December 1971).
87. P. Bijlmer and R. Schliekelmann, The relation of surface condition after pretreatment to bondability of aluminum alloys, SAMPE Quart. (October 1973).
88. W. J. Russell and E. A. Garnis, A study of the FPL process used for preparing aluminum surfaces for adhesive bonding, SAMPE Quart. 7, No. 3:5 (April 1976).
89. W. J. Russell and W. C. Tanner, Component parts assembly with joints adhesive-mechanical, part I: Surface preparation of aluminum for weld bonding, Picatinny Arsenal Technical Report 4610, February 1974.
90. ASTM standard recommended practice for preparation of metal surfaces for adhesive bonding, D-2651.

91. A. W. Bethune, Durability of bonded aluminum structure, 19th Nat. SAMPE Symp., Los Angeles, April 1974.
92. R. F. Wegman, W. M. Bodnar, M. J. Bodnar, and M. J. Barbarisi, The effects of deionized water immersion of prepared aluminum surfaces on adhesive bonding, SAMPE J. (October-November 1967).
93. T. J. Reinhard, Novel concepts for priming metallic adherends for structural adhesive bonding, in *Adhesion 2* (K. W. Allen, ed.), Applied Science Publishers, London, 1978.
94. M. C. Ross, R. F. Wegman, M. J. Bodnar, and W. C. Tanner, Effect of surface exposure time on bonding of 5052-H34 aluminum alloy, SAMPE J. 10, No. 3:5-7 (May-July 1974).
95. M. C. Ross, R. F. Wegman, M. J. Bodnar, and W. C. Tanner, Effect of surface exposure time on bonding of 2024-T3 aluminum alloys, SAMPE J. 10, No. 1 (1974).
96. M. C. Ross, R. F. Wegman, M. J. Bodnar, and W. C. Tanner, Effect of surface exposure time on bonding 6061-T4 aluminum alloys, SAMPE J. 10, No. 2 (1974).
97. T. Smith, Surface treatment for aluminum bonding (STAB), J. Adhesion 9:313-320 (1977).
98. M. C. Locke, W. Scardino, and H. Croop, Non-tank phosphoric acid anodize method of surface preparation of aluminum for repair bonding, in *Materials Review 75*, SAMPE, Azuza, Calif., 1975, pp. 488-504.
99. J. M. Chen, T. S. Sun, J. D. Venables, and R. Hopping, Effects of fluorine contamination on the microstructure and bondability of aluminum surfaces, SAMPE J. (July-August 1978).
100. J. D. Minford, Comparative bondability and joint durability of Alclad 2024-T3, 1100-H14 and 6061-T6 aluminum alloys, unpublished Alcoa Research Report, Alcoa Center, Pa.
101. J. D. Minford, unpublished Alcoa Reports, Alcoa Laboratories, Alcoa Center, Pa.
102. C. V. Cagle, Design criteria and joint design, in *Adhesive Bonding Techniques and Application*, McGraw-Hill, New York, 1968, p. 57.
103. Design considerations, in *Adhesives in Modern Manufacturing*, Soc. Mfg. Engineers, Dearborn, Mich., 1970.
104. D. A. Yurek, Design of adhesive bonded joints, Adhesives Age (December 1965).
105. R. F. Wegman, Durability of some newer structural adhesives, J. Appl. Polym. Sci., Appl. Polym. Symp. 32:1-10 (1977).
106. D. A. Wangsness, Sustained load durability of structural adhesives, J. Appl. Polym. Sci., Appl. Polym. Symp. 32:291-300 (1977).
107. D. R. Askins, Environmental durability testing of structural adhesives, Part I, Air Force Materials Lab. Contract, AFML-TR-78-35, December 1978.

108. D. R. Askins, Environmental durability testing of structural adhesives, Part II, Air Force Materials Lab. Contract, AFML-TR-78-35 Part II, December 1978.
109. R. J. Riel, Corrosion of adhesive bonded aluminum alloy panels, Rohr Corporation Report 24-2047, August 1968.
110. R. J. Riel, Corrosion delamination, SAMPE J. 7, No. 5 (August-September 1971).
111. R. H. Greer, Corrosion resistant adhesive bonding, in *Aerospace Adhesives and Elastomerics*, National SAMPE Technical Conference Series, Vol. 2, 1970, p. 561.
112. Narvel L. Rogers, Corrosion of adhesive bonded clad aluminum, SAE Nat. Bus. Aircraft Meeting No. 720344, Wichita, Kansas, 1972.
113. Narvel L. Rogers, A comparative test for bondline corrosion: clad vs. bare aluminum alloys, Proc. 5th Nat. Tech. Conf., SAMPE, Kiemesha Lake, N.Y., 1973, p. 160.
114. W. T. McFarlen, Environmental testing of adhesive bonded metallic structures, in *Composite Materials in Engineering Design, ASM* (B. R. Noton, ed.), 1973, p. 678.
115. Spot weld through an adhesive? Iron Age, p. 104 (April 24, 1969).
116. Glue-spot welding joins aluminum, stronger than spot, cheaper than fusion, Mater. Eng., p. 14 (January 1971).
117. Weldbond flight component design/manufacturing program, Interim Technical Report First Quarterly Contract No. F33615-71-C-1716, Lockheed-Georgia Co., October 1971.
118. Weldbond flight component design (Manufacturing Program), Interim Technical Report Sixth Quarterly Contract No. F33615-71-C-1716, Lockheed-Georgia Co., January 1973.
119. J. Mahon, C. Vizzi, W. Sisco, and P. Nilsson, Aluminum weld-bonding for 350°F service, in *Materials and Processes for the 70s: Cost Effectiveness and Reliability*, SAMPE, 1973, p. 184.
120. K. C. Wu and B. B. Bowen, Aluminum weldbond manufacturing methods development, Interim AFML Report IR-854-5 (I) Contract No. F33615-75-C-5083, June-November 1975.
121. B. B. Bowen, R. E. Herfert, and K. C. Wu, Development of corrosion resistant surface treatments for aluminum alloys for spot-weld bonding, Contract No. F33615-74-C-5-27, March 1975.
122. White engineers cite truck tractor trends, Metal Progress, p. 50 (February 1974).
123. R. C. Hall, Environmental resistance of a weldbonded joint, SAMPE J. (August-September-October 1974).
124. W. M. Frost, Strength of structural adhesives at temperatures down to minus 424°C, Cryogenic Engineering Laboratory Bureau of Standards, WADC Technical Report No. 260, April 1959.
125. J. Hertz, Cryogenic adhesive evaluation study, Convair (Astronautics) Division, Convair Research Program No. 111-9106, January 1961.

126. J. M. Black and R. B. Blomquist, Metal surface effects on heat resistance of adhesive bonds, Ind. Eng. Chem. 50:918 (June 1958).
127. J. M. Black and R. F. Blomquist, Development of metal bonding adhesive FPL-710 with improved heat resistance properties, Forest Products Laboratory, NACARM 52F19, 1952.
128. J. M. Black and R. F. Blomquist, Development of metal bonding adhesives with improved heat resistant properties, Forest Products Laboratory NACARMS54D01, 1954; Modern Plastics 32:139 (December 1954).
129. J. M. Black and R. F. Blomquist, Metal bonding adhesives for high temperature service, Modern Plastics (June 1956).
130. W. Z. Olson and R. M. Lulling, Shear strength of adhesives in stainless steel and aluminum lap joints at temperatures from -100°F to 1000°F, Forest Products Laboratory, AFML Technical Documentary Report No. ASD-TR61-497, April 1962.
131. R. B. Krieger, Jr., and R. E. Politi, High temperature structural adhesives, in *Aspects of Adhesion*, No. 3 (D. J. Alner, ed.), University of London Press, London 1967.
132. H. W. Eickner, Weathering of adhesive bonded lap joints of clad aluminum alloys, Forest Products Laboratory, WADC Technical Report 54-447, Part I, 1955.
133. H. W. Eickner, Weathering of adhesive bonded lap joints of clad aluminum alloy, Forest Products Laboratory, WADC Technical Report 54-447, Part II, 1957.
134. H. W. Eickner, Weathering of adhesive bonded lap joints of clad aluminum alloy, Forest Products Laboratory, WADC Technical Report 54-447, Part III, 1958.
135. H. W. Eickner, Environmental exposure of adhesive bonded metal lap joints, WADC Technical Report 59-567, Part I, February 1960.
136. R. F. Wegman, W. M. Bodnar, E. S. Duda, and M. J. Bodner, How weathering and aging affect bonded aluminum, Adhesives Age, p. 22 (October 1967).
137. C. I. Hause, W. C. Pagel, and A. G. McKown, The effect of South Florida weather aging on structural adhesive joints, 3M Company Technical Report, October 1965.
138. A. A. Griffith, Phil. Trans., Roy. Soc. 221:163 (1920).
139. G. R. Irwin, Fracture mechanics, in *Structural Mechanics* (Goodier and Hoff, eds.), Pergamon Press, New York, 1960.
140. E. J. Ripling, S. Mostovoy, and R. L. Patrick, Application of fracture mechanics to adhesive joints, in *Adhesion*, ASTM, 1964.
141. S. Mostovoy, P. P. Crosley, and E. J. Ripling, Use of crack-line-loaded specimens for measuring plane-strain fracture toughness, J. Mater. 2, No. 3 (1967).
142. E. J. Ripling and S. Mostovoy, Stress corrosion cracking of adhesive joints, J. Adhesion (1971).

143. E. J. Ripling, S. Mostovoy, and H. T. Corten, Fracture mechanics: A tool for evaluating structural adhesives, J. Adhesion 3: 107 (1971).
144. S. Mostovoy and E. J. Ripling, Fracturing characteristics of adhesive joints, Final Report, Contract NOOO19-71-C-0329, Materials Research Lab., 1972.
145. S. Mostovoy and E. J. Ripling, Factors controlling the strength of composite structures, Final Report, Contract NOOO19-61-C-0231, Materials Research Lab., 1970.
146. S. Mostovoy and E. J. Ripling, Effect of joint geometry on the toughness of epoxy adhesives, J. Appl. Polym. Sci. 15:661 (1971).
147. S. Mostovoy and E. Ripling, J. Appl. Polym. Sci. 13:1083 (1969).
148. S. Mostovoy and E. Ripling, J. Appl. Polym. Sci. 15:641 (1971).
149. S. Mostovoy, H. R. Smith, R. G. Lingwall, and E. Ripling, Eng. Fracture Mech. 3:291 (1971).
150. E. Ripling, S. Mostovoy, and C. F. Bersh, J. *Adhesion* 3:145 (1971).
151. S. Mostovoy and E. Ripling, J. Appl. Polym. Symp. 19:395 (1972).
152. J. A. Marceau and Y. Moji, Development of environmentally stable aluminum adhesive bonds, Document D6-41145, Boeing Company.
153. Common bonding requirements for structural adhesives, Boeing Commercial Airplane Company Process Document BAC 5514.
154. J. A. Marceau and Y. Moji, Application of fracture mechanics testing to process control for adhesive bonding, Document D6-41141, Boeing Company, May 1973.
155. S. Mostovoy and E. J. Ripling, Fracturing characteristics of adhesive joints, Final Report Contract No. NOOO19-74-C-0274, Materials Research Laboratory, January 1974-January 1975.
156. W. D. Bascom, D. O. Timmons, and R. L. Jones, Apparent interfacial failure in mixed-mode adhesive fracture, J. Mater. Sci. 10: 1037-1048 (1975).
157. C. W. Jennings, in *Recent Advances in Adhesion* (L. H. Lee, ed.), Gordon and Breach, New York, 1973, p. 469.
158. D. R. Mulville and R. H. Varshnev, *Army Symposium on Solid Mechanics*, Army Materials and Mechanics Research Center, Watertown, Maine, 1974.
159. R. C. Wilcox and W. A. Jemian, Polym. Eng. Sci. 13:40 (1973).
160. J. R. Huntsberger, J. Polymer Sci. 1A:1339 (1963).
161. J. R. Huntsberger, J. Polymer Sci. 1A:2241 (1963).
162. M. H. Loew, J. M. Fitzgerald, and A. N. Mucciardi, Air Force Materials Laboratory Final Report, AFML-TR-78-206, December 1978.
163. H. T. Clark, Definition and non-destructive detection of critical adhesive bond-line flaws, Air Force Materials Laboratory Final Report, July 1978.

164. W. B. Jones, D. H. Kaelbe, and W. G. Knauss, Investigations of fatigue and crack propagation behavior of adhesives, Air Force Materials Laboratory Report, AFML-TR-72-218, Part I, October 1972.
165. W. J. Jones and D. H. Kaelble, Investigation of fatigue and crack propagation behavior of adhesives, Air Force Materials Laboratory Report, AFML-TR-72-218, Part II, December 1973.
166. W. D. Bascom, R. L. Cottington, and C. O. Timmons, Fracture design criteria for structural adhesive bonding: Promise and problems, Naval Eng. J., pp. 73-85 (August 1978).
167. L. H. Sharpe, Some aspects of the permanence of adhesive joints, Symp. Struc. Adhesives Bonding, Stevens Institute of Technology, Hoboken, N.J., September 16, 1965.
168. L. H. Sharpe, The interphase in adhesion, J. Adhesion 4:51 (1972).
169. H. W. Eickner, Environmental exposure of adhesive bonded metal lap joints, WADC Tech. Report 59-564, Part I, Forest Products Laboratory, 1962.
170. H. W. Eickner, W. Z. Olsen, and R. M. Lulling, Resistance of adhesive-bonded metal lap joints to environmental exposure, Technical Document WADC TR-59-654, Part II, Forest Products Laboratory, October 1962.
171. G. F. Carter, Outdoor durability of adhesive joints under stress, Adhesives Age 10, No. 10:32 (October 1967).
172. G. F. Carter, Durability of adhesive peel joints while stressed in water, ASTM STP 401, p. 28 (August 1966).
173. R. F. Wegman, Durability of adhesive bonded aluminum joints, Picatinny Arsenal Technical Report 4169, June 1971.
174. J. D. Minford, Aluminum joint durability in stress and condensing humidity exposure with varying surface pretreatment, ASM/SAMPE Conf. Specialized Cleaning, Finishing, and Coating Processes, Los Angeles, February 1980.
175. J. D. Minford, Effect of surface preparation on stressed aluminum joints in corrosive saltwater exposure, Adhesives Age, p. 36 (October 1980).
176. J. A. Marceau and W. Scardina, Durability of adhesive bonded joints, Air Force Materials Laboratory Report, AFML-TR-75-3, February 1975.
177. J. C. McMillan, Surface preparation: The key to bondment durability, AGARD Lecture Series No. 102, Wright-Patterson Air Force Base, Dayton, Ohio, October 1979.
178. G. L. Moore, Mechanism of adhesion failure between polymers and metallic substrates, Document D6-40346, Boeing Technical Proposal, January 1972.

179. J. C. Sargent, Research on effects of coupling agents on the failure mechanisms of adhesive bonded joints, Document D6-40347, Boeing Technical Proposal.
180. Raymond B. Krieger, Jr., Evaluating structural adhesives under sustained load in hostile environments, SAMPE 5th Nat. Tech. Conf., Kiamesha Lake, N.Y., October 1973.
181. J. F. Coleman, Durability studies on toughened acrylic resin adhesives, SAMPE 5th Nat. Tech. Conf., Kiamesha Lake, N.Y., October 1973.
182. J. F. Coleman, Modified acrylic adhesives afford durable bonding, Adhesives Age (February 1974).
183. D. A. Yurek, private communication, 3M Adhesives, Coatings and Sealers Division, Test Method No. C-294, June 1972.
184. W. P. Bascom, Stress corrosion of structural adhesive bonds, Adhesives Age, pp. 28-33 (April 1979).
185. T. B. Frazier and A. D. Lajoie, Durability of adhesive bonded joints, Technical Report AFML-TR-74-26, Bell Helicopter Company, March 1974.
186. E. McAbee and D. W. Levi, J. Appl. Polym. Sci. 11:2067 (1967); 13:1899 (1969).
187. E. McAbee and D. W. Levi, Use of a reaction rate method to predict failure times of adhesive bonds at constant stress, Technical Report 4105, Materials Engineering Laboratory, Picatinny Arsenal, December 1970.
188. R. L. Patrick, The use of scanning electron microscopy, in *Treatise on Adhesion and Adhesives*, Vol. 3 (R. L. Patrick, ed.), Marcel Dekker, New York, 1973.
189. P. W. Erickson, A. Volpe, and F. R. Cooper, Mod. Plastics, NOLTR 63-267 (August 1964).

4

Mechanisms of Adhesion Failure Between Polymers and Metallic Substrates: Aluminum 2024-T3 with HT424 Adhesive*

Tennyson Smith and David H. Kaelble

Rockwell International
Thousand Oaks, California

1. INTRODUCTION

1.1. The Purpose

To improve adhesive joints, the investigator concentrates his efforts on the weakest link. If failure occurs primarily in the adhesive, emphasis is placed on a search for a stronger adhesive or means to strengthen a given adhesive. If failure occurs primarily at a weak boundary layer consisting of surface contamination or oxide, emphasis is placed on cleanliness of surface preparations. Since new applications demand stronger and stronger adhesive joints, the tendency will be to continue improvement of the weakest link as it shifts from one locus to another. This study is primarily concerned with the nature of the adherend surface after preparation for bonding and the relationship between the nature of the surface and bond strength and reliability. Two important benefits will follow this type of intensive study: first, new knowledge about the physical and chemical nature resulting from surface preparations will point the way to improved surface preparations; and second, the correlation between surface characteristics and bond strength can be used to develop nondestructive testing (NDT) tools for monitoring assembly line parts that are to be adhesively bonded. Even if it is impractical to monitor all areas of every part, automatic spot-checking of specific areas of a few parts would be valuable with

*This work was conducted under Contract No. F33615-72-C-1641 for the U.S. Air Force Non-Metallic Materials Division, Air Force Materials Laboratory, Wright-Patterson A.F.B., Ohio 45433.

respect to verification that surface preparations and storage procedures are acceptable or not.

1.2. The Problem

All metals, except gold, are thermodynamically unstable at room temperature when in contact with oxygen at atmospheric partial pressure, and hence should tend to form oxide surface layers. The thickness of this oxide may vary from a few atomic layers to many micrometers. Water, as vapor or liquid, is usually available to any surface and many metals such as aluminum and titanium are sufficiently active to displace hydrogen from this water to form hydroxyl groups. However, these reactions often fail to occur at any appreciable rate. The actual film formed on such metals is protective and usually is found to be a thin, nonporous film of insoluble oxide. The Wagner theory [1] of oxidation at the metal-oxide interface is the prevailing theory for describing the formation of protective films on refractory metals. The stoichiometry of the oxide formed is of importance because the relative diffusion rates of the metal or oxygen ions depends on the number of cation or anion vacancies. Studies by Smith [2], Kofstad [3], and others have shown that, for gas-phase oxidation, zirconium and titanium tend toward a slight anion deficiency. Few detailed experiments of this type in the presence of moisture have been made.

In Fig. 4-1, a schematic representation of water and oxide layers on a metal surface is shown. The illustration, taken from a recent review by Bolger and Michaels [4], is based on the results of a number of prior studies of adherend surfaces. The three basic assumptions of the model are (1) that the oxide film is thick enough so that the underlying metal substrate exerts no influence on reactions or electrostatic interactions at the surface; (2) that, while the oxide surface can be dehydrated at elevated temperatures, at normal ambient bonding conditions the outermost surface oxygens hydrate to form a high density of hydroxyl groups; and (3) the hydroxyl adsorbs and retains several molecular layers of water. The spacing and the activity of the hydroxyl groups will depend in general on the spacing and valency of the metal atoms in the oxide. The use of oxidizing acids to clean metal surfaces might be expected to remove these hydrated oxides, oxidize and remove surface impurities, and increase the valence of the surface metal ions. Few detailed studies of surfaces before and after etching are to be found in the literature and, in most instances, the information obtained is inferential rather than direct observation. As discussed next, the structure of the surface layers on aluminum are not well established and it becomes an important part of this program to attempt to clarify this situation.

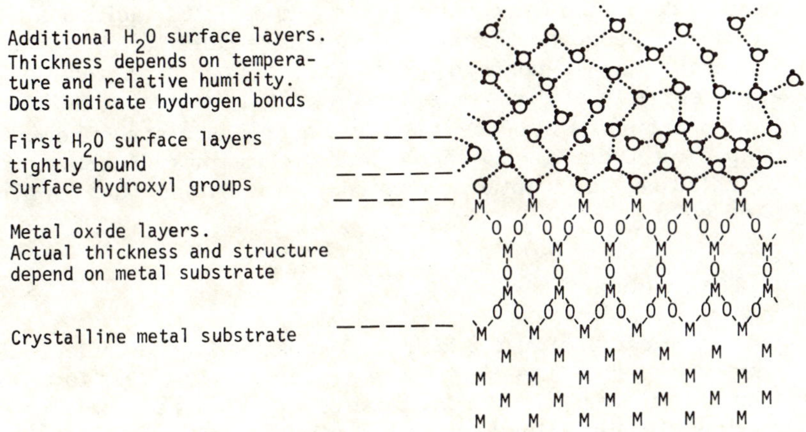

FIG. 4-1 Schematic representation of water and oxide layers on metal surface. M = metal atom, O = oxygen, -O = OH, α = H_2O.

Keller [5] found that the structures of the films formed in sulfuric, chromic, and phosphoric acid electrolytes, after anodizing, are basically similar. The oxide is found to consist of hexagonal cells whose outer and inner surfaces bulge into the solution and metal, respectively. Films formed in concentrated sulfuric acid are believed not to be pure Al_2O_3. For example, films formed in 10% H_2SO_4 at temperatures around 20°C were reported by Liechti and Treadwell [6] to contain 13% SO_3 as well as a few percent of bound H_2O. The films formed in the oxidizing acids often are porous, and have to be sealed to increase corrosion resistance. It is standard practice to seal films formed in sulfuric acid by immersion in boiling neutral water for 20 min. Electron diffraction studies by Harrington and Nelson [7] and Phillips [8] showed that boehmite [α-AlO(OH)] is produced in the film. Since abrasion removed the boehmite pattern, they concluded that the boehmite must have been on the surface. There is considerable controversy in the literature as to the necessity of aluminum metal being present to convert the oxide portion to boehmite.

For adhesive joints with weakest link at the adherend-adhesive boundary, the problem lies in identifying the locus or loci of failure—e.g., in the primer, at the primer-oxide interface, in the oxide, at the oxide-metal interface—and learning enough about the various interface phases to determine the cause of the weakness. In addition to the condition and properties of these phases immediately after surface preparation, one must be concerned with the stability or degradation of these properties in environments expected under factory conditions. That is: What changes will occur due to aging in humid atmospheres? What changes can occur due to inadvertent processing malfunction such as contaminated etching solutions, incorrect formulations, and applications? And so on.

1.3. The Solution

Once having established the locus and cause of failure or degradation, a corrective procedure can be designed into the process to remove the problem, or at least improve the process. In this study we correlate the properties of the metal adherend and adhesive (after a specific preparation but prior to bonding) with subsequent initial bond strength retention for various periods under varying environmental conditions. Characterization of the surface of metal samples is made with a variety of techniques. For example, the thickness and index of refraction of surface films is measured by ellipsometry. Chemical analysis of the surface layers and the chemical profile of films is made with Auger spectroscopy in conjunction with sputter-back etching by ion bombardment. Surface dielectric properties are determined by surface potential difference (SPD) measurements, and surface topography and morphology are determined by optical, electron, and scanning electron microscopy techniques. The sample surfaces and adhesives are characterized by contact angle measurements which delineate the surface tension properties. This allowed predictions of the wettability requirements between the adhesive and adherend to define strong adsorption bonding.

After preparation and characterization of the surfaces of the adherend and the adhesive, joints were fabricated and tested immediately after curing as well as after aging in environmental chambers. Examination of the fracture surfaces were made by optical and scanning electron microscopy and Auger spectroscopy to determine the region of failure (metal-oxide interface, oxide, oxide-adhesive interface, or adhesive) and the chemical constituents at the fracture interface.

Some of the metal samples that have been characterized were subjected to the same aging environments without any adhesive in order to follow the morphologies and chemical changes of the surface films that might be expected if the metal surfaces were exposed for some time or if the bonds were permeable to the environment.

SYMBOLS

Nomenclature for Wettability Parameters

θ	liquid-solid contact angle
γ_{LV}^{d}, γ_{LV}^{p}	respective dispersion (d) and polar (p) parts of liquid surface tension $\gamma_{LV} = \gamma_{LV}^{d} + \gamma_{LV}^{p}$
γ_{SV}^{d}, γ_{SV}^{p}	respective dispersion (d) and polar (p) parts of solid-vapor surface tension $\gamma_{SV} = \gamma_{SV}^{d} + \gamma_{SV}^{p}$
α_L, β_L	respective square roots of γ_{LV}^{d}, γ_{SV}^{p}
α_S, β_S	respective square roots of γ_{SV}^{d}, γ_{SV}^{p}
W_a	nominal value of interfacial work of adhesion
σ_c, σ_I	critical crack propagation stress defined by Griffith theory
E	Young's modulus
c	crack length
γ_G	Griffith surface energy
α_1, β_1	adhesive surface properties
α_2, β_2	surface properties of environment
α_3, β_3	adherend surface properties
S_L	liquid (or adhesive) spreading coefficient
R_0, R, H, K	wettability parameters related to α's and β's

Nomenclature for Surface Tools

Δ	ellipsometry phase shift of the parallel component with respect to perpendicular component of polarized light

ψ	arctangent of the amplitude ration
p	polarizer azimuth
a	analyer azimuth
n_s	real part of the complex refractive index of the substrate
κ_s	imaginary part of the complex refractive index of the substrate
$\hat{n}_s$	complex refractive index of the substrate $\hat{n}_s = n_s(1 - i\kappa_s)$
n_f	real part of the complex refractive index of the film
κ_f	imaginary part of the complex refractive index of the film
$\hat{n}_f$	complex refractive index of the film, $\hat{n}_f = n_f\,(1 - i\kappa_f)$
d	film thickness
λ	light wave length
ϕ	angle of incidence
$\tan \beta_0$	roughness factor

Adhesive Joint Parameters

ϕ'_c	fraction of bond area that failed in the adhesive
ϕ_I	fraction of bond area that failed at the interface
ϕ_v	fraction of bond area that failed in voids
ϕ_g	fraction of bond area that failed in the glass carrier
ϕ_c	fraction of bond area that failed in the adhesive and glass
σ'_c, σ_I, σ_v, σ_q, σ_c	bond strength in region corresponding to the σ values

Auger Electron Spectroscopy (AES)

E	retard voltage
$\frac{dN(E)}{dE}$	derivative of the energy distribution for secondary electrons
APPH	Auger peak-to-peak height in the Auger spectrogram dN(E)/dE vs. E

Surface Potential Difference (SPD)

$\Delta\phi$	change in work function
i	electrical current
R	resistance of the air gap
r	resistance of the electrometer
Γ	adsorbate concentration
μ	dipole moment of adsorbate molecules
μ_0	component of μ normal to the surface
α	polarizability of the adsorbate molecules
q	charge density of adsorbate
F	natural surface field
ξ	dielectric constant of the adsorbate

2. EXPERIMENTAL SECTION

2.1. Materials

2.1.1. Adherends

The metal adherends are bare 2024-T3 aluminum alloy sheet. Specimens were cut from 0.063-in. sheet with dimensions slightly larger than 4 × 1 in. and then milled to precisely 4 × 1 in.

2.1.2. Adhesives

The adhesive was an epoxy phenolic (HT424) from American Cyanamid. This adhesive has a glass fabric carrier and standard weight 0.135 ± 0.005 lb/ft^2. It was shipped in dry ice and stored at -10°C. Unfilled primer with parts A and B were used with the HT424 adhesive. The primer composition was 10 pbw HT424 part A, 10 pbw HT424 part B, and 30 pbw HT424 primer-thinner.

2.1.3. Surface Preparations

As has been shown [9] and as will be seen, every step of the surface preparation may be of significance with respect to bond strength. It is of the utmost importance that procedures be followed in every detail if experimental results are to be compared from laboratory to laboratory. Table A-1 (Appendix A) identifies the various surface preparation treatments for Al 2024-T3. Most of the work on

Al was done with the dichromate cleaning process in 2 of Table A-1 often referred to as the FPL (Forest Products Laboratory) etch. Approximate composition of the aluminum alloy is given in Table A-2.

2.2. *Surface Characterization Techniques*

2.2.1. Ellipsometry

Film thickness and film optical constants can be measured by ellipsometry. A good description of the principles of ellipsometry is given by McCrackin et al. [10]. In brief, when plane-polarized light is reflected from a surface the light becomes elliptically polarized. That is, the surface renders a phase shift (Δ) in the normal and parallel component of the polarized light as well as a change in the ratio of the amplitudes (tan ψ) of these components. The experimental arrangement is shown in Fig. 4-2. Monochromatic light is polarized and then compensated with a quarter-wave plate (compensator) such that after reflection the light is plane-polarized and can be crossed out by the analyzer. Values of Δ and ψ are obtained from the polarizer and analyzer azimuths at which the analyzer extinguishes the light. The experimental parameters measured with the ellipsometer are the polarizer aximuth p and analyzer azimuth a, which are directly related to Δ and ψ; $\Delta = 2p + \pi/2$ and $\psi = a$ where (Δ and ψ) = f(n_s, κ_s, n_f, κ_f, d, λ, θ, tan β_0). Subtracting 360° from Δ does not change its relationship to f( ), e.g., if Δ is 50°, it can just as well be expressed as −310°. The complex refractive index of the substrate is $\hat{n}_s = n_s(1 - i\kappa_s)$ and the film $\hat{n}_f = n_f(1 - i\kappa_f)$, where n_s and n_f are the refractive index and κ_s and κ_f are the absorption index. The other parameters are film thickness d, light wavelength λ, angle of incidence ϕ, and tan β_0 is the mean slope of undulations due to surface roughness. The angle of incidence was 70° and the wavelength of the He-Ne lazer used was 6328 Å. The optical constants for Al at this wave length are approximately $n_s \approx 1.43$, $\kappa_s \approx 5.17$; some measurements were also made at λ = 5461 Å for which $n_s \approx 1.0$, $\kappa_s \approx 6.7$. The computer program searches for solutions of d and n_f for a specified range of κ_f. The search for solutions covered n_f = 1.0 to 3.0, for κ_f from 0 to 1.0 in increments of 0.1. Fenstermaker and McCrackin [12] have shown that surface roughness can seriously affect the ellipsometer parameters; therefore, the p and a or Δ and ψ values are recorded and the computer solutions are considered to be only approximate. Unless two of the parameters d, n_f, κ_f, and tan β_0 are known from independent measurements, an unambiguous solution set (d, n_f, κ_f, tan β_0) cannot be established by ellipsometry. However, order of magnitude values can be obtained and most probable values estimated. Since values of n_f for bulk oxides are known and since bulk oxides are usually transparent ($\kappa \approx 0$) to 6328-Å or 5461-Å light, the most probable solution set is

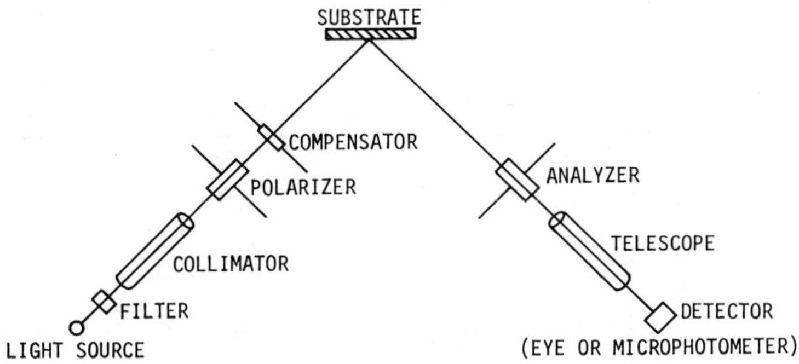

FIG. 4-2 Schematic representation of the optical arrangement of the ellipsometer.

for a minimum value of κ_f and n_f value near the bulk values. The minimum values of κ_f is an effective value which may reflect the unknown roughness parameter tan β_0. The values for n_f for oxides of aluminum [13] are between 1.5 and 1.8.

2.2.2. Auger Spectroscopy

Auger spectroscopy involves the analysis of secondary electrons that are emitted from a surface that has been bombarded with a monochromatic electron beam. The Auger electron spectrogram (AES) reveals the elements that are present in the first three or four atom layers at the surface. A good description of Auger spectroscopy can be found in a paper by Simmons [14]. An example of an AES is given for Al 2024-T3 that had the FPL-etch treatment (no. 2 of Table A-1) in Fig. 4-3. Figure 4-4 shows the AES after ion-sputtering the oxide film away. The abscissa E is the retarding voltage and the ordinate dN(E)/dE is the derivative of the energy distribution curve for secondary electrons. Although progress is being made [15] in the interpretation of the AES with respect to the quantitative chemical composition of a surface, at present only semiquantitative interpretation can be made. Comparison of Auger peak-to-peak height (APPH) for a given peak is significant with respect to changes of surface composition after various surface treatments. The significance of the APPH for different peaks of a given spectrum is not known at this time.

2.2.3. Chemical Profiles

Monitoring the chemical composition of a surface as the film is sputter-etched by ion bombardment is a very powerful tool. However, at present the interpretation of these profiles are only semiquantitative.

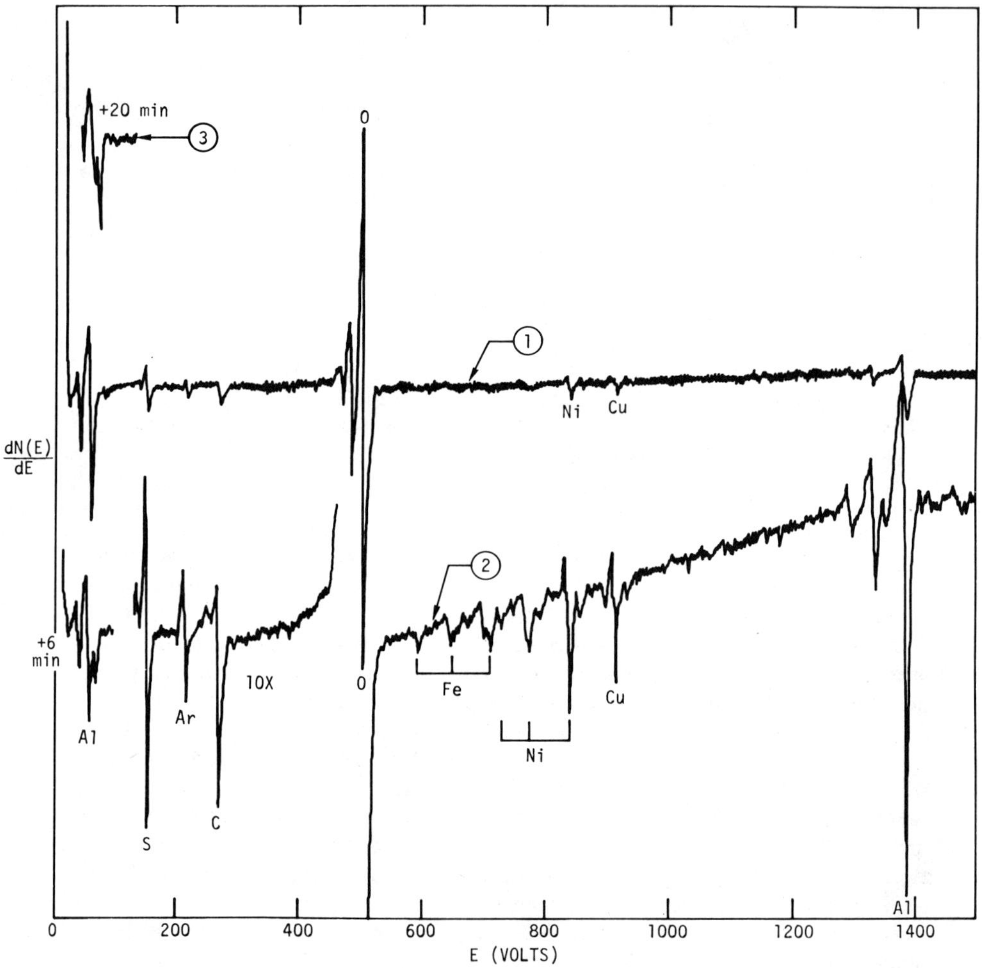

FIG. 4-3 An Auger electron spectrogram of an FPL-etched Al 2024-T3 sample.

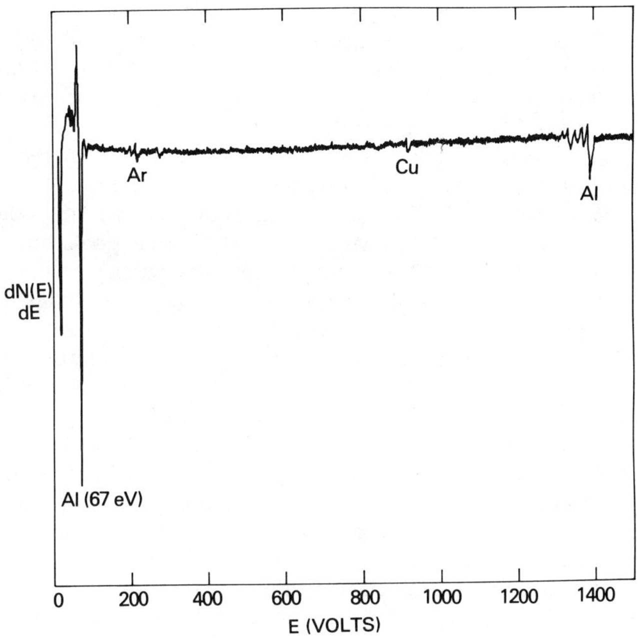

FIG. 4-4 An Auger electron spectrogram of an FPL-etched 2024-T3 sample after ion-sputter-etching through the oxide film.

To yield quantitative chemical profiles much has yet to be learned. As a first approximation the APPH of the derivative dN(E)/dE, for a particular Auger transition, is proportional to the concentration of the particular element, subject to the validity of the following assumptions:

1. The mean electron escape depth is small compared to the film thickness.
2. APPH is directly proportional to the Auger current.
3. The electron beam does not change the atomic ratios at the surface.
4. The ion beam does not change the atomic ratios at the surface.
5. The electron beam diameter is small compared to the ion beam diameter.

The mean escape depth for metals and oxides [16] range from 5 to about 30 Å for electron energies to about 2 KeV. Therefore, AES monitors the outer four or five monolayers and assumption 1 is good for films $>\sim 100$ Å. Tarng and Wehner [17] give a detailed analysis of the relationships and parameters involved.

Although assumption 2 has generally been accepted, Grant et al. [18] have shown that in some cases assumption 2 is not valid due to the shape of the Auger derivative peaks. They show that this problem can be overcome by integration of the AES and measuring the currents directly from the area under the integrated peaks.

We have discovered assumption 3 to be invalid for electron currents usually used for AES (20-100 μA). This is demonstrated in Fig. 4-3 for curves 1, 2, and 3. Curve 1 is for the initial spectrum which took about 6 min. It will be noted that the low-energy aluminum peak in curve 1 is at 55 eV. This 55-eV peak is associated [19] with a transition between oxygen and aluminum and therefore is related to aluminum oxide and not pure aluminum. After 6 min of electron gun current (at 90 μA) the 55-eV peak has diminished and the 67-eV peak appears. The 67-eV peak is associated [19] with pure aluminum and not the oxide. After 20 min of electron current, curve 3 shows that the 55-eV peak has further diminished and the 67-eV peak has increased. Experiments on pure single crystal α-alumina showed the same trends. It can only be concluded that the electron beam is transforming the outer layers of the oxide to pure aliminum. This effect was not discovered until after most of the Auger work reported in this report was done. Experiments have shown that the use of gun currents between 1 and 10 μA greatly reduces this problem. Although some loss in sensitivity results, Auger measurements with minimum electron exposure should increase the validity of assumption 3 to an acceptable level. Assumption 3 should be approximately valid for AES results near zero sputter time in the profile figures since the surfaces had been exposed to very little electron current.

Shimizu et al. [20] have shown that unless the ion yield is the same for each element, some changes will occur in the composition of the outer layers during ion sputter-etching. We have observed that the ion sputter yields are much closer for metal and oxygen than the electron beam yield; however, this problem may be more important than the electron yield problem because one cannot lower the ion current and still etch through the film. Very little is known about the validity of assumption 3 at this time.

Assumption 5 is easily made valid since the ion beams can be large, >1 cm diameter, and the electron beams very small, <0.05 cm diameter. In our system the electron beam was ~0.05 cm diameter and the ion beam ~1 cm diameter and the position of the beam on the sample could be observed with a telescope mounted on the ultra high vacuum (UHV) system.

In spite of the problems mentioned, the AES is a very powerful tool for semiquantitative analysis of surfaces and will become increasingly more powerful as more is learned about Auger processes and sputter-etching.

2.2.4. Surface Potential Difference

The surface potential difference (SPD) is the potential between a reference electrode and the sample to be studied. It arises from the work function difference $\Delta\phi$ of the two surfaces. If the reference electrode is inert, any changes in SPD will be due to changes that occur on the sample surface. Any process that will change the work function of the sample surface will affect SPD. Therefore, the SPD is related to the structure and dielectric properties of the surface film. Due to the large number of property changes that can affect SPD, its usefulness is quite limited if measured by itself. However, if the SPD is measured in conjunction with other measurements such as ellipsometry, AES, scanning electron microscopy, etc., it can add valuable information for the interpretation of the structural and chemical properties of films.

Figure 4-5 is a schematic current diagram used for the measurement of SPD. The current i flowing in the circuit can be expressed by Ohm's law:

$$i = \frac{\Delta\phi}{R + r} \tag{1}$$

where $\Delta\phi$ is the SPD, R is the air gap resistance between the sample surface and reference electrode, r is the electrometer resistance, and e is the electrometer voltage reading. A radioactive substance sealed behind a metal foil on the reference electrode is used to provide ionization of the gas between the electrodes, thus reducing R such that $R << r$. In this case,

$$i = \frac{\Delta\phi}{r} = \frac{e}{r} \tag{2}$$

and $\Delta\phi = e$. That is, the electrometer yields SPD directly. To demonstrate that $R << r$, e is measured with increasing distance d between the electrodes; as long as e is independent of d, $R << r$. Our reference electrode has Am^{241} sealed behind gold foil and e proves to be constant with distance between 8 and 3 mm (see Fig. 4-6). Our measurements were made with about 5 mm between the reference electrode and sample. The gold foil was sprayed with a thin layer of Teflon to maintain its inertness.

The most comprehensive analysis of the relationship of the measured values of the surface potential difference $\Delta\phi$ to the molecular adsorbed state has been given by MacDonald and Barlow [21,22]. Elementary electrostatics applied to a continuous dipole sheet leads to the Helmholtz equation [23]:

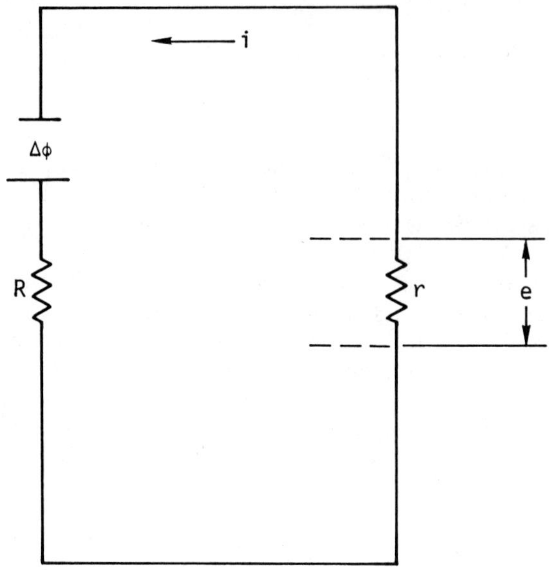

FIG. 4-5 Schematic equivalent circuit diagram of the ionization technique for measuring the surface potential difference (SPD).

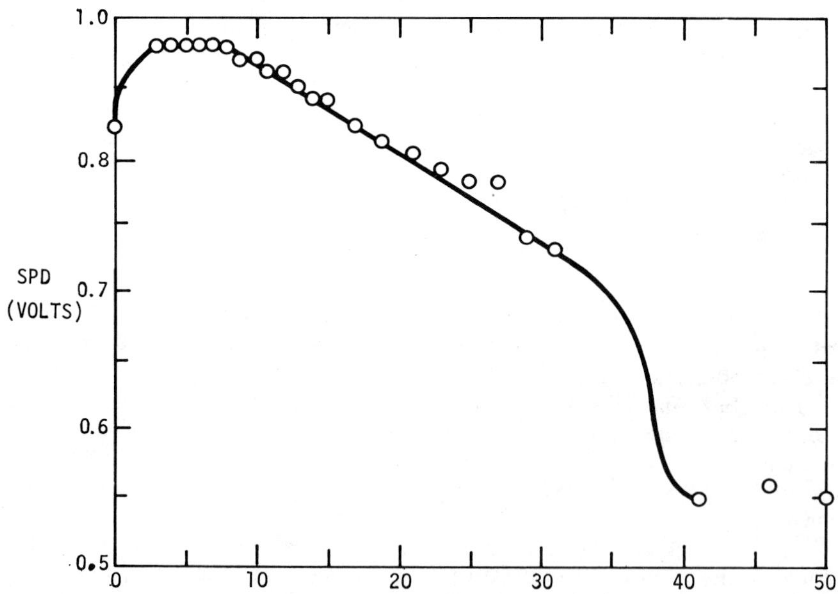

FIG. 4-6 Variation of SPD with electrode distance from an FPL-etched aluminum sample.

$$\Delta\phi = \pm 4\pi\mu\Gamma \tag{3}$$

where μ is the dipole moment and Γ is the adsorbate surface concentration. This equation was modified to account for the polarizability of real dipoles and their discrete nature. A depolarizing field at the position of a given dipole arises from all surrounding dipoles. The modification to Eq. (3) is

$$\mu \approx \frac{\alpha(4\pi q + F) + \mu_0}{\xi} \tag{4}$$

where q is the charge density of the adsorbate, F is the natural surface field, α is the polarizability of the adsorbate, μ_0 is the component of the permanent dipole moment normal to the surface, and ξ is the dielectric constant of the adsorbate film. The sign is positive if the positive end of the dipole is pointed away from the surface. The effect of oxide films on SPD is discussed by Boggio and Plumb [24].

Because the SPD is relative to the reference electrode, the absolute values reported here are not of significance. However, the values of SPD for one surface treatment relative to another is of significance, the change being due to differences in the work function of the two samples.

2.2.5. Microscopy

Optical, scanning electron, transmission electron, and interference microscopy are used in this study. These microscopy studies allow morphology, topography, and structure of films to be revealed. The electron microscope is a Philips Model 300, the SEM is an ETEC autoscan, and the interference microscope is a Zeiss.

2.2.6. Contact Angles

This section will introduce and briefly discuss the concepts currently applied to the analysis of liquid-solid contact angle measurements. Zisman and coworkers [25] developed detailed procedures for experimental measurement of liquid-solid contact angles θ_{LS}. Zisman showed that a plot of cos θ vs. liquid surface tension γ_{LV}, for a homologous liquid series against a common solid tends to produce a linear curve of cos θ vs. γ_{LV}. Extrapolation of cos θ vs. γ_{LV} curve to cos $\theta = 1.0$ defines a critical surface tension for wettability γ_C for the solid with the prediction that when $\gamma_{LV} \leqslant \gamma_C$, the liquid will spontaneous wet and spread (with $\theta = 0$). A number of refinements in the analysis of contact angle data have been more recently introduced to define surface energetic criteria for adhesive bonding and fracture. Recent reviews of this subject by Kaelble [26,27] provide details of the surface energy analysis applied to the wettability measurements conducted in this study.

Present theory provides an equation for work of adhesion which states [28]:

$$\frac{W_a}{2} = \alpha_L\alpha_S + \beta_L\beta_S \tag{5}$$

where

$$\gamma_{LV} = \gamma_{LV}^d + \gamma_{LV}^p = \alpha_L^2 + \beta_L^2 \tag{6}$$

$$\gamma_{SV} = \gamma_{SV}^d + \gamma_{SV}^p = \alpha_S^2 + \beta_S^2 \tag{7}$$

$$W_a = \gamma_{LV}(1 + \cos\theta) \qquad \theta \geqslant 0 \tag{8}$$

A new basis for designing wettability experiments which becomes evident by rearranging Eq. (5) as follows:

$$\frac{W_a}{2\alpha_L} = \alpha_S + \beta_S\frac{\beta_L}{\alpha_L} \tag{9}$$

Equation (9) points out that a plot of $W_a/2\alpha_L$ vs. (β_L/α_L) defines α_S as the intercept at $\beta_L/\alpha_L = 0$ and β_S as a slope. Thus by selecting test liquids with varied values of surface tension γ_{LV} and (polar/dispersion) $= \beta_L/\alpha_L$ character a means becomes available to isolate two solid-vapor surface tension properties as defined by Eq. (7).

Based on the detailed knowledge of solid surface properties, as defined by Eq. (7), new and more detailed surface-energetic criteria for adhesive bonding become available. The concept of the adherend "wettability envelope" was introduced in our work to define the equilibrium conditions under which an adhesive phase will spontaneously wet and spread over an adherend surface to displace the ambient air phase and eliminate interfacial cavities. The wettability envelope represents a critical function of γ_{LV}^d and γ_{LV}^p which can be generated by introducing polar coordinates T and θ defined by

$$T = \left[\left(\gamma_{LV}^d\right)^2 + \left(\gamma_{LV}^p\right)^2\right]^{1/2} \tag{10}$$

$$\tan\theta = \frac{\gamma_{LV}^p}{\gamma_{LV}^d} \tag{11}$$

Substituting Eqs. (10) and (11) into the equation for the liquid (or adhesive) phase spreading coefficient,

$$S_L = W_a - 2\gamma_{LV} \tag{12}$$

it follows that for spontaneous spreading to occur where $S_L \geqslant 0$ the wettability envelope for the adherend surface is defined by the region of γ_{LV}^d vs. γ_{LV}^p bounded by the following function:

$$T(\theta) = \left[\frac{(\gamma_{SV}^d \cos\theta)^{1/2} + (\gamma_{SV}^p \sin\theta)^{1/2}}{\cos\theta + \sin\theta}\right]^2 \tag{13}$$

If a combined value of γ_{LV}^d and γ_{LV}^p for the liquid (or adhesive) phase falls within the region enclosed by Eq. (13), that adhesive is predicted to spread. Conversely, if the surface energy properties of the adhesive are outside the wettability envelope, nonspreading and weak bonding with interfacial cavities is predicted.

Environmental effects on adhesive joint strength have recently been defined in terms of the critical Griffith surface energy γ_G for crack propagation. The new definition of γ_G appears in the following special relation for the critical stress σ_c for crack opening under normal stress loading:

$$\sigma_c = \left(\frac{2E\gamma_G}{\pi c}\right)^{1/2} = \left(\frac{2E}{\pi c}\right)^{1/2}(R^2 - R_0^2)^{1/2} \tag{14}$$

where

$$R_0^2 = 0.25[(\alpha_1 - \alpha_3)^2 + (\beta_1 - \beta_3)^2] \tag{15}$$

$$R^2 = (\alpha_2 - H)^2 + (\beta_2 - K)^2 \tag{16}$$

$$H = 0.50(\alpha_1 + \alpha_3) \tag{17}$$

$$K = 0.50(\beta_1 + \beta_3) \tag{18}$$

where the subscripts relate respectively to 1 = adhesive, 2 = environmental immersion phase, and 3 = adherend, and $\alpha = (\gamma^d)^{1/2}$ and $\beta = (\gamma^p)^{1/2}$ of these phases. Equation (14) provides a means of predicting the suppression of the critical Griffith stress σ_c which may be induced by modifying the surface energy properties of the environment as defined by $\alpha_2 = \beta_2 = 0$ for dry air to $\alpha_2 = 4.67\ (\text{dyn/cm})^{1/2}$ and $\beta_2 = 7.14\ (\text{dyn/cm})^{1/2}$ for water immersion [34].

2.3. *Adhesive Primer Procedures*

2.3.1. Applications and Cure Cycles

The priming procedure for Al 2024-T3 with HT-424 A&B primer is as follows: Freshly treated coupons are spray-primed to an approximate thickness of 0.001 in. using an atomizer operated on low-pressure, clean, dry (argon) gas. The primer has the following composition:

10 pbw	HT-424	part A
10 "	"	part B
30 "	"	primer-thinner

The primed coupons are dried for 30 min under ambient laboratory conditions. The primed coupons are then oven-dried for 60 min at 66°C. This priming procedure is recommended [29] by the manufacturer.

2.4. *Bonding Procedures*

2.4.1. Bonding Fixture

A 1 × 0.5 in. section of adhesive is tacked onto the treated and primed end of a 4.000 × 1.000 in. coupon to be bonded. This coupon is then positioned against one end of a 1.000 × 7.500 in. hole cut in a 0.126-in.-thick aluminum sheet (see Fig. 4-7). Six such sections have been cut in this sheet allowing up to six bonds to be prepared at once. At the other end of the cutout section a spacer is positioned as in Fig. 4-8. Then the other half of the bond which is also 4.000 in. long and 1.000 in. wide is placed against the other end of the cut-out section on top of the spacer. From Figs. 4-7 and 4-8 we see that by this arrangement the two coupons making up the bond are prevented from moving horizontally and the overlap between the two bonding coupons is maintained at the desired 0.500 in. Measurement of the

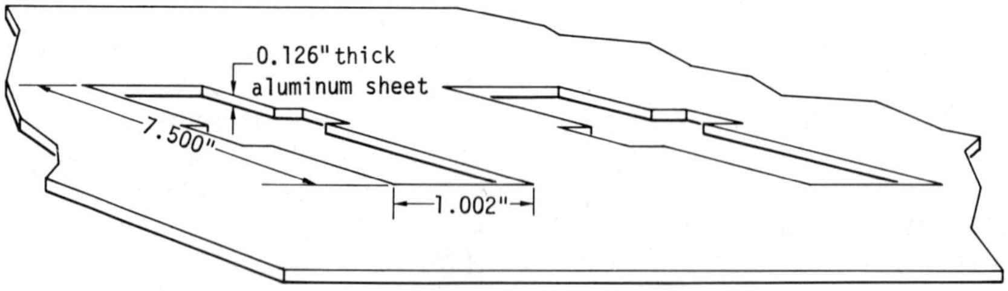

FIG. 4-7 Schematic drawing of cut-out sections for the bonding jig.

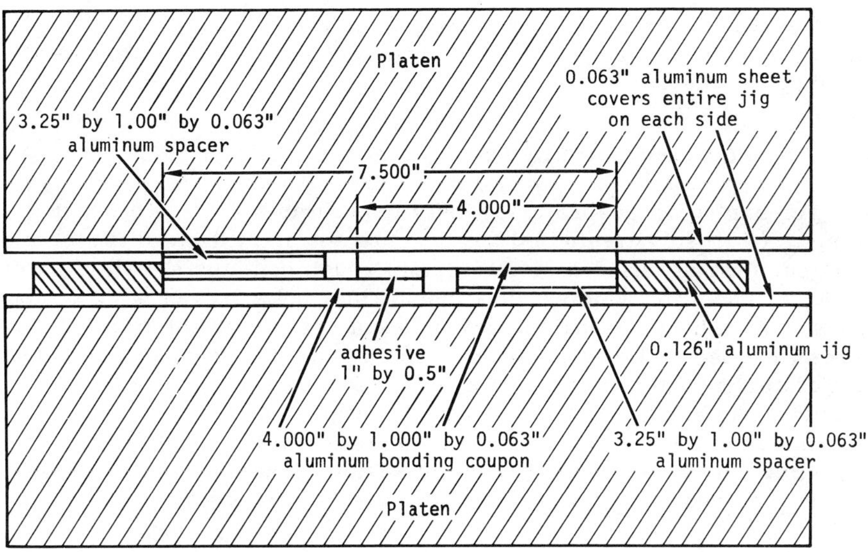

FIG. 4-8 Schematic drawing of bonding jig and press.

glue line thickness showed that the glass carrier in the HT-424 adhesive maintains the glue line thickness at 0.008 in. As shown in Fig. 4-7, a 1/8-in. overcut around the bonding area allows for excess adhesive to flow out without bonding to the jig. A 0.063-in.-thick aluminum sheet with an aluminum foil cover to prevent bonding to it by excess adhesive is then taped over the jig on each side. This assembly is then placed in a press at ~50 psi during the cure cycle.

2.4.2. Cure Cycles

The cure cycle for the HT424 system was that recommended by the manufacturer [29]:

1. Raise temperature to 171°C (340°F) in 60 min.
2. Cure for 60 min at 171°C (340°F).
3. Cool down.

2.5. Tensile Lap Shear Testing

1. 1-1/2 × 1 × 0.063 in. aluminum shims are bonded to the inside (bonded side) ends of the bonded pieces.
2. Bonded pieces are clamped in 0.25-in. jaws of floor model Instron testing machine leaving 4.5 in. jaw-to-jaw distance (2 in. from each glue line).

3. Bonds are sheared at crosshead rate = 0.01 in./min.
4. Strip chart recorder records stress in lbs vs. strain (crosshead displacement).

2.6. Surface-aging Effects

For surface exposure time (SET) experiments 4 × 1 × 0.063 in. samples of freshly treated Al 2024-T3 are set in an 8-in.-diameter glass chamber (usually used as a desiccator) which contains a dish of saturated K_2SO_4. The 1 in. wide samples just fit into holes in a ceramic bottom plate, which hold them in approximately a vertical orientation. The lid is then secured on with tape, as no stopcock grease is used. The desiccator is then placed in an oven at 55°C and the saturated salt solution provides a 95% RH condition. Samples are then withdrawn at various SETs. For contact angle measurements 4 × 1 × 0.063 in. samples were cut into 4 × 0.12 × 0.063 in. strips and placed in the chamber. After SET these strips were then cut into 0.5 in. lengths.

2.7. Bond-joint-aging Effects

Bond exposure time (BET) experiments at 54°C and 95% RH are also carried out in the glass chamber with a saturated K_2SO_4 solution to provide constant humidity.

3. EXPERIMENTAL RESULTS AND DISCUSSION

3.1. Surface Character vs. Joint Strength

3.1.1. Ellipsometry

The average values of Δ and ψ for 22 different samples with the FPL etch are 110.5 ± 2.5 and 37.6± 0.3, respectively. The wavelength was 6328 Å and angle of incidence 70°. Nomographs of Δ vs. ψ are plotted in Figs. 4-9 and 4-10 for various film thickness and optical constants. The average values of Δ and ψ for the FPL etch on 2024-T3 aluminum fall outside the curves for reasonable values of n_f between 1.5 and 3.0 and d between 0 and 2000 Å for transparent films ($\kappa_f = 0$). The values of Δ and ψ for the FPL etch fall within reasonable limits if the effective value of κ_f is assumed to be ~0.3, as in Fig. 10. The best estimate of film thickness is obtained as ~200 Å if $n_f \approx 1.7$, as for bulk oxide.

According to the literature [30], anodized aluminum in borate solutions produces oxide films ~14 Å/V with optical constants similar to bulk α-alumina ($n_f = 1.7$, $\kappa_f = 0$). To establish the reliability of our

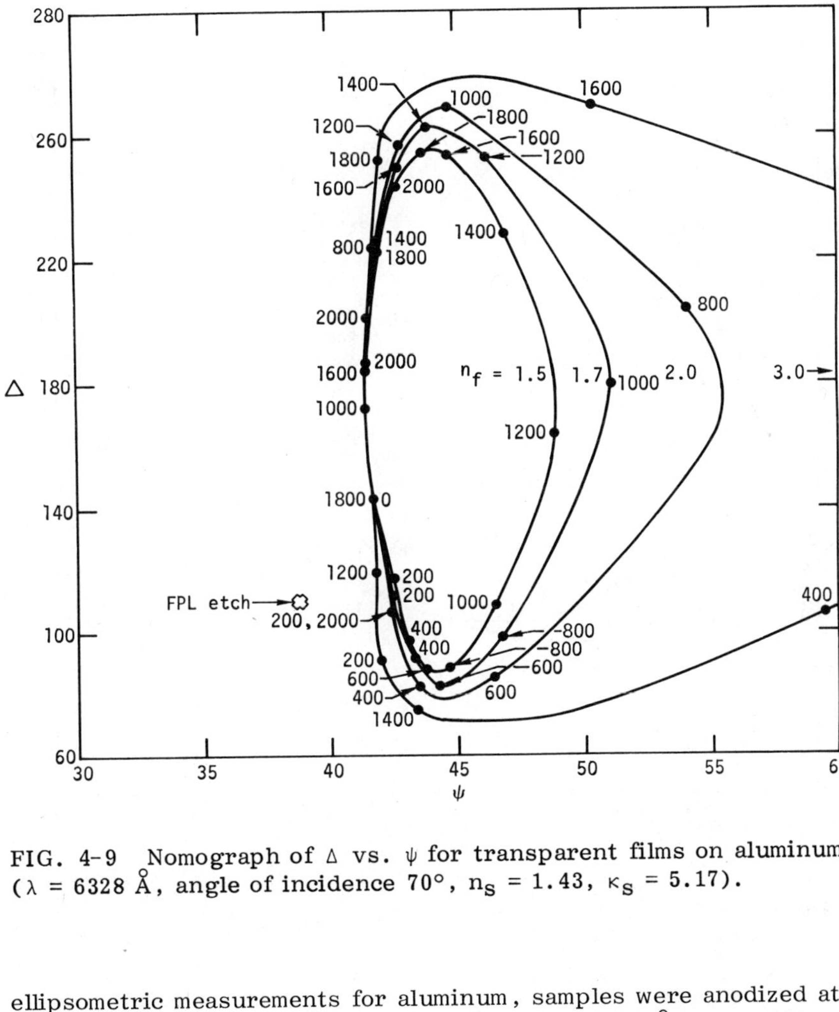

FIG. 4-9 Nomograph of Δ vs. ψ for transparent films on aluminum (λ = 6328 Å, angle of incidence 70°, n_s = 1.43, κ_s = 5.17).

ellipsometric measurements for aluminum, samples were anodized at various voltages to yield films between 50 and 1000 Å. Our ellipsometric results were within 5% of the values predicted from the literature (i.e., 14 Å/V, n_f = 1.7, κ_f = 0). The thickness of oxide on FPL-etched aluminum has been estimated by another independent technique, i.e., photoelectron emission. Photoelectron currents from aluminum that had been anodized to thicknesses between 50 and 1000 Å (with light of λ = 2500 Å, $h\nu \approx$ 5 eV), yielded a calibration curve. The photoemission current from an FPL-etched sample corresponded to 210 Å on the calibration curve.

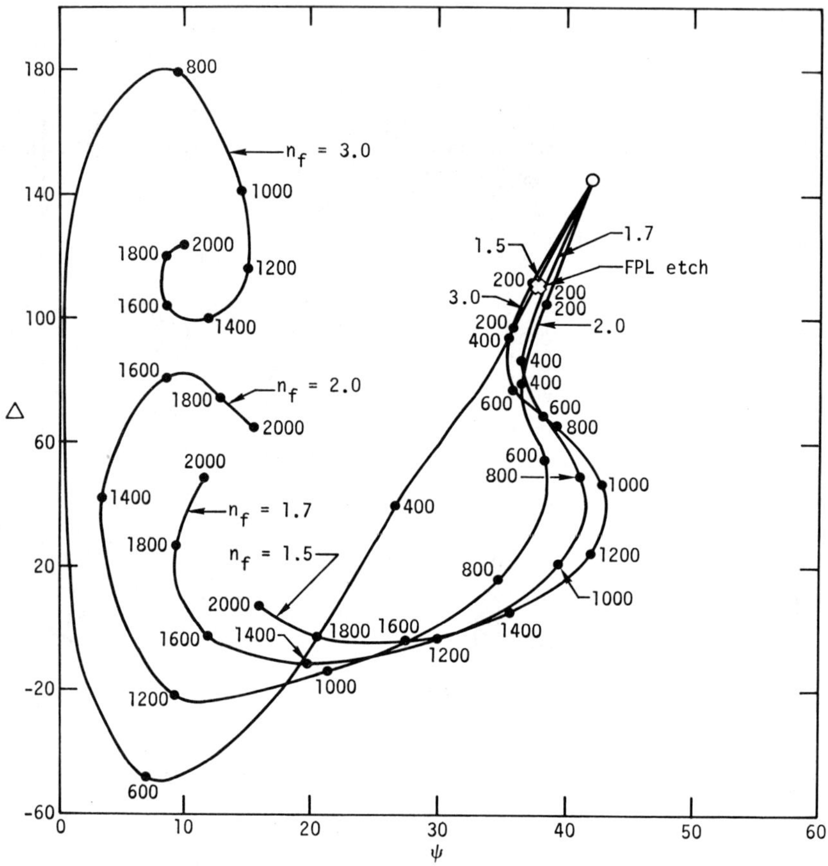

FIG. 4-10 Nomograph of Δ vs. ψ for an absorbing ($\kappa_f = 0.3$) film on aluminum.

Attempts to remove FPL-etch films by the methods of Hamilton and Lyerly [31] were not successful because the films appeared to grow in thickness. Lyerly (personal communication) confirmed they had observed the apparent film growth problem and report films of thicknesses from 650 μm to 4500 μm.

It should be noted in Fig. 29 of the Picatinny report [31] that alkaline-cleaned 2024-T3 aluminum produced a thick cracked film estimated to be about 1000 μm thick from the protruding layer. One might expect that films 4500 μm thick from the FPL etch might also crack and be revealed by scanning electron microscopy (SEM). However, none of the SEM pictures in the Picatinny report or in our SEM pictures reveal such films. It is believed that FPL-etch films were grossly thickened by the removal process reported in the Picatinny report [31] and that oxide films left on aluminum from the FPL etch are very thin (~200 Å).

The estimate of thickness of the oxide film on vapor-degreased 2024-T3 aluminum is 280 Å and $\kappa_f \approx 0.5$ if $n_f = 1.7$, and Table 4-1 shows ellipsometric data for plasma-treated 2024-T3 aluminum. These experiments were carried out in a glass chamber in order to control the atmosphere during and after the plasma treatment. The purpose of the plasma treatment was to remove surface layers prior to bonding. In order to make ellipsometric measurements through the glass chamber, it was necessary to provide birefringence-free windows. The first two rows of data in Table 4-1 indicate that the presence of the windows had a very small effect on Δ and ψ. For these experiments the angle of incidence was 60° and the wavelength 5461 Å. A 1/2-hour plasma treatment removed approximately 23 Å of the film and the next half-hour only removed about 2 Å more. The surface pretreatment was an alkaline-clean process. A similar small decrease in film thickness and saturation effect is noted in Table 4-2 for the vapor-degrease pretreatment, whereas continued plasma exposure continues to remove film from an FPL-etch-treated surface. For the experiments reported in Table 4-2, the samples were removed from the plasma chamber to make the ellipsometric measurements. The plasma etching rates were much lower than had been hoped for, but it is expected that outer surface contamination is removed by the process. Values for estimated film thickness and optical constants for the various surface treatments are reported in Table 4-3.

TABLE 4-1 Ellipsometer Study of Plasma-Treated Al 2024-T3 (λ = 5461 Å, ϕ = 60°)[a]

Time in plasma (min)	Power (W)	Δ	ψ	n_f	κ_f	Thickness d (Å)	δd (Å)
0	No windows	116.8	34.50	1.8	0.6	310	0
0	Windows	117.0	34.85	1.7	0.6	315	0
10	50	117.5	34.65	1.8	0.6	301	14
20	100	117.9	34.72	1.9	0.6	296	19
30	150	119.4	35.3	1.8	0.6	282	23
60	200	120.6	36.0	1.8	0.6	280	25

[a]See no. 3 of Table A-1.

TABLE 4-2 Surface Potential and Ellipsometric Study of Plasma Treated Al2024-T3 (Measurements in Air)[a]

Plasma treatment time (hr)	Initial treatment	Average surface potential difference (V)	Average thickness (Å)	δd (Å)
0	Vapor degrease	0.31	250	0
0.25	" "	0.15	217	33
1.25	" "	0.25	217	33
2.91	" "	0.30	217	33
0	FPL etch	0.26	250	0
0.25	" "	0.02	220	30
1.25	" "	-0.01	210	40
2.91	" "	0.03	195	55

[a]See nos. 1 and 2 of Table A-1.

3.1.2. Auger Spectroscopy

The initial Auger peak-to-peak height (APPH) for various elements on the aluminum surface after the vapor degrease (no. 1 of Table 4-1) and the FPL etch (no. 2 of Table 4-1) are also recorded in Table 4-6. No Auger results were obtained for the plasma-treated surface. Relative to the FPL etch, the vapor-degrease treatment leaves considerably more Mg, Si, C, Cl, and S at the outer surface. The aluminum peak on the vapor-degreased sample is masked by the contamination. Other than aluminum and oxygen, Fe, Ni, Cu, C, and S were detected on the FPL-etch sample if the sample was placed in the vacuum chamber within 1 hr of treatment. It should be noted that no chromium is observed. Bowen [32] reports observation of Cr on the surface immediately after placing the sample in the vacuum system but that it is removed by the vacuum system in a few hours. Our samples were not measured with Auger for about 12-24 hr after being placed in the vacuum system and this may be why no Cr was observed.

3.1.3. Chemical Profiles

Figure 4-11 shows chemical profiles of two identically treated 2024-T3 aluminum samples with the FPL etch which had been exposed to laboratory air for some hours. The solid points are for one sample and the open circles for the other. The reproducibility is excellent for the APPHs. The Ar^+ ion-sputtering was done at about 10^{-5} torr,

TABLE 4-3 Surface Characterization of Al 2024-T3 Samples and Bond Strengths for Various Surface Preparations (Adhesive HT-424)

Surface preparation	Surface characterization: Ellipsometry: Film thickness (Å)	n	κ	SPD ΔV (V)	γ_{SV} dyn/cm	Auger spectroscopy APPH: Al	Cu	Mg	Ni	Fe	Si	C	O	Cr	Cl	S	Bond strength: tensile shear strength (psi)
Vapor degrease	280	1.7	0.5	0.56	45	5	0	29			14	65	100	0	12	6	2367 ± 50
FPL etch	200	1.7	0.3	0.48	80	38	4	0	4	0.7	0	0.5	87	0	0	0.5	3080 ± 250
Plasma	296	1.9	0.6	—	64												3130 ± 263

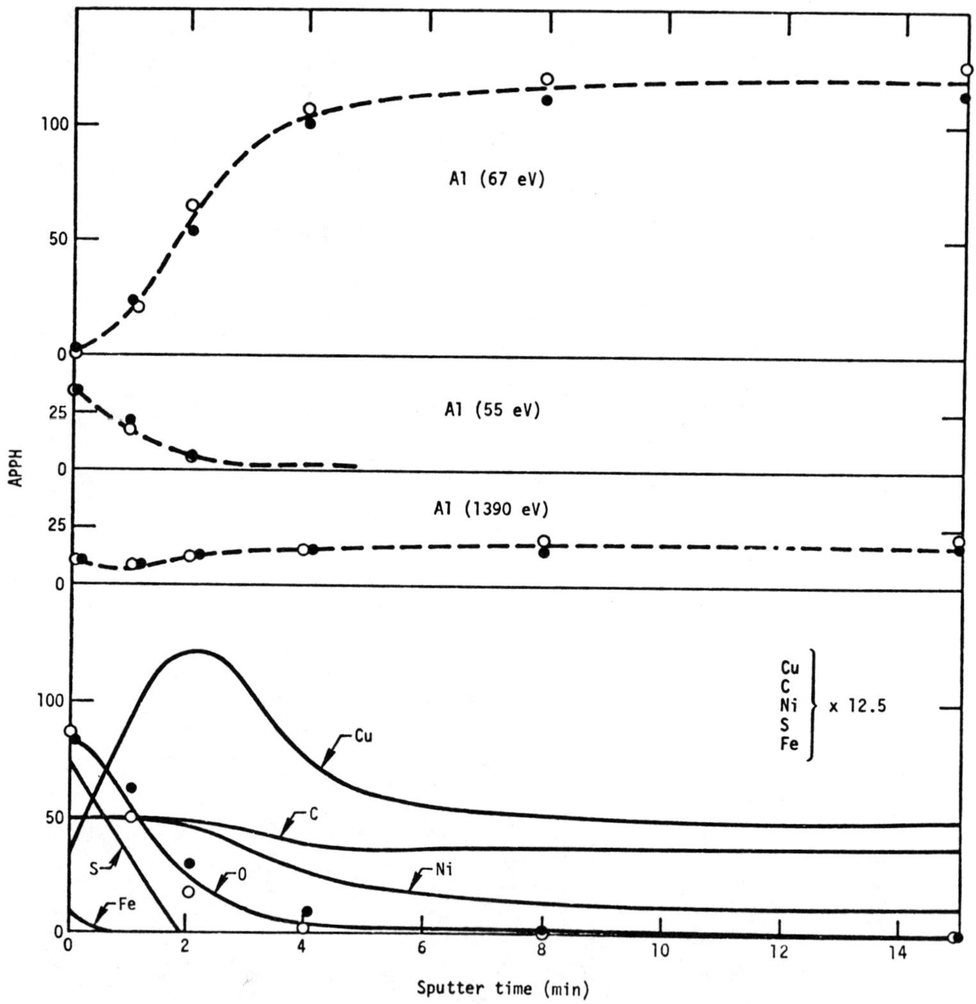

FIG. 4-11 Chemical profiles of oxide films formed on Al 2024-T3 by the FPL etch.

2 kV, ~100 μa/cm^2 over an area of ~0.25 in. diameter. As pointed out previously, the interpretation of these profiles with respect to actual atomic concentrations is too difficult at present. The value of the profiles lies in qualitative and semiquantitative analysis. The APPHs for Cu, C, Ni, S, and Fe have been expanded by a factor of 12.5 with respect to the APPH for Al and oxygen. The alloy elements Al, Cu, and Fe (see Appendix A, Table A-2) are observed in the oxide film and the substrate metal. The Cu has segregated to higher concentration in the oxide than in the metal and Fe is in the outer atom layers only. Within the resolution of the spectrometer, the alloy elements Mg, Si, Zn, Cr, and Mn are not observed, whereas Ni is observed. Carbon contamination is probably from laboratory air or from adsorption of CO in the UHV system.

Figure 4-12 shows the profile of the back side of a sample that had the FPL etch and then exposed to 55°C and 95% RH for 48 hr, bonded, and fractured. The low-energy Al (55-eV and 67-eV) peaks were not resolved in these measurements, the curve-labeled Al (63-eV) reflects mostly the Al (67-eV) peak. Initial carbon contamination was much greater for the 48-hr SET sample (e.g., APPH ≈ 34) as compared to the fresh sample (APPH ≈ 50/12.5 = 4). Cu, Mg, and Fe were observed, but no Ni was observed, and S and N in addition to C was observed. Since the electron beam diameter was small (~0.04 cm^2), an attempt was made to profile the adhesive side of the fractured couple. Figure 4-13 is a profile in an area where interfacial failure occurred. The striking feature of Fig. 4-13 is the high APPH value for Si in the failure area. This may indicate that interfacial failure (weak oxide) is associated with areas that were contaminated by the carrier glass fibers or silicones from the atmosphere. Figure 4-14 shows profiles as the electron beam was moved slightly into the adhesive region. No Si is observed and as expected the C peak increased dramatically. It should also be noted that the sputter rate has been decreased by a factor of 10 for the adhesive side (Figs. 4-13 and 4-14).

3.1.4. Surface Potential Difference

If the reference electrode changes over a period of time, comparison of SPDs becomes difficult. It is therefore useful to compare SPD values for a series of measurements made at one time—for example, as in Table 4-2. The plasma treatment decreased SPD from 0.31 to 0.15 V and then increased it to 0.3 again for the vapor-degrease pretreatment. Plasma treatment of the FPL-etch sample decrease SPD from 0.26 to ~0 V where it remained.

In a previous set of experiments SPD for a vapor-degreased sample was 0.56 V as compared to 0.48 V after the FPL etch, a decrease of 0.08 V. This decrease can be compared with a decrease of 0.31 - 0.26 = 0.05 V in Table 4-2.

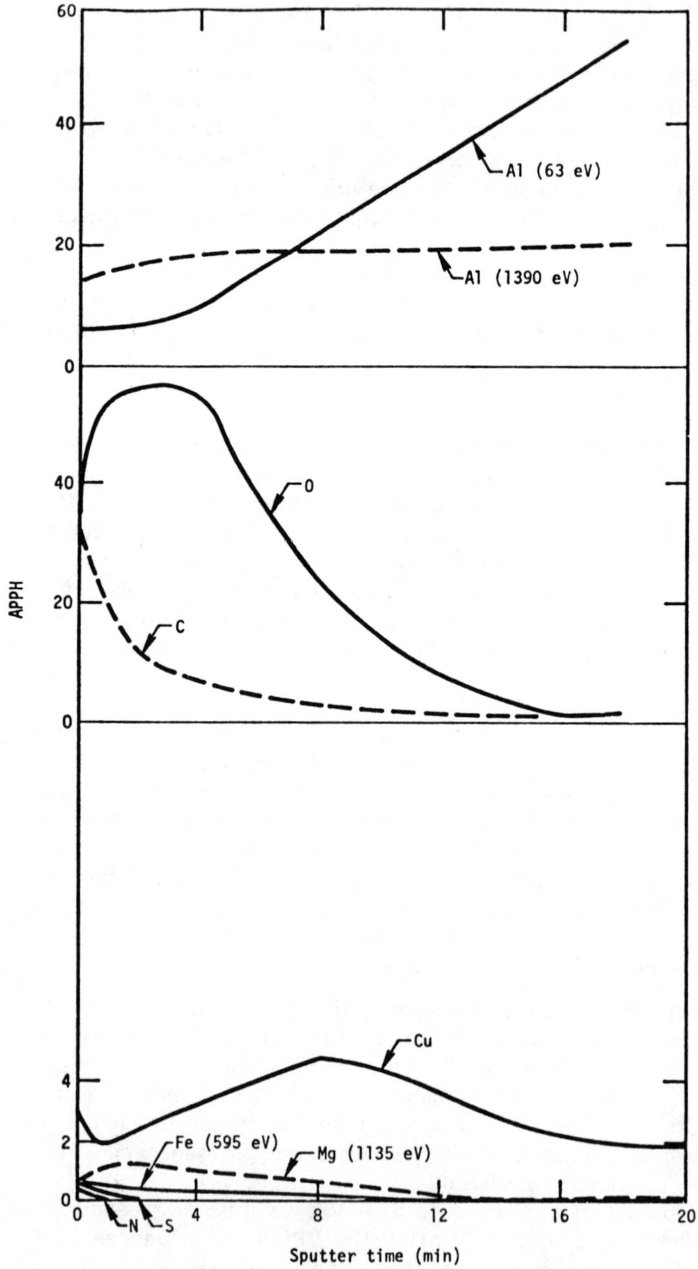

FIG. 4-12 Chemical profile for the unbonded side of a sample with the FPL etch and SET = 48 hr at 54°C and 95% RH.

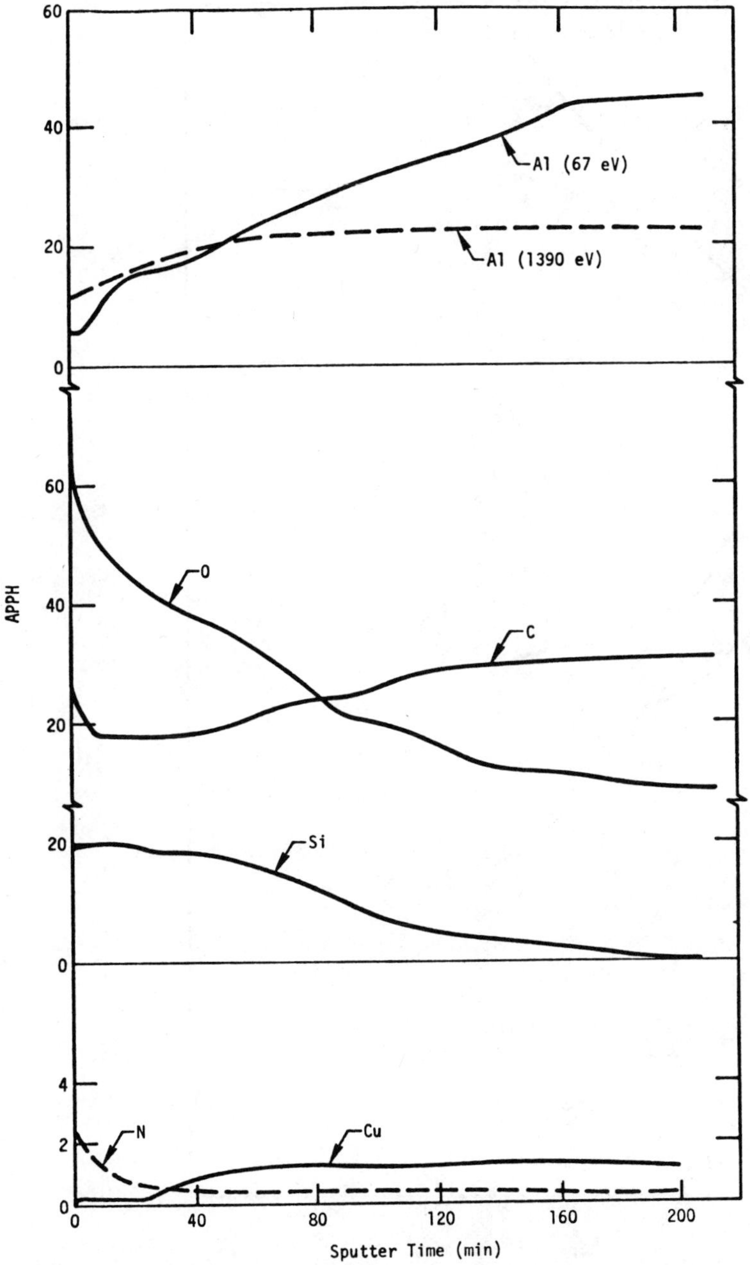

FIG. 4-13 Chemical profile in the interfacial failure area of an Al 2024-T3--HT424 joint. SET = 48 hr, 54°C, 95% RH.

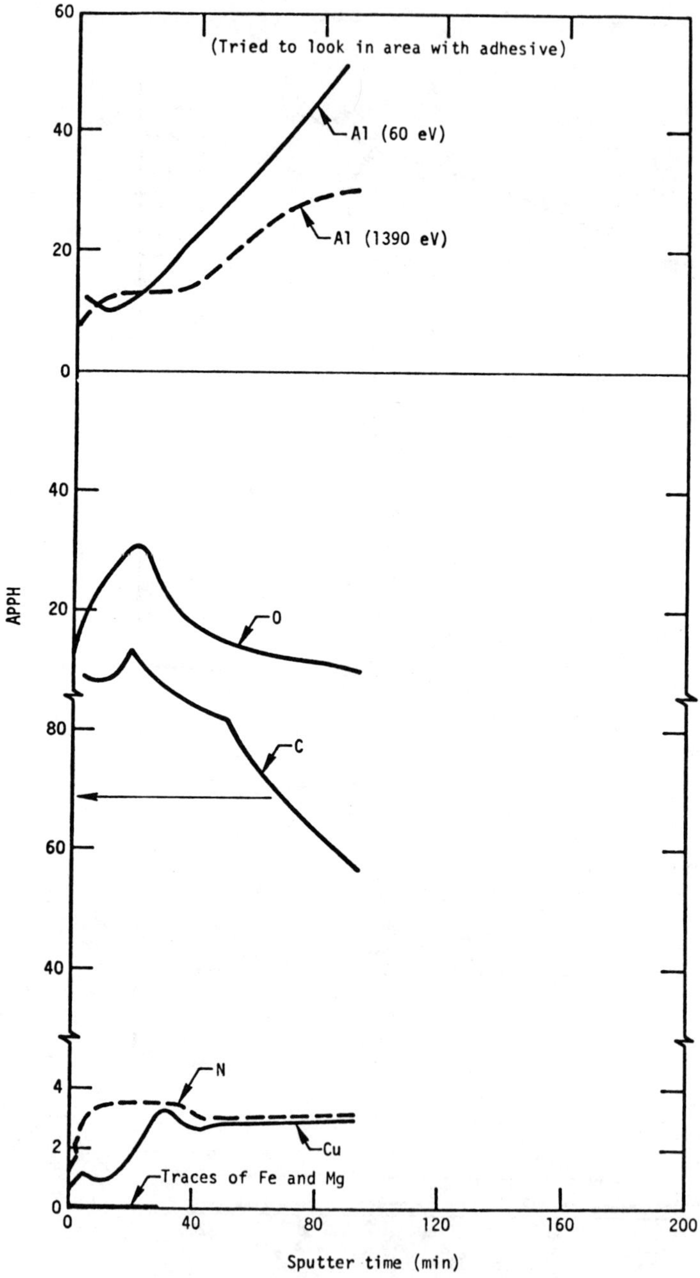

FIG. 4-14 Chemical profile in a region of cohesive failure for Al 2024-T3--HT424. SET - 48 hr, 54°C, 95% RH.

3.1.5. Microscopy

Figure 4-15 shows SEM photomicrographs of a 2024-T3 aluminum surface after the FPL etch. The pitted nature of the surface is similar to that revealed by SEM in other laboratories. For example, Fig. 4-15 can be compared with Figs. 31 and 32 of the Picatinny report [31] and Figs. 10 and 11 of a paper by Patrick [33]. Figures 4-16 and 4-17 are optical microscope pictures of typical fractured bonds for 2024-T3 aluminum with HT-424 adhesive. The photographs are much more difficult to interpret than observation under the microscope. Low-power microscopic observation reveals three failure modes in the 0.5 × 1 in. bond area. First the glass carrier network usually fails, with about half the fiber bundle on one adherend and half on the other, leaving the fiber net observed in Figs. 4-16 and 4-17. Second, most of the bond appears to fail at metal-adhesive interface, leaving matching surfaces. The light areas between the fiber net is the metal interface, the darker, more dull areas are the matching adhesive surface. Third, cohesive failure occurs in the adhesive, revealing a large amount of natural porosity caused by water vapor bubbles formed during the cure. Details of the cure process with respect to this porosity can be found in a paper by Kaelble and Cirlin [34]. The failure mode is related to the stress distribution of the lap shear joint during the tensile test. The stress distribution is schematically represented in Fig. 4-18 (by Bolger [33]). The predicted pattern (Fig.

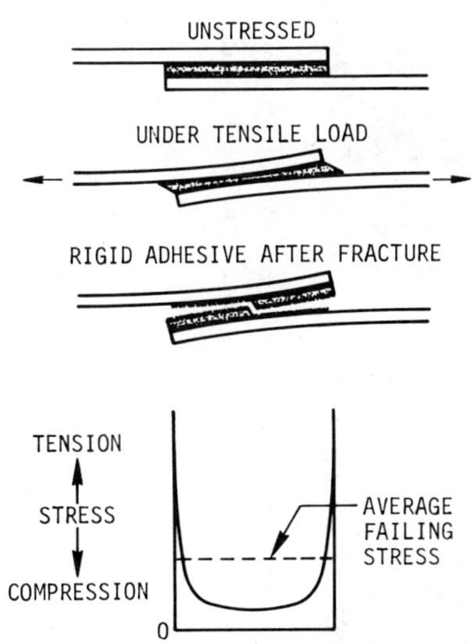

FIG. 4-18 Failure region in lap shear joint due to stress distribution.

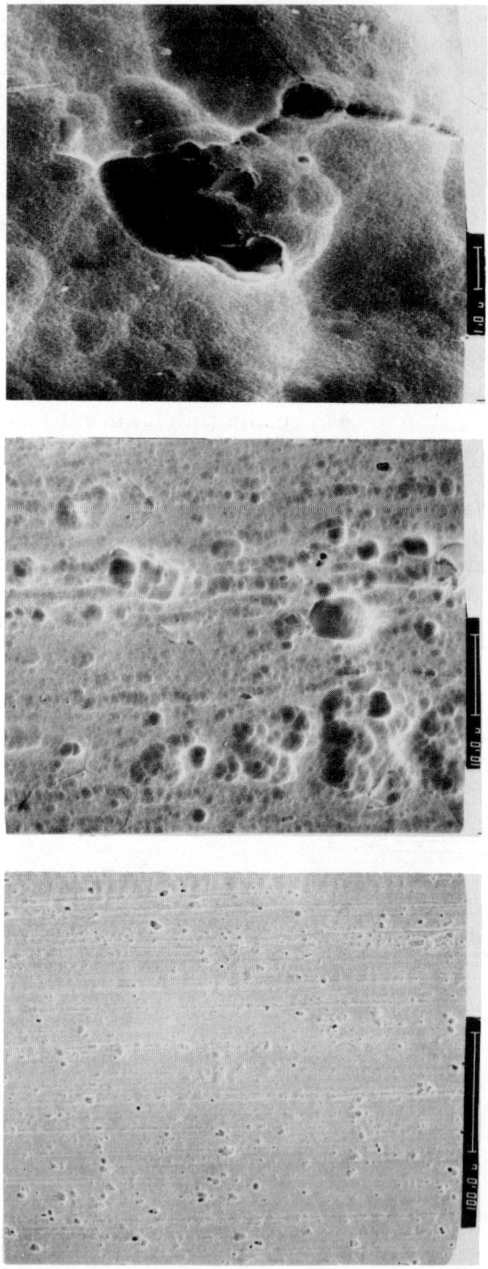

FIG. 4-15 Scanning electron photomicrographs of an FPL-etched Al 2024-T3 surface at three magnifications: (left) 240×; (center) 1600×; (right) 8000×.

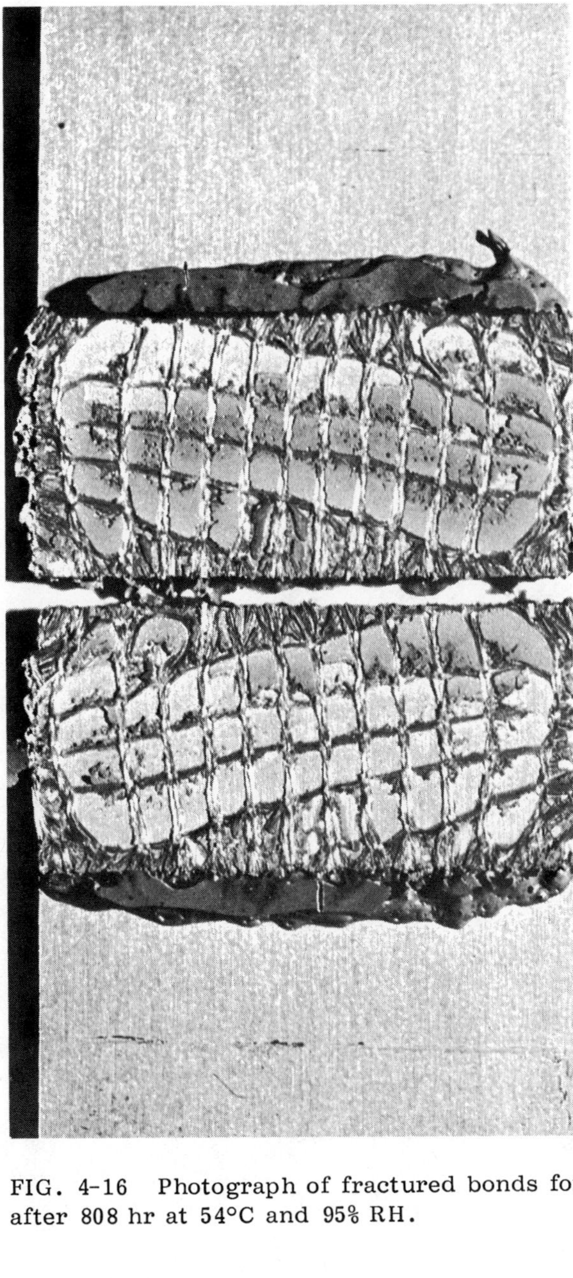

FIG. 4-16 Photograph of fractured bonds for Al 2024-T3, HT424 joint after 808 hr at 54°C and 95% RH.

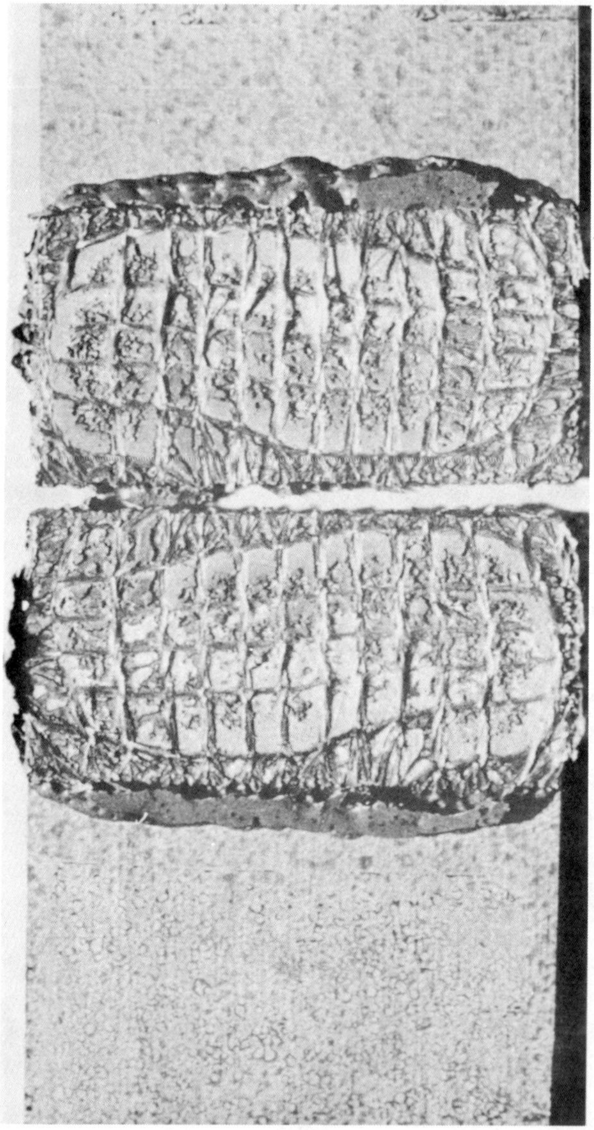

FIG. 4-17 Photograph of fractured bonds for Al 2024-T3, HT424 adhesive after 1023 hr at 54°C and 95% RH.

4-18) of interfacial failure, in the region farthest from the end of the metal coupon, was observed in general and is somewhat revealed in Figs. 4-16 and 4-17. Cohesive failure was generally more prevalent at the periphery of the bond area, as in Fig. 4-16. Cohesive failure spread toward the center of the bond area for the bond with longer aging time (Fig. 4-17). In some instances failure occurred almost entirely at the metal-adhesive interface, as shown in Fig. 4-19 for aging in 100% relative humidity.

It is of interest to investigate regions of different failure modes in more detail. Figures 4-20a and b are low-magnification SEM photomicrographs for a bond that had not been aged. The bright areas in these photomicrographs are caused by removal of the metal overlayer used in the SEM process and charging of the dielectric substrate. Figure 4-20a shows a region of the adhesive surface that had mostly failed at the metal interface. An enlargement of the central part of Fig. 4-20a is seen in Fig. 4-20c. The fractured adhesive in Fig. 4-20c indicates that void regions in Fig. 4-20a penetrate to the opposite interface. There is considerable cohesive and interfacial failure in the region of Fig. 4-20a. Figure 4-20b shows interfacial failure in the top half of the picture and cohesive failure in the lower half. The enlargement of a region of the lower half of Fig. 4-20b, in Fig. 4-20d shows cohesive failure parallel to the metal surfaces in contrast to interfacial failure and cohesive failure perpendicular to the plane of the metal in Fig. 4-20c. The large pore regions revealed at the bottom of Fig. 4-20b are presumably also beneath the adhesive surface in the top of Fig. 4-20b.

Figure 4-21a is a higher magnification of the adhesive surface in Fig. 4-20a. Figure 4-21b is a magnification of a region near the metal surface and reveals a region of primer layer that apparently did not adhere to the adhesive. The thin transparent nature of this region can be compared with the opaque pore region of Fig. 4-20d which is adhesive. Closer scrutiny of the edge of the primer film in Fig. 4-21c reveals small holes in the primer which will be shown are due to penetration of primer into holes in the FPL-etched surface (see Fig. 4-15). The region in the lower portion of Figs. 4-21b and c show that what might be interpreted from low-power optical micrographs as interfacial failure includes a great deal of failure in the primer layer. The small voids within the primer (lower part of Fig. 4-21c) are elongated in a direction approximately perpendicular to the metal surfaces and probably nucleate and grow during fracture.

Figures 4-22, 4-23, and 4-24 are for a fractured bond that had been aged for 1000 hr at 54°C and 95% RH. Figure 4-22a is an optical micrograph of the adhesive surface and Fig. 4-22b is an SEM of the mating metal surface. Figure 4-22c is an SEM of a fiber bundle, the top of the adhesive surface (top left) and metal surface (bottom left).

FIG. 4-19 Photograph of fractured bonds for Al 2024-T3, HT424 after aging at 54°C ~100% RH.

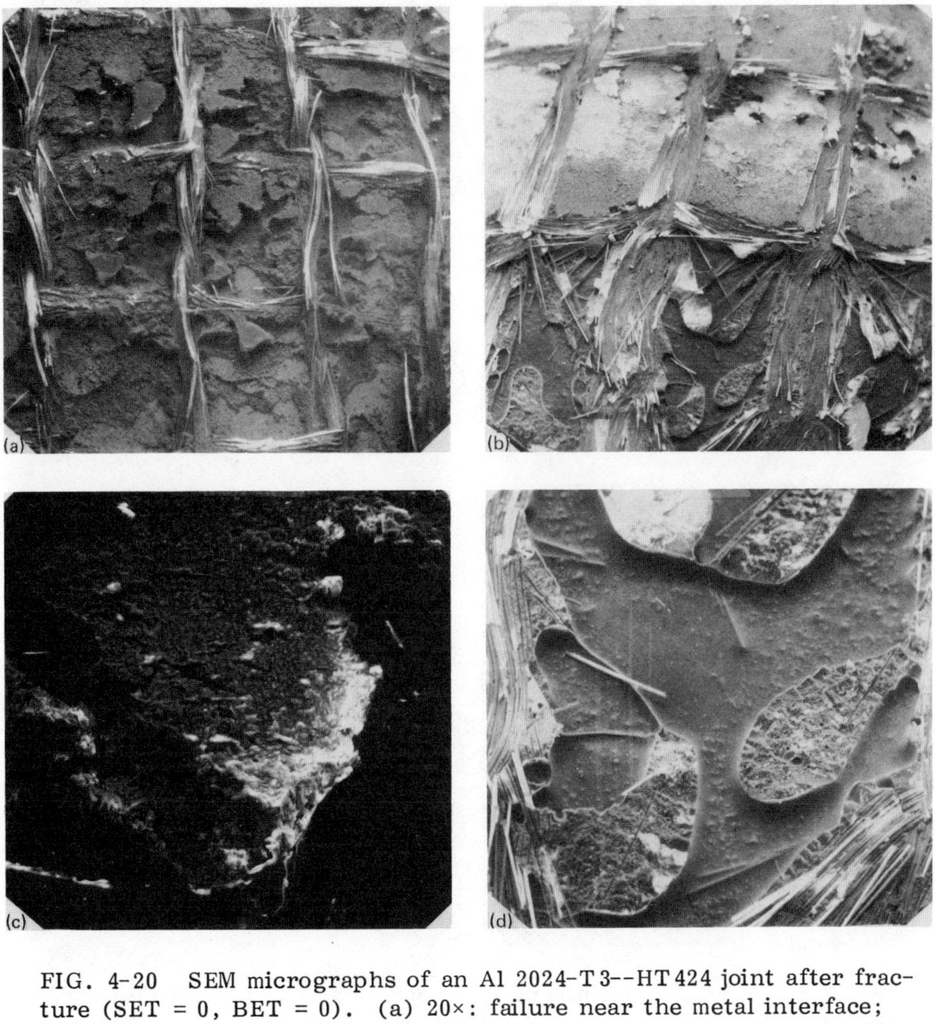

FIG. 4-20 SEM micrographs of an Al 2024-T3--HT424 joint after fracture (SET = 0, BET = 0). (a) 20×: failure near the metal interface; (b) 20×: failure interfacial (top of picture) and cohesive (lower half); (c) 200×: enlargement of central part of (a); (d) 60×: enlargement of lower part of (b).

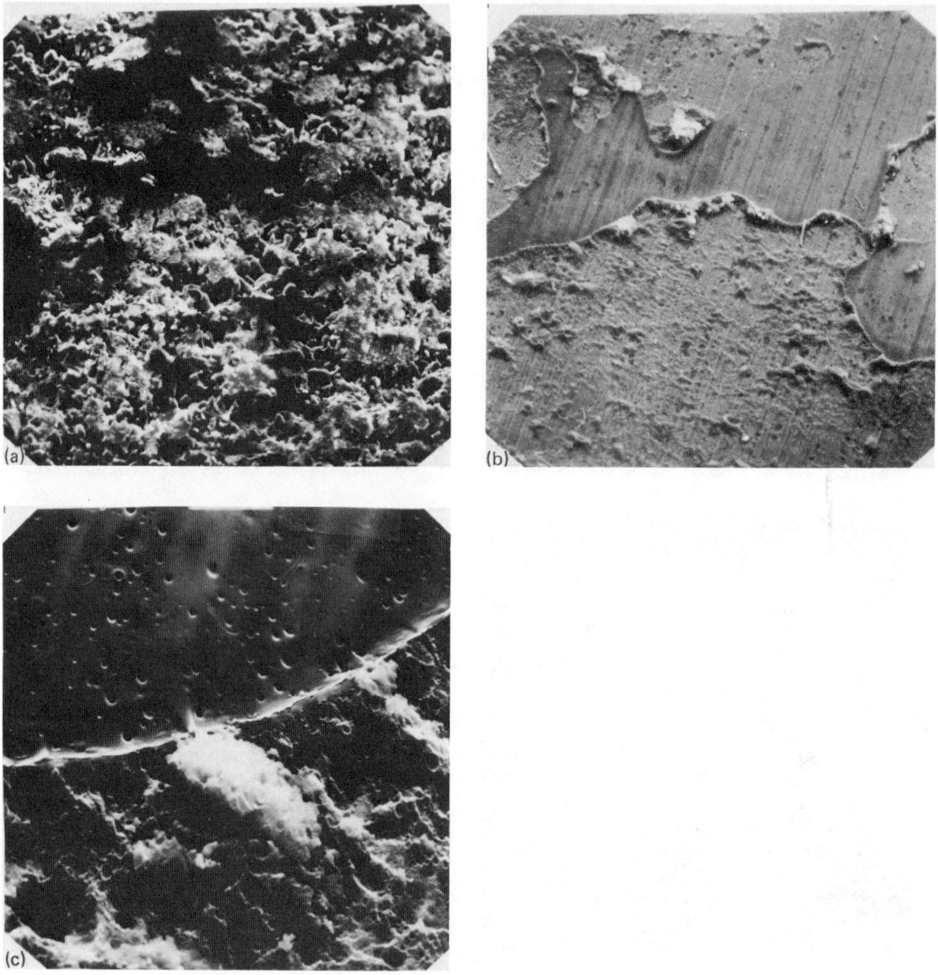

FIG. 4-21 SEM micrographs of an Al 2024-T3--HT424 joint after fracture (SET = 0, BET = 0). (a) 200×: higher magnification of Fig. 4-20a; (b) 180×: region near metal surface; (c) 1900×: edge of primer film.

FIG. 4-22 (a) Optical micrograph (10×) of Al 2024-T3--HT424 joint after fracture; (b) 10×, (c) 200×, and (d) 100×: SEM micrographs. SET = 0, BET = 1000 hr at 54°C, 95% RH.

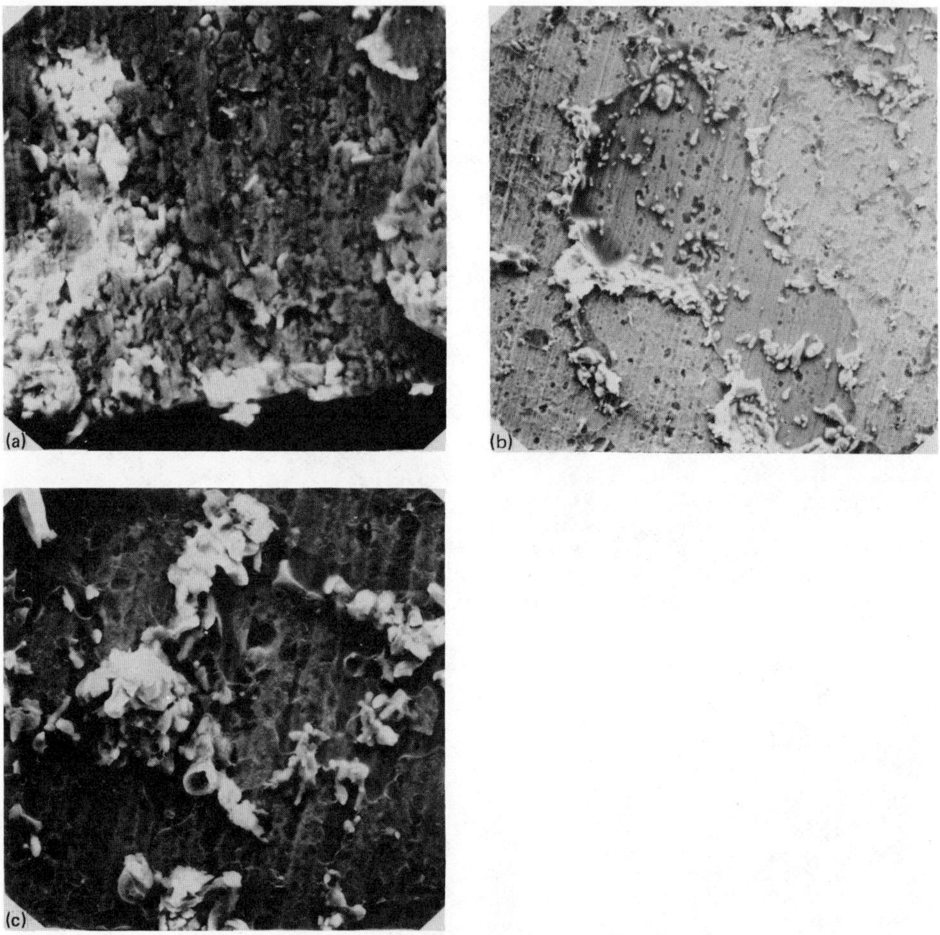

FIG. 4-23 SEM micrographs of an Al 2024-T4--HT424 joint after fracture (SET = 0, BET = 1000 hr at 54°C, 95% RH). (a) 500×: enlargement of adhesive surface after 1000-hr BET; (b) 200×: region of primer layer; (c) 600×: enlargement of a region in (b).

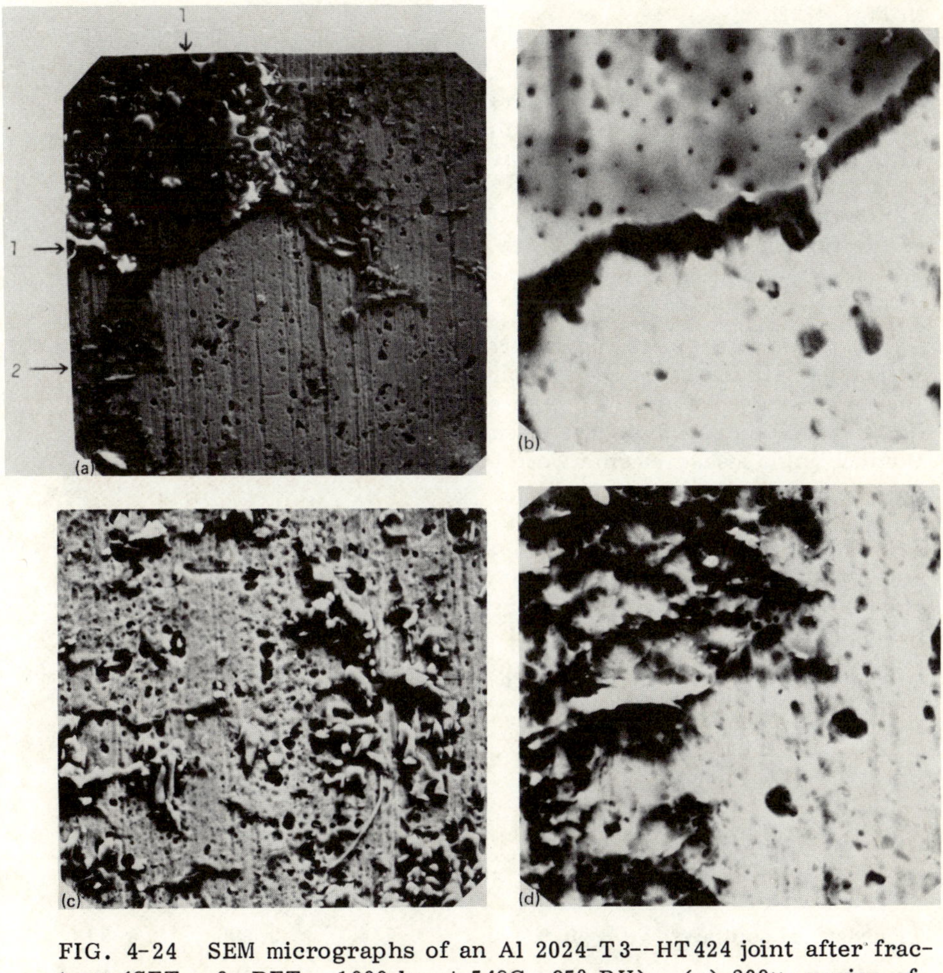

FIG. 4-24 SEM micrographs of an Al 2024-T3--HT424 joint after fracture (SET = 0, BET = 1000 hr at 54°C, 95% RH). (a) 200×: region of adhesive pore surface (top left), cohesive failure (bottom left and top center) and interfacial failure (bottom right); (b) 1400×: enlargement of region with coordinates 1,1; (c) 500×: enlargement of region of interfacial and cohesive failure; (d) 1000×: enlargement of region with coordinates 2,2 in (a).

Figure 4-22d shows the metal area (left side) and adhesive pore area (right side) near a glass fiber bundle. Figure 4-23a is an enlargement of the adhesive surface after the 1000-hr BET and can be compared with Fig. 4-21a for 0-hr BET. Figure 4-23b shows a region of primer layer as for Fig. 4-21b and Fig. 4-23c is an enlargement of a region in Fig. 4-23b.

Figure 4-24a shows a region of adhesive pore surface (top left), regions of cohesive failure (bottom left and top center), and regions of interfacial failure (bottom right). Figure 4-24b is an enlargement of the region designated by coordinates marked 1 in Fig. 4-24a. Holes in the primer can be associated with penetration of primer into holes in the metal substrate. Figure 4-24c is an enlargement of the region of interfacial and cohesive failure, and Fig. 4-24d is an enlargement of the region with coordinates 2 in Fig. 4-24a.

3.1.6. Contact Angles

In order to evaluate the several methods of surface energy analysis, wettability studies were carried out on panels of Al 2024-T3 given the vapor degrease and FPL-etch surface treatments. The results of liquid-solid contact angle measurements are summarized in Table 4-4. Contact angle measurements were conducted at 22°C in the B-100

TABLE 4-4 Experimental Values of Liquid Surface Tension (γ_{LV}) and Cosine of Liquid-Solid Contact Angle (cos θ) at 22°C

Al 2024-T3 surface treatment		VD	FPL etch	Plasma
Test liquid	γ_{LV} (dyn/cm)			
Water[a]	72.8	-.643 (~0.94)	0.998	1.00
Glycerol	64.0	-0.052	0.998	0.98
1-Br-naphthalene	44.6	0.616		0.84
Tricresyl phosphate	40.9	0.574	0.996	0.99
Polyglycol PG-1200	31.3	0.829	0.994	0.99

[a]All liquid-solid interactions display time-invarient contact angles from $t \leqslant 10$ sec to $t \geqslant 10$ min except water on surface treatments 1, 2, and 3, where $t \leqslant 10$ sec values are without parentheses and $t \geqslant 10$ min values are in parentheses.

TABLE 4-5 Analysis of Liquid-Solid Interactions

Al 2024-T3 surface treatment			$W_a/2\alpha_L$ (dyn/cm)$^{1/2}$		
γ_L (dyn/cm)	$2\alpha_L$ (dyn/cm)$^{1/2}$	β_L/α_L	VD	FPL	Plasma
72.8	9.34	1.53	2.78	15.6	—
64.0	11.66	0.94	5.20	11.0	10.9
44.6	13.36	0	4.39		6.1
40.9	12.52	0.21	5.14	6.50	6.5
31.3	9.90	0.53	5.78	6.31	6.3

environmental chamber of the NRL contact angle apparatus (Rame-Hart, Inc.) using sessile drops. All liquids displayed stable advancing contact angles over the time range $t \leqslant 10$ sec to $t \geqslant 10$ min except water on surface treatments 1, 2, and 3, as noted in Table 4-4. For the special case of water on surfaces 1, 2, and 3 one may suspect surface modification by interaction with water. The data of Table 4-4 may be plotted as cos θ vs. γ_{LV}. When the data display a linear variation of cos θ vs. γ_{LV}, an extrapolation to cos $\theta = 1.0$ identifies the critical surface tension γ_c for the solid surface, as shown by Zisman and coworkers [25].

These data may be further analyzed by a more detailed model for interface interactions discussed in Sec. 2.2.6. To apply this model the measured values of γ_{LV} and cos θ reported in Table 4-4 are introduced into the standard Young equation $W_a = \gamma_{LV}(1 + \cos\theta)$ to provide nominal values for the work of adhesion W_a. Table 4-5 organizes these data into the format defined by Eq. (9) of Sec. 2.2.6, which describes a linear function of $W_a/2\alpha_L$ vs. β_L/α_L whose intercept at $(\beta_L/\alpha_L) = 0$ is $\alpha_S = (\gamma_{SV}^d)^{1/2}$ and whose slope is $\beta_S = (\gamma_{SV}^p)^{1/2}$.

Computed average values γ_{SV}^d, γ_{SV}^p, and γ_{SV} in conjunction with standard deviations $\pm\gamma^d$, $\pm\delta^p$, and $\pm\delta$ are summarized in Table 4-6 for the surface treatments. The calculations exclude the parenthesized data of Tables 4-4 and 4-5. Quantitative conclusions concerning the effects of surface treatment on Al 2024-T3 wettability are automatically restricted by the magnitudes of the standard deviations shown in Table 4-6. Qualitative ranking would suggest the FPL etch to be the most wettable, followed by plasma treatment and then the vapor degrease when γ_{SV} is used as a wettability index. The results of this initial wettability experiment showed that the surface energy analysis discussed in Sec. 2.2.6 was applicable to surface-treated Al 2024-T3 plates.

TABLE 4-6 Calculated Solid-Vapor Surface Tension Properties for Al 2024-T3 Surface Treatments by Application of the Determinant Method

Surface treatment	$\gamma_{SV}^{d} \pm \delta^{d}$	$\gamma_{SV}^{p} \pm \delta^{p}$	$\gamma_{SV} \pm \delta$
FPL etch	14.8 ± 5.2	65.5 ± 14.0	80.3 ± 9.4
Plasma	28.8 ± 6.0	35.5 ± 18	64.3 ± 12.7
VD	40.3 ± 6.8	4.3 ± 2.1	44.6 ± 8.8

Wettability studies combined with surface energy analysis were subsequently carried out for the surfaces which form the interfacial bond between HT424 adhesive and sulfuric-acid-chromate-etched Al 2024-T3. Films of HT424 primer were deposited from solution on glass microscope slides. Uncured primer films were dried, following the manufacturer's recommendations, as follows [29]:

1. 30 min dry at 23°C
2. 60 min dry at 66°C

Primer films were dried and then cured as follows:

1. 30 min dry at 23°C
2. 60 min dry at 66°C
3. 30 min cure under dry N_2 at 171°C

Steps 1 and 2 above follow the adhesive manufacturer's recommended procedure for drying of primer films. Step 3 above correlates with curing conditions recommended for HT424 adhesive bonding. This last step is conducted under dry N_2 to minimize oxidative and hydrolytic degradation of the primer surface. Assuming the curing kinetics of HT424 primer follow results of prior curing studies [34] of HT424 structural adhesive film, it follows that steps 1 + 2 complete solvent removal without substantial crosslinking and that step 3 substantially completes the chemical curing of the HT424 primer films. Liquid-solid contact angle measurements were conducted at 20°C in the B-100 Environmental Chamber of the NRL contact angle apparatus (Rame-Hart, Ind.) using sessile drops. The results of contact angle measurements, reported in terms of the cosine of the advancing liquid-solid angle θ, are summarized in Table 4-10. When the data of Table 4-10 are plotted as $\cos\theta$ vs. γ_{LV}, a linear variation of $\cos\theta$ vs. γ_{LV} and extrapolation to $\cos\theta = 1.0$ identifies the critical surface tension γ_c of the adhesive film [25].

TABLE 4-7 Cosine of Liquid-Solid Contact Angles (cos θ) of Test Liquids on Dried and Cured Films of HT424 Primer

Test liquid	γ_{LV} (dyn/cm)	Dried cos θ	Cured cos θ
Water	72.8	0.276	0.326
Glycerol	64.0	0.515	0.469
Ethylene glycol	48.3	0.743	0.669
Tricresyl phosphate	40.9	0.987	0.982
Polyglycol P-1200	31.3	0.981	0.982

These data have also been further analyzed by a more detailed model for interface interactions discussed in Sec. 2.2.6. To apply this model the measured values of γ_{LV} and cos θ reported in Table 4-7 are converted to values of nominal work of adhesion by application of the Young equation. Table 4-8 presents these data in the format described by Eq. (9) of Sec. 2.2.6.

Computed values of γ_{SV}^d, γ_{SV}^p, and γ_{SV} in conjunction with standard deviations $\pm\delta^d$, $\pm\delta^p$, and $\pm\delta$ obtained by a method of determinant calculations are summarized in Table 4-9. Both Table 4-8 and 4-9 indicate that the HT424 primer displays a well-resolved polar component γ_{LV}^p to its total surface tension $\gamma_{SV} = \gamma_{SV}^d + \gamma_{SV}^p$. It is also evident that the dried and cured forms of HT424 primer have essentially equivalent wettability and surface tension properties. Figure 4-25 illustrates the wetting data in terms of cos θ vs. γ_{LV}, while Fig. 4-26 provides the experimental variation of $W_a/2\alpha_L$ vs. β_L/α_L described by Eq. (9) of Sec. 2.2.6. Earlier surface energy studies of HT424 structural adhesive film provide reference [28] values of $\gamma_{SV}^d \pm \delta^d =$

TABLE 4-8 Analysis of Liquid-Solid Interactions

HT-424 primer			$W_a/2\alpha_L$ $(dyn/cm)^{1/2}$	
γ_L (dyn/cm)	$2\alpha_L$ $(dyn/cm)^{1/2}$	β_L/α_L	Dried	Cured
72.8	9.34	1.53	9.94	10.33
64.0	11.66	0.94	8.32	8.07
48.3	10.83	0.81	7.77	7.44
40.9	12.52	0.21	6.49	6.47
31.3	9.90	0.53	6.26	6.26

TABLE 4-9 Calculated Solid-Vapor Surface Tension's Properties by the Determinant Method

HT424 primer	Dried	Cured
$\gamma_{SV}^{d} \pm \delta^{d}$	28.7 ± 3.5	25.5 ± 3.8
$\gamma_{SV}^{p} \pm \delta^{p}$	10.3 ± 2.6	11.7 ± 2.4
$\gamma_{SV} \pm \delta$	39.0 ± 1.3	37.2 ± 1.5

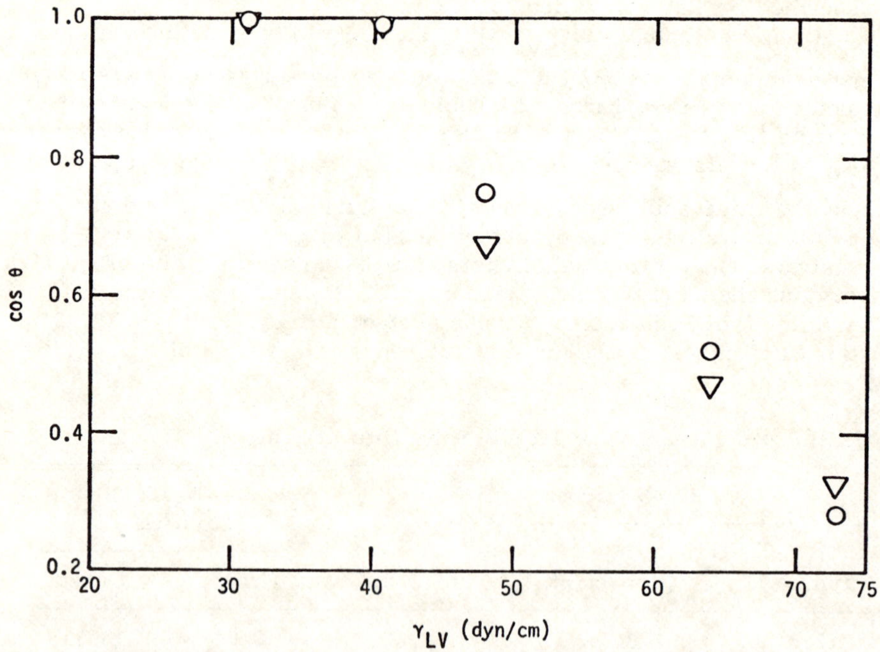

FIG. 4-25 Zisman plot of cos θ vs. γ_{LV} for dried (o) and cured (▽) films of HT424 primer.

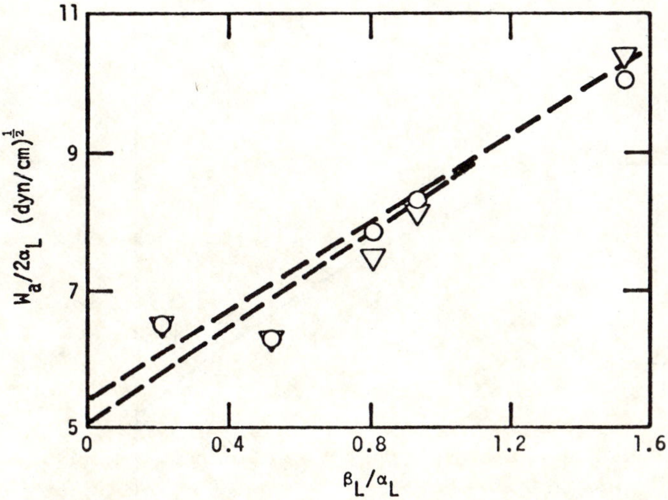

FIG. 4-26 Variation of $W_a/2\alpha_L$ vs. β_L/α_L for dried (o) and cured (∇) films of HT424 primer.

30.3 ± 4.5 dyn/cm, $\gamma^p_{SV} \pm \gamma^p$ = 12.0 ± 2.6 dyn/cm, and $\gamma_{SV} \pm \delta$ = 42.3 ± 2.3 dyn/cm. A close similarity between these literature values $\gamma_{SV} = \gamma^d_{SV} + \gamma^p_{SV}$ for HT424 adhesive and the values reported for HT424 primer films in Table 4-9 is evident.

The results of the surface energy analysis for HT424 primer and the FPL-etch 2024-T3 aluminum can be combined in a "wettability envelope" diagram as defined by Eq. (13) of Sec. 2.2.6. The diagram in Fig. 4-27 reveals that the HT424 primer and adhesive fall well within the envelope for freshly etched aluminum and therefore should, by this criterion, form strong adhesive bonds.

3.1.7. Joint Strengths

The last column of Table 4-3 gives average (of 6 bonds) bond strengths and the mean deviation from the mean, for comparison with surface preparation and surface character. The weakest bond results from the degrease-only surface preparation. This preparation had the highest value of SPD, the lowest value of γ_{SV} (lowest wettability), and the most contamination. These results suggest that the low bond strength for the vapor-degrease treatment results from a nonwettable weak boundary layer at the oxide-primer interface. The bond strength for the FPL-etch and plasma-treated surfaces are nearly the same.

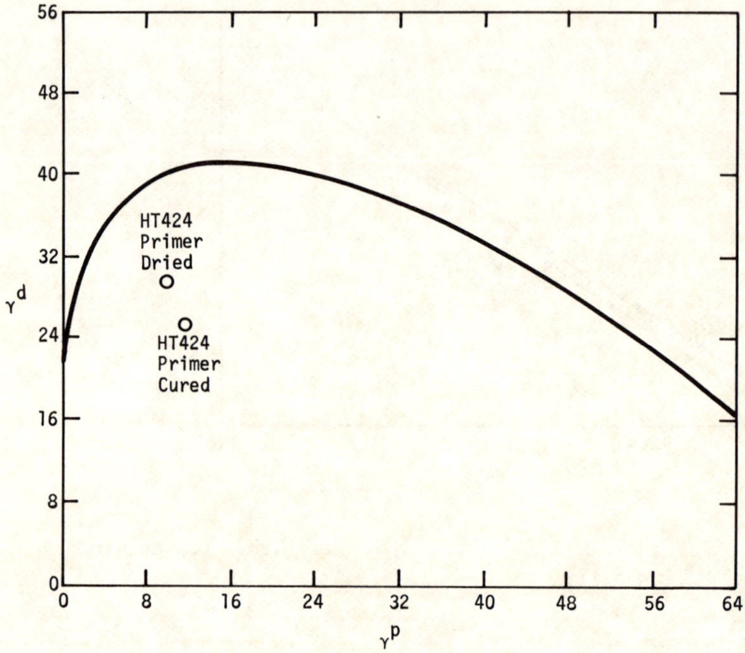

FIG. 4-27 Wettability envelope for freshly etched (FPL) Al 2024-T3.

Although AES measurements were not made on the plasma-treated surface, ion bombardment is shown to remove contamination (chemical profile section), and surface contamination is expected to be low as for the FPL etch. There does not appear to be any correlation between film thickness and bond strength for these thin films. However, it will be shown that for thicker films a correlation does exist.

3.1.8. Loci of Fracture

The optical and SEM micrographs reveal that for each particular bond, failure occurs in a nonuniform manner. Failure appears to occur in the adhesive, in the glass carrier bundles, in the primer, and at the metal surface. The spatial distribution of the areas of different failure modes are determined by the stress profile during fracture and the exposure condition of the metal surface prior to bonding or exposure condition of the bond.

It is apparent that the surface after the FPL etch is not uniform in character and that surface aging accentuates this nonuniformity. Mapping of surfaces after surface treatment and aging reveals this to be true, as will be seen in the next section. It is clear that since failure occurs by different modes at different loci on the surface, an investigation of the mechanisms of bond failure requires identification of these various regions prior to and after bond fracture. Identification prior to bonding is done by mapping the surface with respect to ellipsometry and SPD. Identification after fracture is done with a low-power optical microscope. The area of regions that appeared to fail interfacially (at the metal-adhesive interface) are estimated by taking low-power micrographs as in Fig. 4-16 but with about a factor of 10 magnification. A cellophane sheet with a grid (20 lines/in.) was placed over the photographs in order to keep track of each small area in the grid. An estimate of the fraction of each grid square that failed interfacially was recorded, then the total fraction of the bond area that failed interfacially was computed and designated ϕ_I. The complimentary fraction is designated ϕ_c, where $\phi_I + \phi_c = 1$. The average bond strength corresponding to these regions are designated σ_I and σ_c. Alternative techniques for estimating ϕ_I and ϕ_c were investigated and are reported in Appendix B.

We have referred to interfacial failure as failure at the metal-adhesive interface. However, one cannot determine by visual observation whether fracture occurred in the primer very near the oxide surface, at the primer-oxide interface, in the oxide, or at the metal-oxide interface. To determine which of these is the locus of failure for the FPL-etch surfaces, ellipsometric measurements were performed on the samples prior to bonding and in areas of interfacial failure after bond fracture. If the interfacial failure occurred in the primer near the oxide interface or at the oxide interface, the effective film thickness should be greater or approximately the same as before bonding. The results in Table 4-10 indicate that in each case the thickness of the oxide after fracture was substantially less than before. It is concluded that the aged FPL-etched surfaces with thick oxide layers ~500 Å fail in the oxide. It should be noted that solutions of n_f, κ_f, and d could not be found for the Δ and ψ values near $n_f = 1.7$, and $\kappa_f \simeq 0.3$ as for unaged FPL-etch films. The values reported in Table 4-10 are the minimum values of n_f and κ_f that correlate with the Δ and ψ values. This indicates that aging has caused a structural as well as thickness change in the oxide.

3.2. Surface Aging Effects

3.2.1. Effect of Humidity

For the aging experiments, samples were placed upright in a glass chamber which contained saturated K_2SO_4. The glass chamber was placed in an oven that had temperature control. However, it was observed that some condensation occurred at the side of the desiccator

TABLE 4-10 Oxide Thickness before and after Fracture SET and BET were at 54°C, 95-100% RH, λ = 5461 Å, Angle of Incidence 70°

Bond history		Δ	ψ	n_f	κ_f	Thickness (Å)
FPL etch SET (24 hr) BET (264 hr)	Before bonding	9.6	21.2	2.4	0.3	623
	After fracture	87.2	35.5	2.9	0.3	195
FPL etch SET (24 hr) BET (0)	Before bonding	108	51.9	2.9	0.4	576
	After fracture	6.8	19.8	2.9	0.3	433
FPL etch SET (24 hr) BET (0)	Before bonding	90.5	16.7	2.5	0.7	550
	After fracture	88.0	35.9	2.5	0.3	273

nearest the door of the oven. This suggested that the relative humidity was more near 100% for some of the samples than others. To check this hypothesis, samples were placed side by side as couples at specific distances from the front of the desiccator to obtain a systematic gradient of RH and left for 23.6 hr at 54°C. After SET the samples were characterized with respect to surface potential difference (SPD) and ellipsometry, then bonded and fractured.

The results in Table 4-11 show the effect of relative humidity (position in desiccator) on the surface characteristics as well as ϕ_c and the total bond strength σ_b. In Table 4-11, A and B refer to the two Al samples that were adjacent to each other at the various positions. The results in Table 4-11 also show that systematic trends in the surface characteristics exist as a function of relative humidity (position in the desiccator) between R.H. ~95 and 100%. Figure 4-28 shows that a definite correlation exists between the surface characteristics, SPD, and film thickness, and the fraction of the surface that failed cohesively. The solid points in Fig. 4-28 are for the A sample of the couple, and the open circles are for the B sample. The samples in position 1 yielded ellipsometric values quite different from the other samples. No computer solution was found for sample A in position 1. The optical constants for sample B prove to be larger than for the other samples. This is probably due to the fact that condensation was observed to occur on the samples in position 1.

Figure 4-29 shows that except for the couple in positions 1 and 5 an excellent correlation exists between σ_b and ϕ_c. The values of σ_I and σ_c obtained from the intercepts of the line through the points are 1700 and 3800 psi, respectively. σ_c is close to the maximum value

TABLE 4-11 Effect of Humidity on Surface Characteristics and Bond Strengths of 2024-T3 Al after SET of 23.6 hr 54°C

	Higher RH		Sample position in desiccator								Lower RH	
	1		2		3		4		5		6	
	A	B	A	B	A	B	A	B	A	B	A	B
SPD (V)	0.17	0.16	0.22	0.21	0.24	0.22	0.28	0.25	0.27	0.26	0.28	0.27
Δ (deg)	143.8	90.5	81.7	83.5	85.2	84.8	86.5	88.8	89.3	89.3	94.4	93.6
ψ (deg)	14.9	16.7	41.8	39.1	35.6	35.5	35.6	35.1	34.2	35.0	34.2	34.3
Film thickness (Å)	—	550	350	350	277	282	260	—	240	258	216	224
n	—	2.47	1.8	1.8	2.04	2.0	2.1	—	2.2	1.96	2.2	2.1
κ		0.7	0.2	0.2	0.4	0.4	0.5	—	0.5	0.5	0.6	0.6
ϕ_c (Visual, with stds)	0.1		0.45		0.55		0.65		0.65		0.75	
σ_b (psi)	2620		2640		2860		3050		2770		3290	

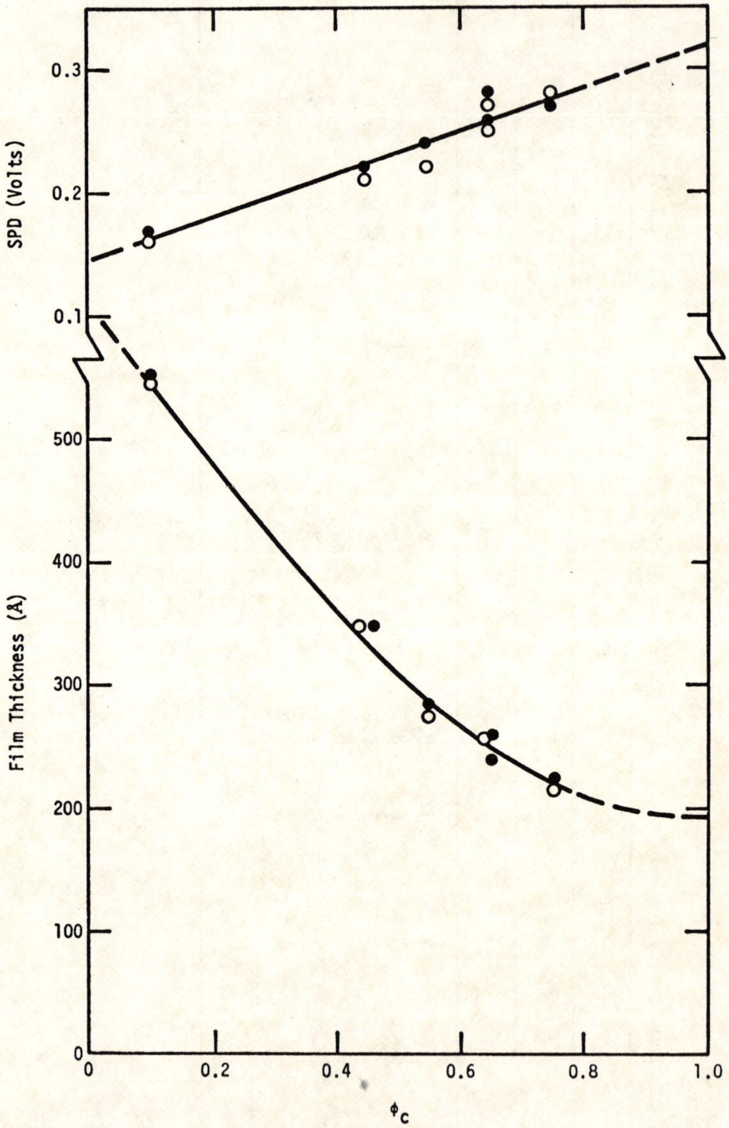

FIG. 4-28 Plot of film thickness and SPD vs. ϕ_c.

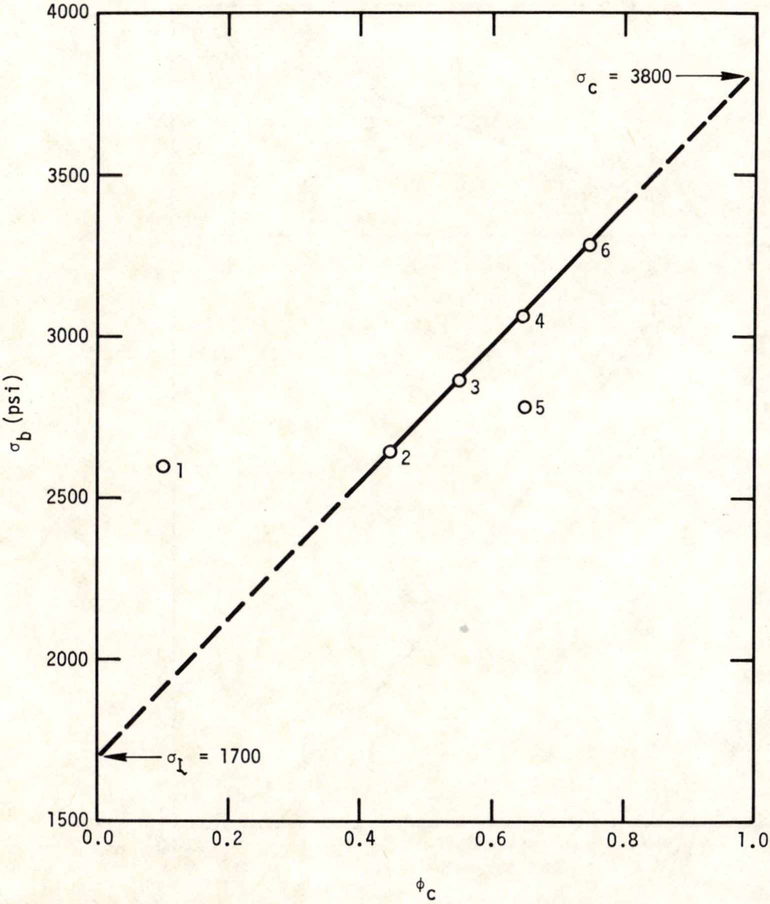

FIG. 4-29 Plot of σ_b vs. ϕ_c for samples at various positions in the humidity chamber

reported for the adhesive (3950 psi) by the manufacturer. Figures 4-28 and 4-29 show that a direct correlation exists between the surface characteristics and ϕ_c and thus the bond strength. The thicker the film, the higher is the probability that fracture will occur at the interface. Figure 4-30 is a plot of SPD and film thickness vs. σ_b, where σ_b is obtained from Fig. 4-29.

Table 4-11 shows that except for bond 1, n_f and κ_f tend to increase as the film becomes thinner. The values of the optical constants are related to film structure and may indicate stronger films for higher

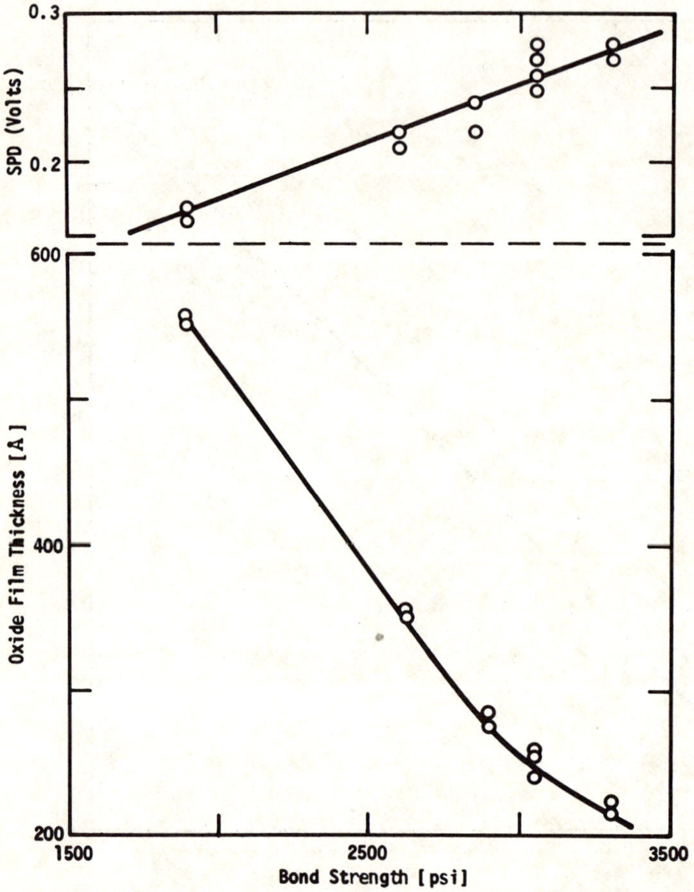

FIG. 4-30 Oxide film thickness and SPD vs. bond strength.

values of n_f and κ_f. Even though samples from position 1 have thicker films and would thus, from the line in Fig. 4-29, be predicted to fail at lower σ_b, the values of n_f and κ_f are larger, indicating stronger structure. This may account for the larger value of σ_b for position 1. The reason for the deviation from the line for the point for position 5 is not known.

The straight-line relationship in Fig. 4-29 indicates that films of approximately the same optical properties (structure) fail at the interface with about the same specific bond strength σ_I. However, the fraction of the bond area that fails at the interface ϕ_I is related to film thickness. These results are in accord with the results of Table

4-10: for those areas that fail interfacially, fracture probably occurs in the oxide film rather than at the oxide-primer interface or in the adhesive very near to the interface.

3.2.2. Effect of Glow Discharge

Plasma (glow discharge) treatment of surfaces may remove surface contamination and weak boundary layers. Experiments were carried out to discover if plasma-treated surfaces would produce better adhesive bonds. In order not to expose the plasma-treated surfaces to air, the sealed plasma-treating tube was opened in a glove box with controlled atmosphere. Five experiments were performed as follows:

1. Three samples (three bond couples) were degreased (tricresylphosphate), primed, and bonded.
2. Twelve samples were plasma-treated and immediately primed in the glove box.
3. Six samples were plasma-treated then exposed to 1 atm of pure oxygen in the plasma tube for 20 hr, then primed in the glove box.
4. Six samples were plasma-treated, then exposed exclusively to water vapor for 20 hr by opening a valve between the plasma tube and a bulb containing water but no air.
5. Six samples were plasma-treated, then exposed to a mixture of pure oxygen and water vapor.

The precise steps for these treatments are found in Table A-1. The results in Table 4-12 are plotted in Fig. 4-31. Comparison of the average values of σ_b for each treatment reveals that plasma-only yields the highest and plasma + H_2O + O_2 yields the lowest. Note that at $\theta_c \approx 0.1$ values of σ_b are above the line at the same position as for Fig. 4-29. That is, the σ_b and ϕ_c values are the same for samples exposed to water condensation at 54°C for 23 hr and samples exposed to pure oxygen at retention time (RT) after the plasma treatment. It can be concluded from Fig. 4-31 that except for the plasma + O_2 (open circles at lowest ϕ_c value) the strength of the oxide is $\sigma_I \approx 1700$ psi and the strength of the adhesive is $\sigma_c \approx 3800$ Å, as for the experiments recorded in Fig. 4-29.

Figures 4-32 and 4-33 show plots of bond strength σ_b vs. ϕ_c for various SET aging. When these experiments were performed, the variation in humidity (see Fig. 4-29) within the glass chamber had not been discovered. Consequently the RH probably varied between ~95 and 100%, but the position of the samples was not recorded. For the set of experiments presented in Fig. 4-32, interfacial failure (ϕ_I) increases and bond strength decreases as SET increases to 24 hr, then ϕ_c and σ_b increase again. The fact that the data in Fig. 4-32 cluster around a straight line indicates that for this set σ_c is approximately constant

TABLE 4-12 Effect of Plasma Treatment on Bond Strength, ϕ_I and ϕ_c

Bond couple	Surface treatment	ϕ_I	ϕ_c	σ_b (psi)
1	TCE	0.40	0.60	3000
2	only	0.50	0.50	2650
3		0.70	0.30	2710
				avg 2790 ± 140
1	Plasma only	0.25	0.75	3320
2	" "	0.35	0.65	2940
3	" "	0.30	0.70	3430
4	" "	0.45	0.55	2830
5	" "	0.50	0.50	2830
6	" "	0.20	0.80	3430
				avg 3130 ± 263
1	Plasma + O_2	0.90	0.10	2540
2	" "	0.90	0.10	2560
3	" "	0.80	0.20	2490
				avg 2530 ± 27
1	Plasma + H_2O vapor	0.60	0.40	2630
2	" " "	0.60	0.40	2960
3	" " "	0.50	0.50	2770
				avg 2787 ± 116
1	Plasma + H_2O + O_2	0.85	0.15	2140
2	" " "	0.75	0.25	2330
3	" " "	0.75	0.25	2370
				avg 2280 ± 93

(~3800 psi) for cohesive failure and σ_I is approximately constant but much lower (~800 psi) than for the plasma experiments reported in Fig. 4-31 or those reported in Fig. 4-29. The experiments reported in Fig. 4-33 do not cluster around a straight line, indicating that for these experiments, although σ_c may approximate a constant, σ_I does not.

3.2.3. Surface Mapping

Thus far we have reported average values for ϕ_c and surface properties averaged over a small area of the sample. No attention has been paid to nonuniformity over the surfaces to be bonded. However, as pointed out in Sec. 3.1.1, the micrographs taken after fracture reveal that the mode of fracture is not uniform over the surface. The tendency to fail by a particular mode in a particular area must be built into the surface during the sample history, perhaps prior to bonding. It was thought that the scatter of the data in Fig. 4-33 may be related to nonuniformity. It follows that a map of the surface with respect to ellipsometry and SPD may reveal the areas of interfacial failure prior to bonding. To test this the following experiments were performed. Two Al samples that were contaminated with dichromate solution during aging at 95% RH at 54°C revealed stains at various positions across the areas to be bonded. Two other samples were aged with RH between 95 and 100% at 54°C for 24 hr. Each of these samples were bonded with a freshly prepared sample so that the interfacial fracture would only occur on the stained and aged samples. Prior to bonding the aged and and stained samples were mapped in the bond region with ellipsometry and SPD. The samples were then bonded and fractured and ϕ_I measured by photographing each map position under a low-power microscope and

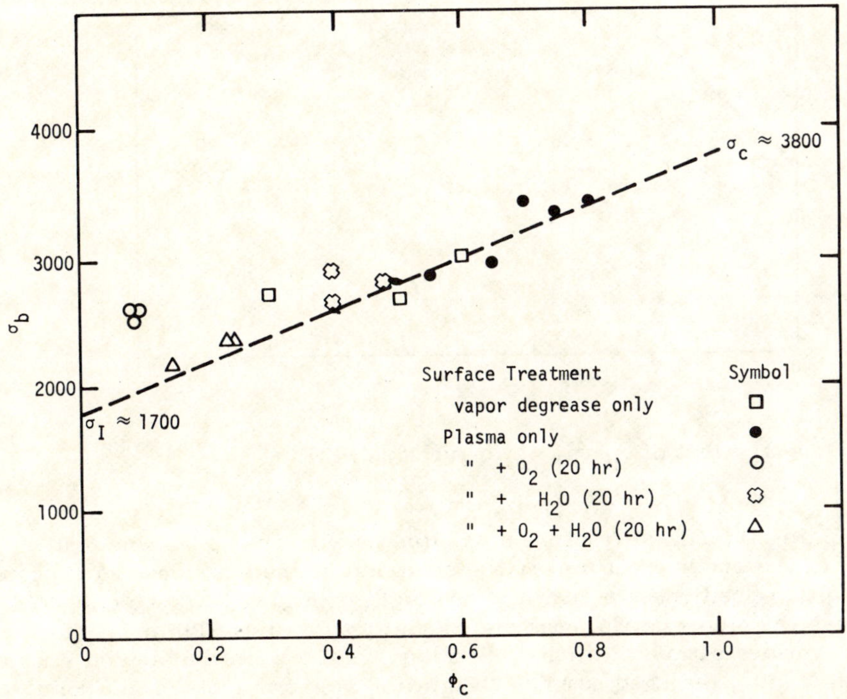

FIG. 4-31 Plot of σ_b vs. ϕ_c for plasma-treated surfaces of Al 2024-T3.

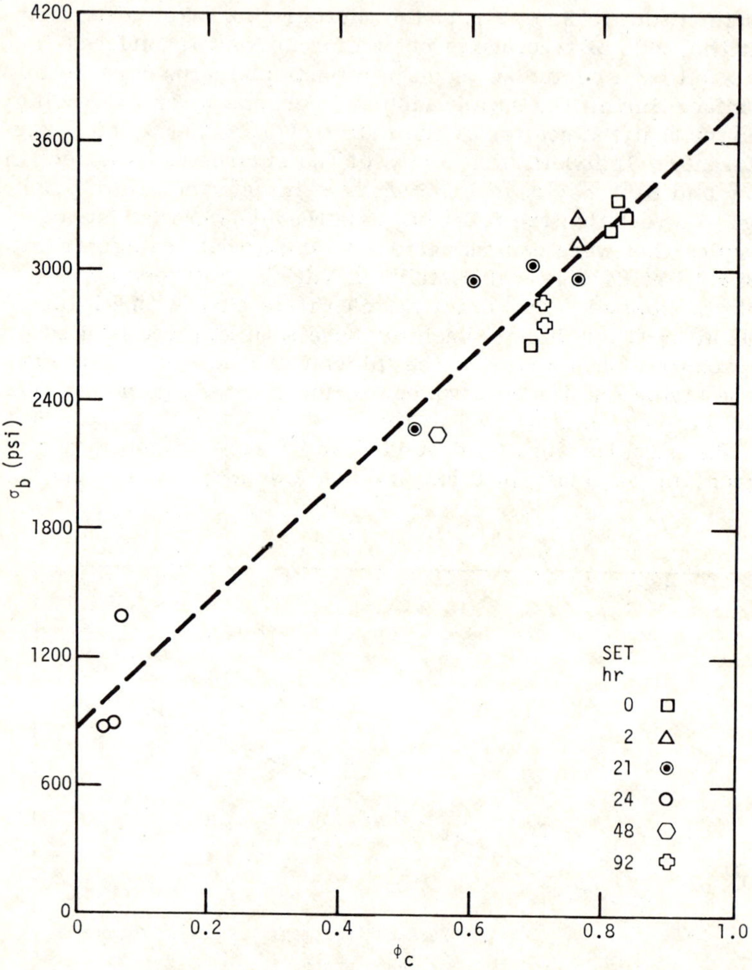

FIG. 4-32 Plot of σ_b vs. ϕ_c for various SET.

counting interfacial fracture area squares with the cellophane grid. All of the bonds fractured at the aged and stained samples, as expected. The map positions are shown at the right of Fig. 4-34. Table 4-13 gives the ellipsometric parameter Δ and ψ as well as SPD and ϕ_I for the various sample positions. Samples 1 and 2 were the stained samples and 3 and 4 the aged samples. Figure 4-34 shows plots of ϕ_I (solid lines) and Δ (dashed lines) as a function of position. At the right of Fig. 4-34 estimates of film thicknesses are given for each sample.

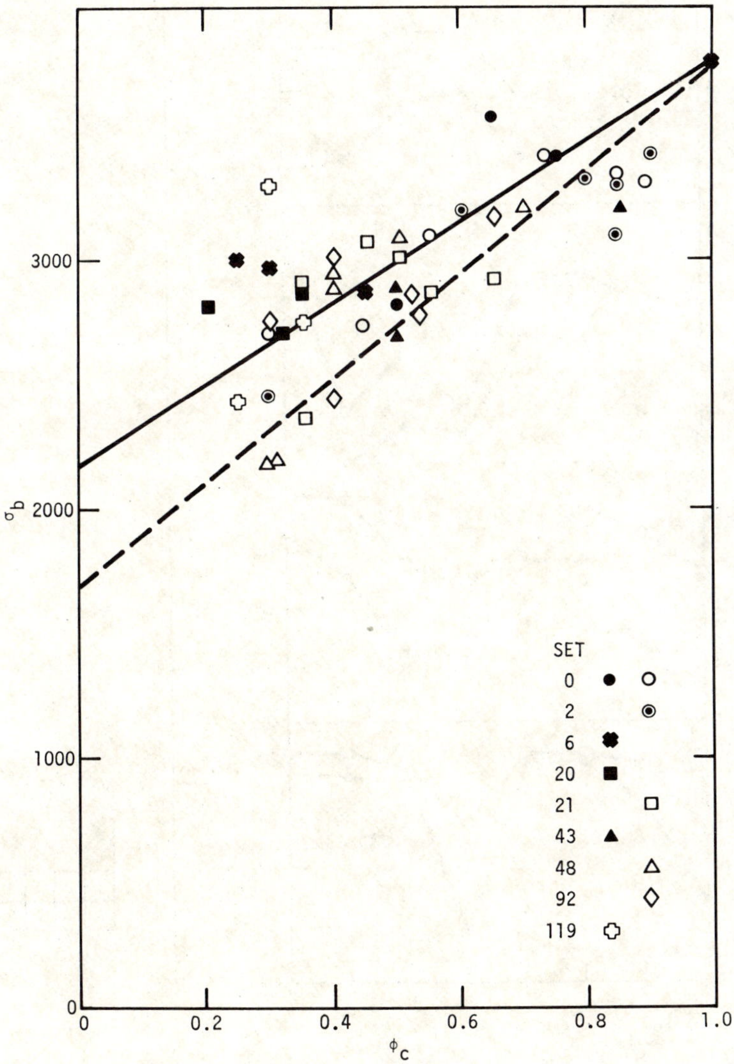

FIG. 4-33 Plot of σ_b vs. ϕ_c for Al 2024-T3 after SET at 54°C and 95-100% RH.

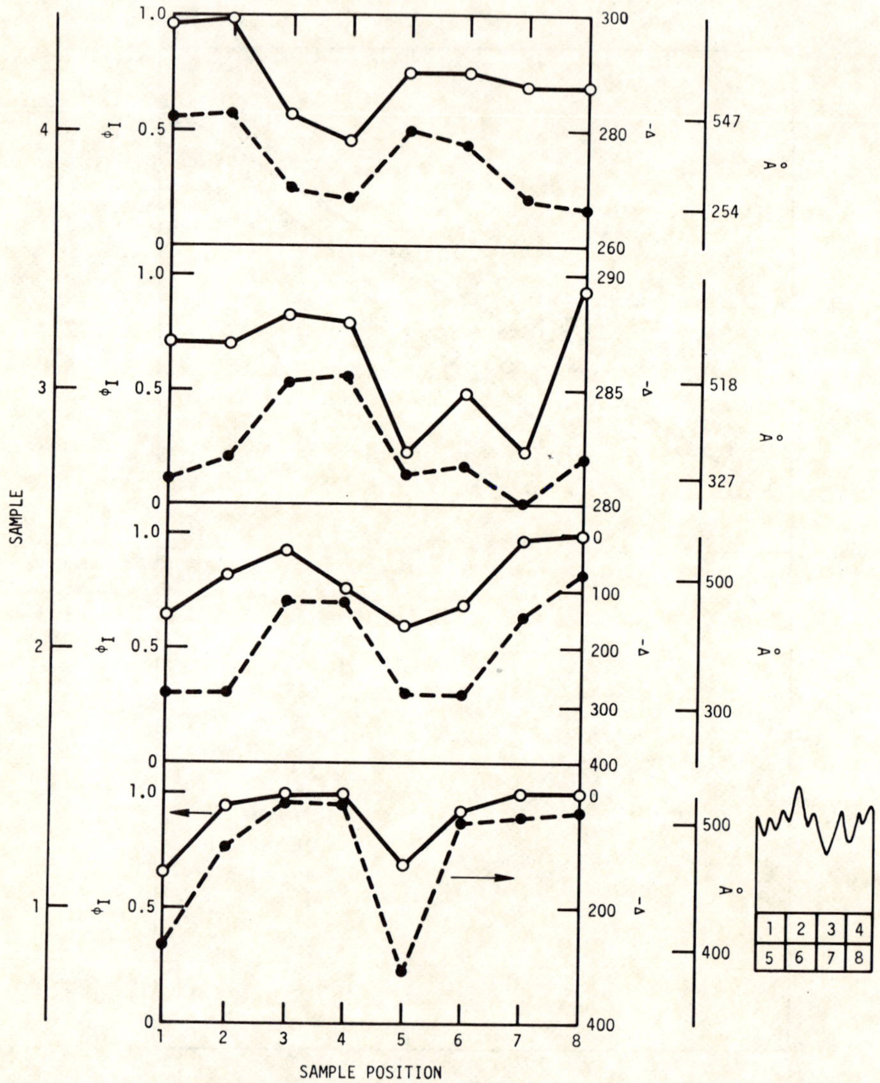

FIG. 4-34 Maps of the ellipsometric parameters Δ for positions on Al 2024-T3 samples prior to bonding, compared with ϕ_I measured after bonding and fracture. Estimates of film thickness corresponding to Δ are given on the right ordinate.

TABLE 4-13 Mapping of Surface Characteristics and Comparison with ϕ_I

Position	Δ	ψ	SPD (V)	ϕ_I	σ_b (psi)
Sample #1					
1	-260.8	58.4	-0.16	0.65	
2	-98.8	50.6	-0.14	0.88	
3	-18.0	11.5	+0.02	0.97	
4	-22.2	14.2	+0.01	0.98	
5	-313.2	70.2	-0.16	0.68	
6	-52.4	30.9	-0.20	0.84	
7	-47.1	14.3	-0.05	0.99	
8	-32.0	19.1	-0.05	0.97	
				avg 0.87	2040
Sample #2					
1	-274.8	39.0	-0.07	0.66	
2	-275.8	46.0	-0.07	0.83	
3	—		+0.02	0.93	
4	-107.6	71.8	+0.06	0.77	
5	-275.8	40.0	-0.04	0.59	
6	-278.0	45.5	-0.04	0.68	
7	-145.0	6.4	-0.01	0.97	
8	-70.0	38.0	+0.04	0.99	
				avg 0.80	2500
Sample #3					
1	-281.2	41.6	+0.23	0.71	
2	-282.0	45.1	+0.11	0.70	
3	-285.3	46.8	+0.12	0.82	
4	-285.6	41.8	+0.06	0.78	
5	-281.4	39.9	+0.04	0.23	
6	-281.6	44.9	-0.12	0.49	
7	-280.0	44.5	-0.11	0.22	
8	-282.0	46.8	-0.13	0.92	
				avg 0.61	2770
Sample #4					
1	-282.4	41.9	-0.27	0.96	
2	-283.1	44.1	-0.28	0.98	
3	-270.2	38.4	-0.29	0.56	
4	-268.4	39.0	-0.26	0.47	
5	-279.8	40.3	-0.24	0.74	
6	-276.0	38.6	-0.27	0.74	
7	-267.8	38.4	-0.28	0.66	
8	-266.4	37.6	-0.27	0.65	
				avg 0.72	2110

Figure 4-34 dramatically demonstrates that mapping of the surfaces by ellipsometry prior to bonding allowed prediction of the trend in ϕ_I in every case. Again it is observed that ϕ_I increases with film thickness. Although the range of film thickness for all of the samples is about the same (300-500 Å), the large differences in Δ and ψ for samples 1 and 2 as compared to 3 and 4 are due to differences in optical constants for the two sets. The optical constants for samples 1 and 2 are about $n_f \approx 2.8$, $\kappa_f \approx 0\text{-}0.3$, whereas for samples 3 and 4, $n_f \approx 1.8$, $\kappa_f \approx 0.1\text{-}0.3$. Note that the negative values of Δ in Table 4-13 can be expressed as positive values by adding 360°. Table 4-13 shows that a good correlation exists between SPD and ϕ_I for samples 1 and 2. The correlation is not as good for 3 and 4.

3.2.4. Effect of Time

Some of the aged samples were used to measure SPD and ellipsometry, others were used to measure contact angles θ for water and tricresylphosphate (TCP), and others were used to make adhesive joints with HT424. Figure 4-35 is a plot of Δ and ψ, SPD and θ_{H_2O}, θ_{TCP} with SET.

It is apparent from Fig. 4-35 that all parameters level off in about 20-30 hr. The shape of curves SPD, ψ, and θ_{TCP} indicate that a different process is occurring in the first 10 hr than is occurring after 10 hr. A clue as to the process during the first 10 hr is given by the curve for ψ and SPD. Computer calculations show that growth of an oxide film of constant optical properties ($\kappa_f \sim 0.1$) will cause both Δ and ψ to decrease. The increase in ψ with SET can be interpreted in two ways: ψ can increase if the oxide is being transformed to one with smaller absorption coefficient (κ_f) and/or organic surface contamination is forming. The value of SPD in air 26 min after the dichromate treatment was 0.3 V and 0.27 V after 43 min. These two values are shown at SET = 0 in Fig. 4-35 and indicate that the initial process began before exposure to the humidity chamber. The results in Fig. 4-35 are best interpreted as oxide film transformation during the first 6 hr, followed by organic contamination that continues to 20-30 hr. The curve labeled "organic contamination" is obtained from the ellipsometric data if one assumes that the total change in Δ and ψ after 6 hr is due to formation of an organic film with refractive index 1.5 on the transformed oxide film, 200 Å thick with optical properties $n_f = 1.7$ and $\kappa_f = 0.4$. At about 30 hr the contamination reaches about 25 Å. The data between SET = 0 and 120 hr can also be interpreted as transformation alone if the oxide changes from one of 195 Å with $n_f = 1.4$, $\kappa_f = 0.5$ to 223 Å with $n_f = 1.7$, $\kappa_f = 0.3$. The data cannot be interpreted in terms of organic contamination alone.

To estimate σ_I, it is assumed that σ_c is a constant and that the relation between σ_b, σ_I, σ_c, ϕ_c, and ϕ_I is

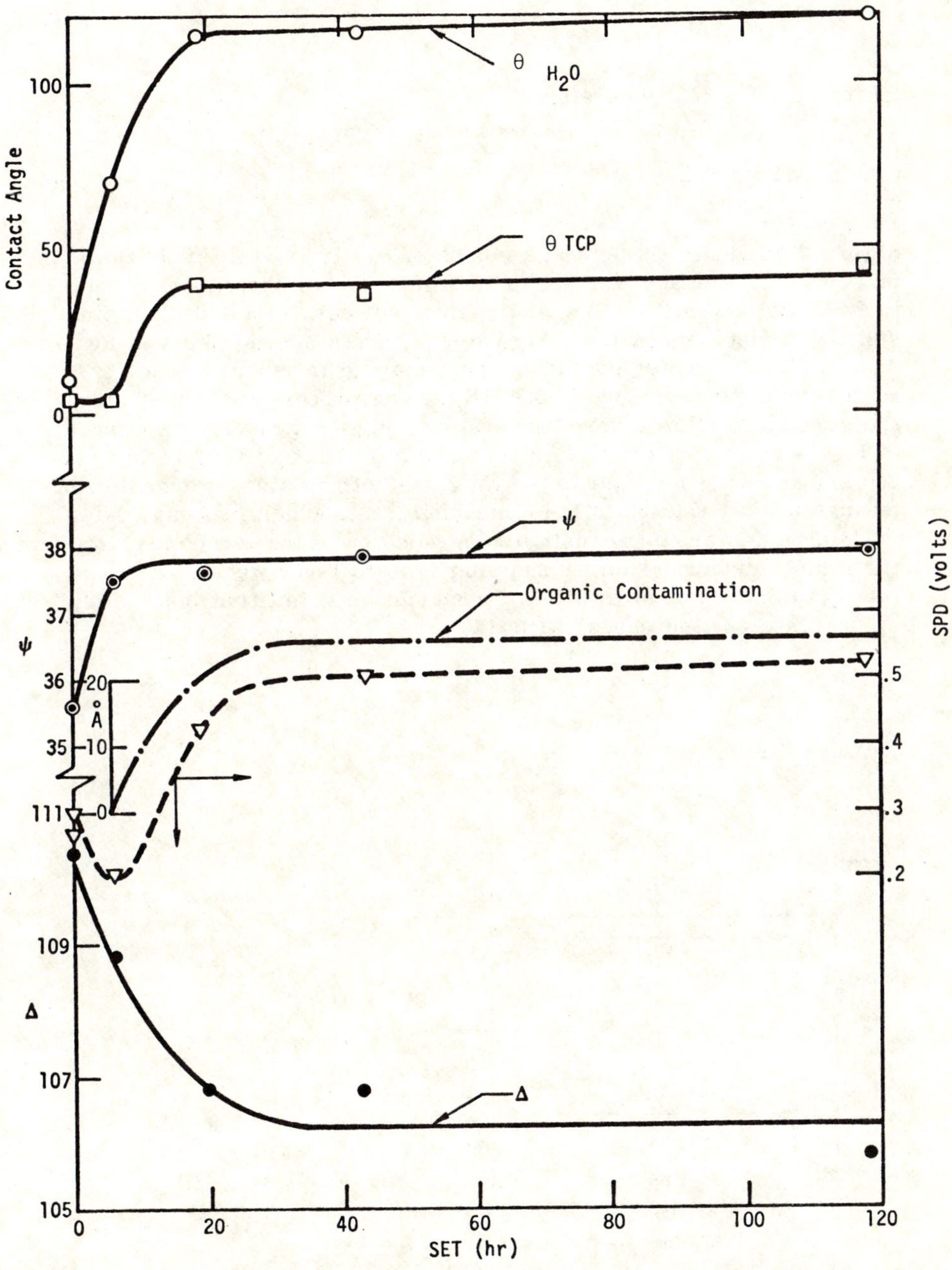

FIG. 4-35 Plot of Δ, ψ, SPD, θ_{H_2O}, and θ_{TCP} vs. SET for 2024-T3 at 54°C, 95% RH.

$$\sigma_b = \phi_c \sigma_c + \phi_I \sigma_I \tag{19}$$

where $\phi_I = 1 - \phi_c$. Therefore,

$$\sigma_I \approx \frac{\sigma_b - \phi_c \sigma_c}{1 - \phi_c} \tag{20}$$

Figure 4-36 shows the average values of σ_b, σ_I, and ϕ_c plotted vs. SET. As for surface properties in Fig. 4-35, all parameters level off in 20-30 hr, indicating the relationship between the bond character and the surface character. Figure 4-37 shows average curves for Δ, ψ, SPD, and water contact angle for aging in laboratory air at room temperatures (~23°C, ~50% RH). The processes are probably the same as for the higher-temperature, higher-humidity experiment in Fig. 4-36 but slower. Note that ψ and SPD are still increasing at 300 hr, as would be expected if they are both related to organic contamination. At this point it is uncertain as to which process, oxide transformation or contamination, degrades the adhesive bond. However, the results discussed under mapping tend to favor weak oxides and therefore oxide transformation. It should be noted that both σ_I and ϕ_c decrease (ϕ_I increases) with SET.

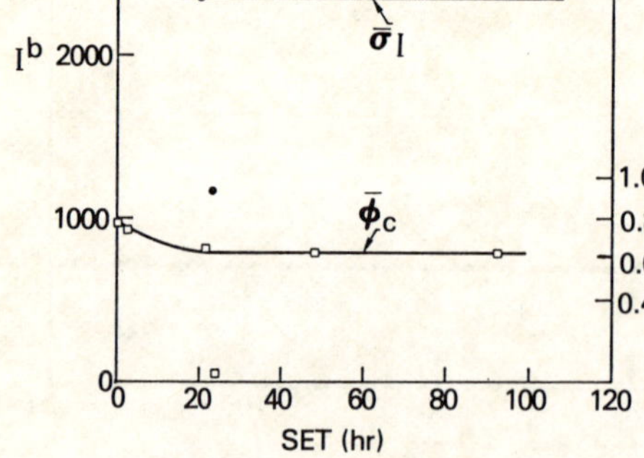

FIG. 4-36 Plot of σ_b, σ_I, and ϕ_c vs. SET at 54°C, 95% RH.

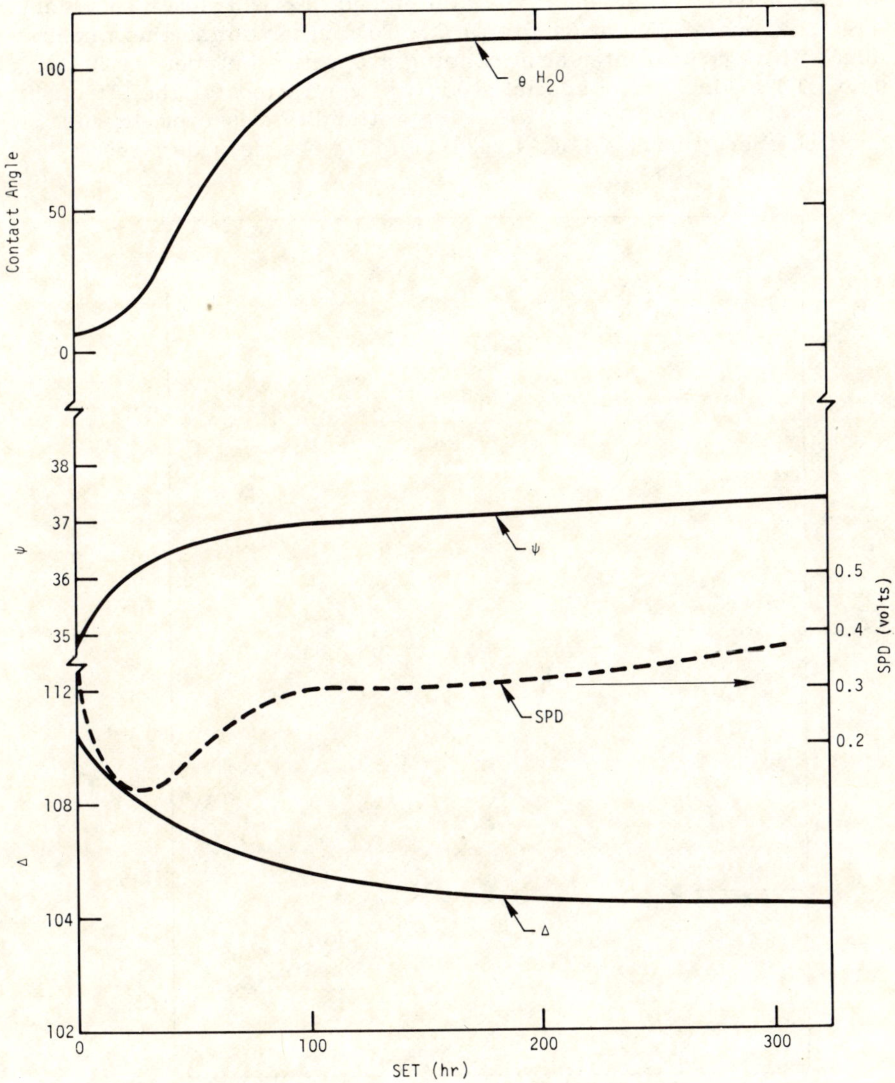

FIG. 4-37 Plot of Δ, ψ, SPD, and θ_{H_2O} vs. SET at 23°C and 50% RH.

3.2.5. Effect of SET on Contact Angles

Wettability measurements and analysis identified notable variations in the surface properties of the FPL-etched 2024-T3 aluminum which are functions of SET at 24 and 54°C. To further define this phenomenon with regard to interfacial bonding properties between Al 2024 and HT424 adhesive, a special study was conducted for the effects of SET at 23°C and 50 ± 5% RH. The wettability measurement and surface energy analysis of Al 2024-T3 follow the methods already

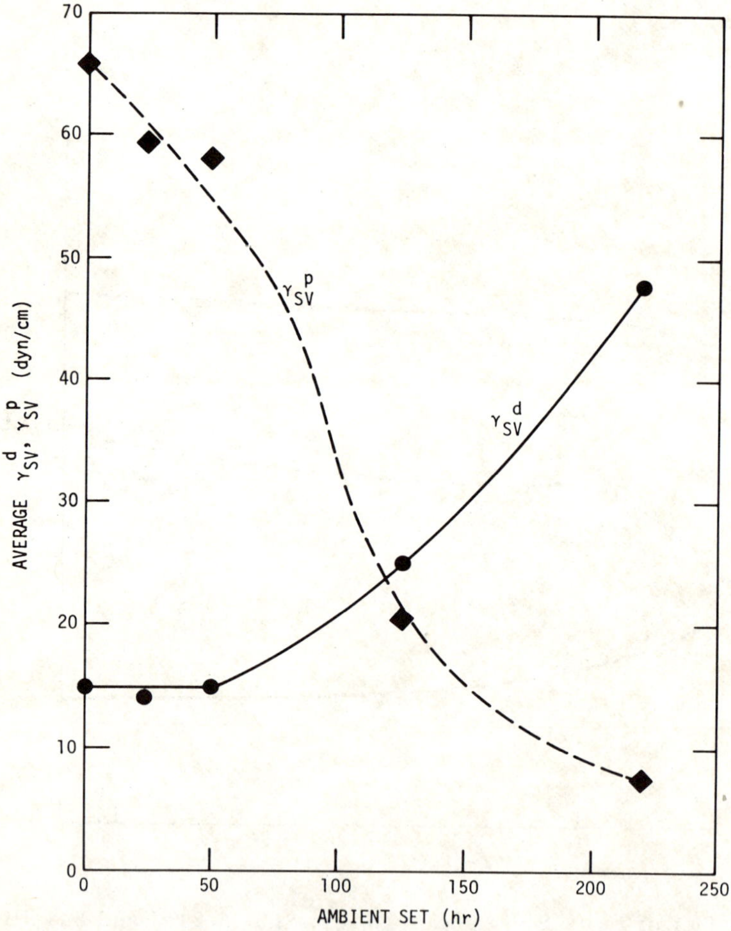

FIG. 4-38 Effect of SET at 23°C and 50 ± 5% RH on calculated values for γ_{SV}^d and γ_{SV}^p for FPL-etch Al 2024-T3.

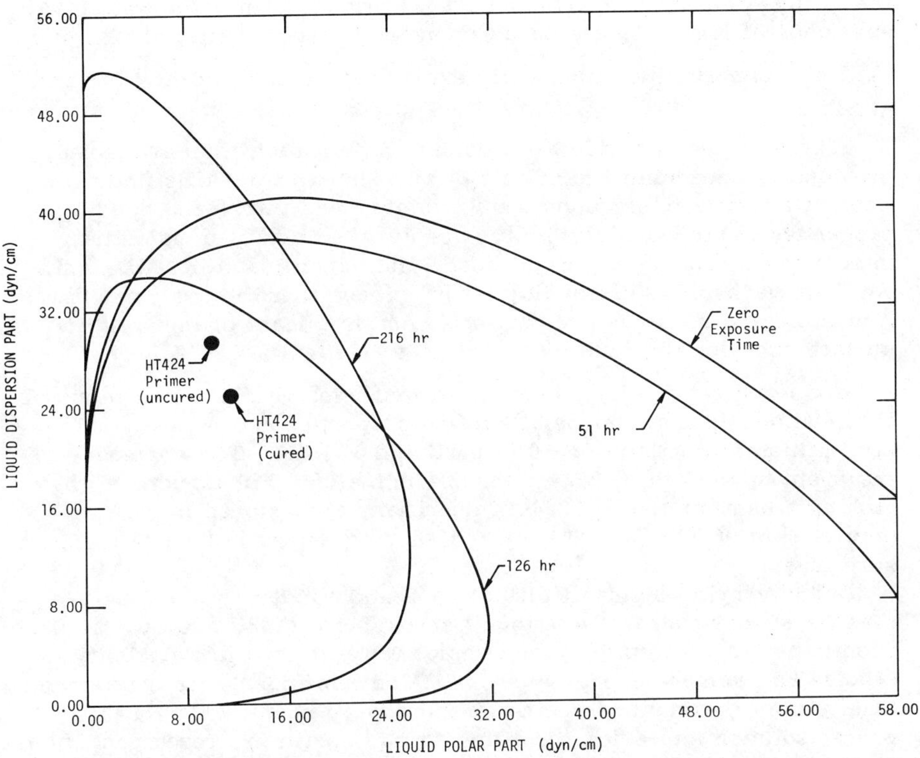

FIG. 4-39 Wettability envelopes for Al 2024-T3 with FPL etch and SET at 23°C, 50-60% RH.

described for HT424 primer. Subsequent to the etching treatment the Al 2024-T3 specimens were stored under ambient laboratory air at 23°C and 50 ± 5% RH while protected from dirt and dust by tissue paper covers. Curves of average values of γ^d_{SV} and γ^p_{SV} versus ambient SET are plotted in Fig. 4-38.

FPL-etch-treated Al 2024-T3 can be combined on a "wettability envelope" diagram, as defined by Eq. (13) of Sec. 2.2.6, and shown in Fig. 4-39. The distance of γ^d_{LV} and γ^p_{LV} extending from the origin $\gamma^d = 0$, $\gamma^p = 0$ to a particular wettability envelope curve describes a condition of spreading $\cos\theta = 1.0$ for the liquid on the solid surface. The average values of γ^d and γ^p for HT424 primer are represented in Fig. 4-39. It may be noted that all the wettability envelopes for FPL-etched Al 2024-T3 at different ambient SET conditions enclose the surface tension properties of HT424 primer. Since the wettability envelopes of Fig. 4-39 are based upon the average values of γ^d_{SV} and γ^p_{SV}, the uncertainties inherently expressed by $\pm\delta^d$ and $\pm\delta^p$ are not illustrated.

These studies provide more detailed information than available previously concerning the effects of surface exposure time under ambient air atmosphere upon the wettability and interfacial bonding properties of etched Al 2024-T3. Despite the obvious complexities introduced by microroughness, adsorption layers, and possible ionic-covalent mechani5ms of bonding, it is evident that the analysis summarized in Sec. 2.2.6 is capable of resolving significant changes in the surface energies γ^d_{SV} and γ^p_{SV} surface-treated Al 2024-T3.

A special set of integrated experiments were implemented to examine the relationship between metal surface properties and adhesive joint strength as a function of SET at 54°C and 95% RH. The surface treatment applied to Al 2024-T3 is the FPL etch. Subsequent to the etching treatment the Al 2024-T3 specimens were stored in a closed environment at 54°C and 95% RH for varying degrees of surface exposure time.

After varying levels of SET at 54°C and 95% RH the Al 2024-T3 samples were withdrawn to ambient atmospheric (23°C and 50 + 5% RH) conditions. A portion of these samples were utilized in wettability studies and surface energy analysis. The remainder were conditioned briefly under ambient atmosphere, then spray-coated with HT424 primer solution and dried according to manufacturer's recommendations as follows:

1. 30 min dry at 23°C
2. 60 min dry at 66°C

The average thickness of the dried primer layer is 0.001 in. These primer-coated surfaces were subsequently bonded into standard lap shear joints using HT424 structural adhesive film (0.135 lb/ft^2) and cured under 50 psi pressure for 60 min at 171°C.

Liquid-solid contact angles on the Al 2024-T3 samples after SET at 54°C and 95% RH were conducted at 20°C in the B-100 Environmental Chamber of the NRL contact angle apparatus (Rame-Hart, Inc.) using sessile drops.

The values of γ^d_{SV} and γ^p_{SV} are plotted vs. SET in the curves of Fig. 4-40. The curves of Fig. 4-40 display an initial rapid increase in γ^d_{SV} and concurrent decrease in γ^p_{SV} over the first 8-hr period of SET at 54°C and 95% RH similar to that for longer set at 23°C and 50± 5% RH (see Fig. 4-38). A direct correlation of these two experiments is obtained by the time ratio:

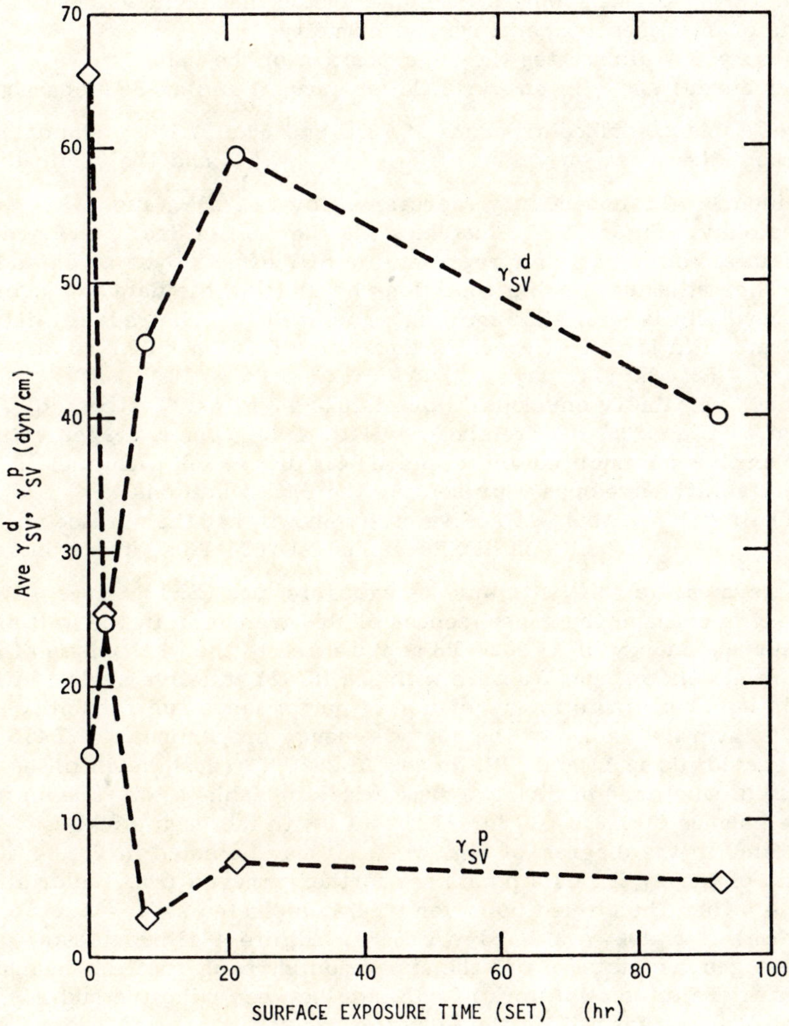

FIG. 4-40 Effect of SET at 54°C, 95% RH on calculated average values of γ_{SV}^d and γ_{SV}^p.

$$a_{TC} = \frac{\text{SET}(23°\text{C},\ 50\%\ \text{RH})}{\text{SET}(54°\text{C},\ 95\%\ \text{RH})} \approx 58.5 \tag{21}$$

where a_{TC} is a dimensionless time shift factor that includes both the effects of increased temperature and humidity.

Figure 4-41 illustrates the superposition of the values of γ_{SV} (upper curve) and γ_{SV}^d and γ_{SV}^p (lower curves) and an SET characteristic of ambient surface exposure at 23°C and 50 ± 5% RH. The data represent the averaged values of γ_{SV}, γ_{SV}^d, γ_{SV}^p, and the vertical bars identify the uncertainty represented by $\pm\delta$, $\pm\delta^d$, and $\pm\delta^p$, respectively. Figure 4-41 illustrates that Eq. (5) of Sec. 2.2.6 provides an effective basis for representing SET effects over an extended time range at ambient aging conditions by shifting the data to longer times characteristic of slow aging at ambient SET. The reduced data represented in Fig. 4-41 covers a time range 0.25 hr ≲ SET (23°C, 50% RH) ⩽ 5250 hr (218 days).

The "wettability envelopes" for Al 2024-T3 with FPL etch and SET at 54°C and 95% RH are shown in Fig. 4-42. The averaged values of γ^d and γ^p for uncured and cured HT424 primer fall within all of the wettability envelopes representing the SET conditions.

The results of this wettability study shows that the surface energy $\gamma_{SV} = \gamma_{SV}^d + \gamma_{SV}^p = \alpha_S^2 + \beta_S^2$ of FPL-etched Al 2024-T3 surfaces undergoes a systematic shift with surface exposure time (SET). It is of interest to examine the consequences of this transition in the wettability and surface energy of Al 2024-T3 as it relates to the mechanisms of interfacial bonding and fracture with the HT424 adhesive system utilized in lap shear tests for evaluation of performance and reliability.

The average values of the surface energy properties for HT424 (designated phase 1) and FPL-etched Al 2024-T3 (designated phase 3) at varied conditions of SET are summarized in Table 4-14. The shift in the balance of α_3 vs. β_3 for Al 2024-T3 with increasing SET is illustrated in the diagram of Fig. 4-43. It may be noted in Fig. 4-43 that all of the α_3, β_3 data points are farther removed from the origin $\alpha = \beta = 0$ than the circled point for HT424, indicated as $\alpha_1 = 5.21$ $(\text{dyn/cm})^{1/2}$ and $\beta_1 = 3.31$ $(\text{dyn/cm})^{1/2}$. Figure 4-43 provides a compact graphic statement of the experimental result that the calculated values of the interfacial work of adhesion between adhesive and adherend W_{13}, as defined by the following relation,

$$W_{13} = 2(\alpha_1\alpha_3 + \beta_1\beta_3)\ \text{erg/cm}^2 \tag{22}$$

exceeds the work of cohesion W_{11} for HT424 which is defined by the relation

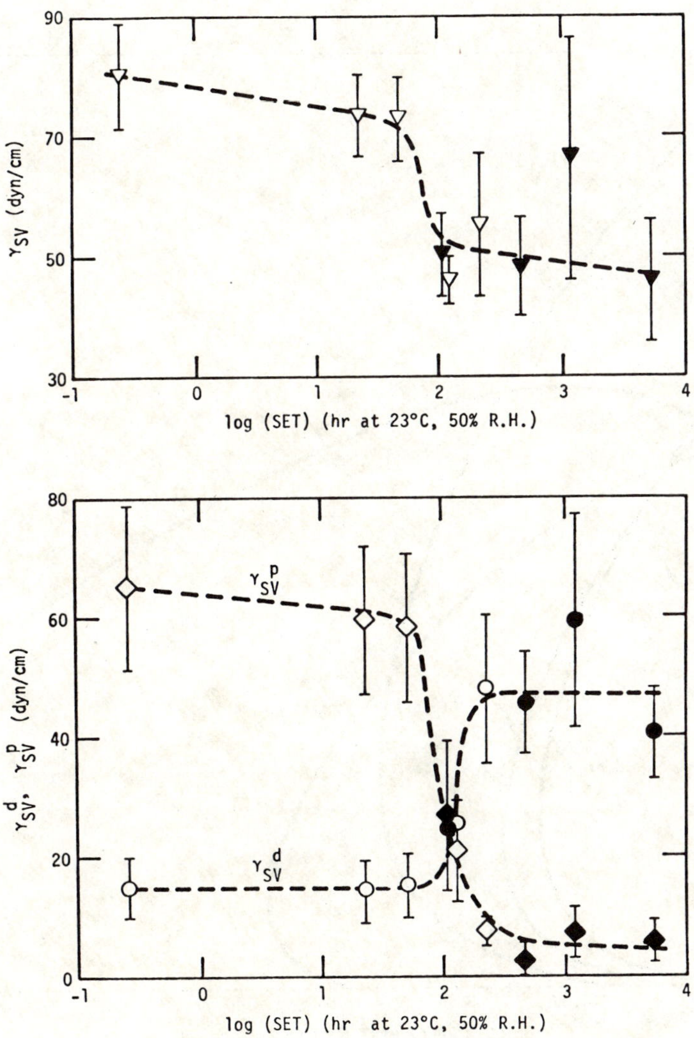

FIG. 4-41 Superposition of SET with $\gamma_{SV} = \gamma^d_{SV} + \gamma^p_{SV}$ at 23°C, 50% RH (open points) and SET 54°C, 95% RH (solid points) by the line shift factor $a_{TC} = 58.5$.

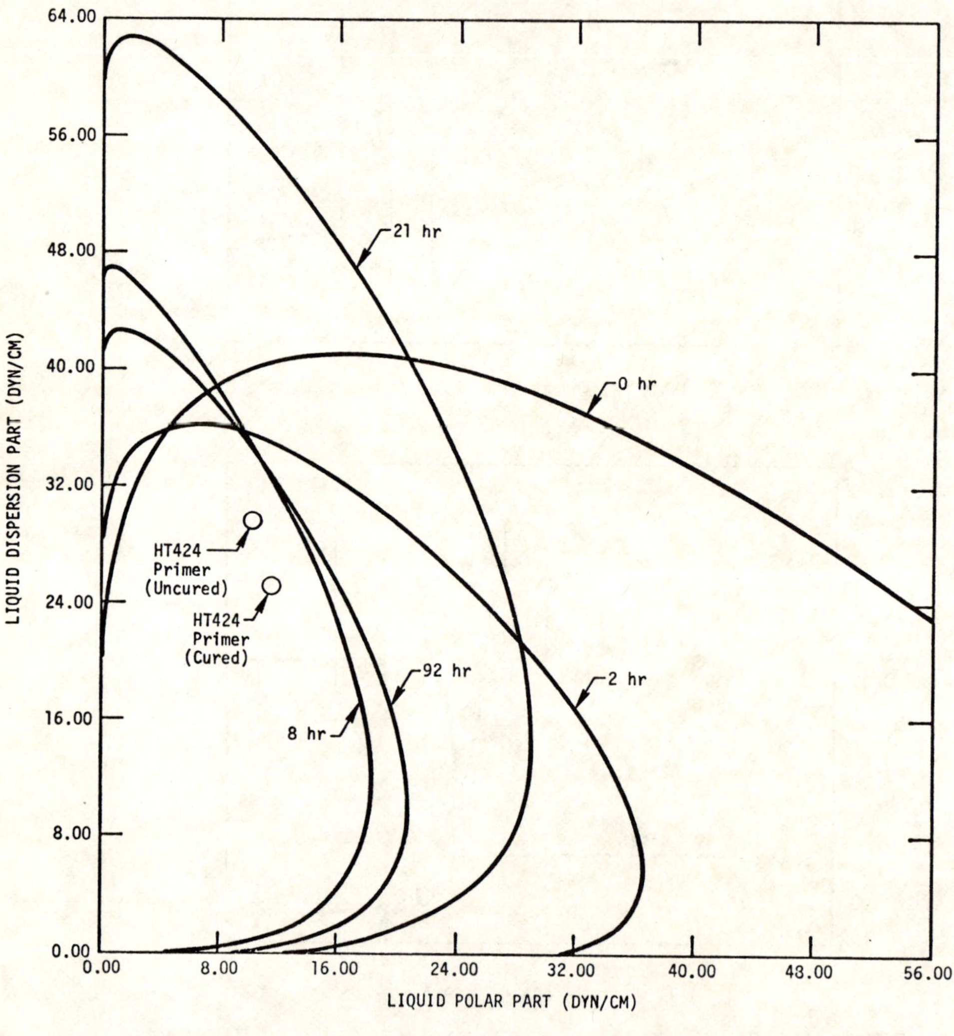

FIG. 4-42 Wettability envelopes for Al 2024-T3 with FPL etch and SET at 54°C, 95% RH.

TABLE 4-14 Surface Energy Analysis and Work of Adhesion between HT424 (Phase 1) and Sulfuric-Chromate-Etched Al 2024-T3 (Phase 3) at Varied Surface Exposure Times (SET)

A—Adhesive surface properties

Material	α_1 $(dyn/cm)^{1/2}$	β_1 $(dyn/cm)^{1/2}$
Uncured HT424	5.36	3.21
Cured HT424	5.05	3.42
Averaged values:	α_1 = 5.21	β_1 = 3.31

B—Adherend surface properties and work of adhesion W_{13}

SET (hr) (23°C, 50% RH)	Reduced SET (hr) (54°C, 95% RH)	α_3 $(dyn/cm)^{1/2}$	β_3 $(dyn/cm)^{1/2}$	W_{13} (erg/cm^2)
0.25 ± .25	0.0043	3.85	8.09	93.7
23	0.39	3.74	7.69	89.9
51	0.87	3.86	7.62	90.7
126	2.15	5.03	4.52	82.3
219	3.74	6.91	2.72	90.0
SET (hr) (54°C, 95% RH)				
0	≈ 0 (0.0043)	3.85	8.09	93.7
2	2	4.97	5.06	85.3
8	8	6.75	1.61	81.0
21	21	7.68	2.68	97.8
92	92	6.33	2.37	81.6

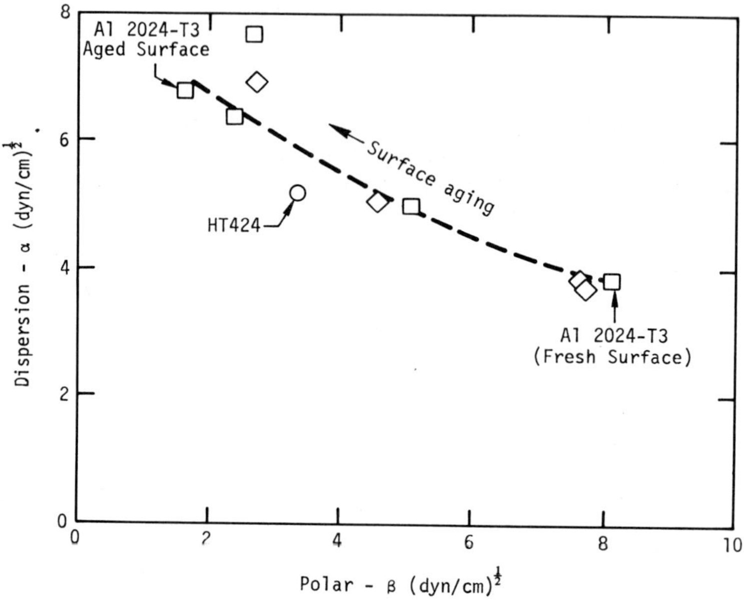

FIG. 4-43 Dispersion (α) and polar (β) components of the solid-vapor surface tension $\gamma_{SV} = \alpha_S^2 = \beta_S^2$ for HT424 primer (phase 1) and Al 2024-T3 adherend (phase 3).

$$W_{11} = 2(\alpha_1^2 + \beta_1^2) = 76.2 \text{ erg/cm}^2 \tag{23}$$

Calculated values of W_{13} are tabulated in the lower right columns of Table 4-14 and graphed as a function of SET at 54°C and 95% RH in the upper curve of Fig. 4-44. The reduced SET time scale for the upper curve of Fig. 4-44 combines previously reported data for SET at 23°C and 50% RH and SET at 54°C and 95% RH. One notes that the calculated values of W_{13} diminish in a monotonic fashion with increasing SET over the first 10-hr period and then display an apparent time invariance. The lower curve of Fig. 4-44 plots experimental data for the averaged values of lap shear bond strength σ_b as a function of SET time between FPL-etch treatment and bond formation. It may be noted that the variation in σ_b with SET shows a close correspondence with the variation in W_{13} with SET, as was noted in Figs. 4-35 and 4-36 for other surface properties.

Based upon the correlation demonstrated in Fig. 4-44 between the nearly equivalent effects of surface exposure time (SET) upon interfacial work of adhesion W_{13}, the lap shear bond strength σ_b, and the interfacial bond strength σ_I, the propositions outlined by Eqs. (14)

through (18) of Sec. 2.2.6 were applied. While the modified Griffith failure criterion defined by Eq. (14) represents a highly idealized boundary failure criterion, it makes a useful simplified basis for correlating fracture mechanics and surface energetic arguments.

The averaged values of $\alpha_1 = 5.21\ (\text{dyn/cm})^{1/2}$ and $\beta_1 = 3.31\ (\text{dyn/cm})^{1/2}$ define the surface properties of HT424 primer as summarized in Table 4-9. For surface-aged Al 2024-T3, surface energy

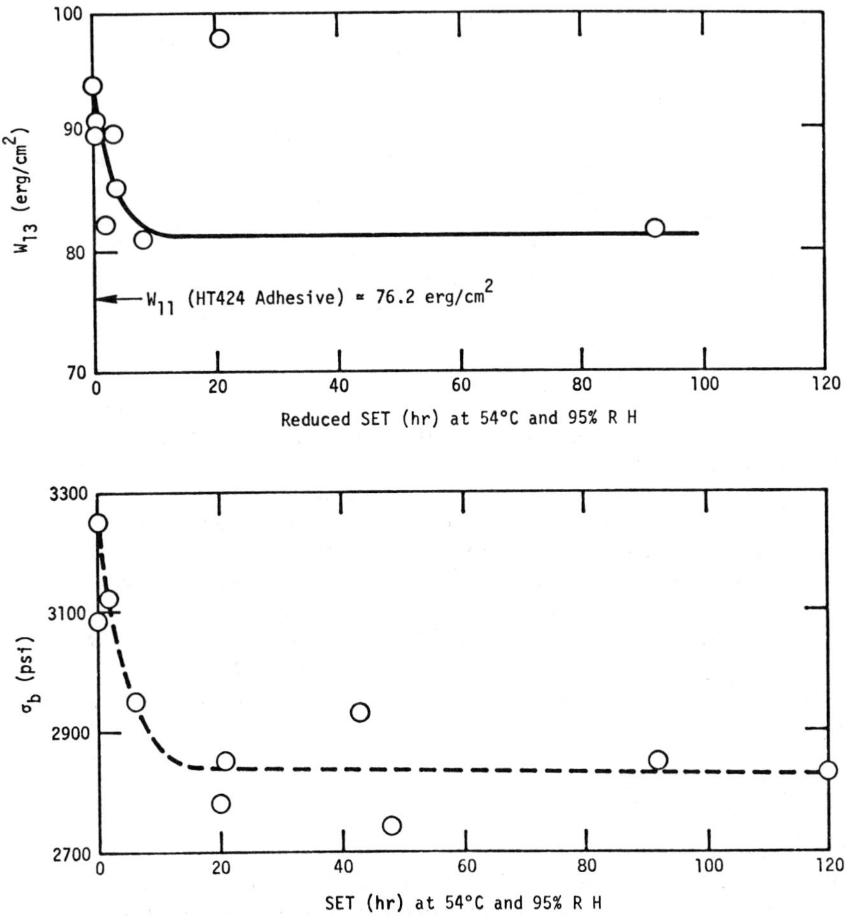

FIG. 4-44 Dependence of interfacial work of adhesion W (upper curve) and lap shear bond strength σ_b (lower curve) at varies SETs.

analysis has defined the following values as representing a fully aged surface at high SET:

SET (hr) At 54°C and 95% RH	α_3 $(dyn/cm)^{1/2}$	β_3 $(dyn/cm)^{1/2}$
21	7.68	2.68
92	6.33	2.37
Averaged values:	α_3 = 7.00	β_3 = 2.52

These surface energy parameters for HT424 and etched Al 2024-T3 are indicated in the α vs. β diagram of Fig. 4-45 (left view). The points H = 6.11 and K = 2.92 $(dyn/cm)^{1/2}$ defined by Eqs. (17) and (18) of Sec. 2.2.6 represent the locus of the vectors R_0, R(air), and R(H_2O), where for air $\alpha_2 = \beta_2 = 0$ and for H_2O $\alpha_2 = 4.67$ $(dyn/cm)^{1/2}$ and $\beta_2 = 7.14$ $(dyn/cm)^{1/2}$ for interfacial failure, with $\phi_I = 1 - \phi_c = 1.0$ between HT424 and Al 2024-T3. The right view of Fig. 4-45 presents the function of $\gamma_G^{1/2} \propto (\alpha_c)$ vs. R which may be generated from Eq. (14) for the case where E and c are assumed constant. The prediction from the Griffith surface energy analysis is that

$$\frac{\sigma_I(H_2O)}{\sigma_I(\text{air})} = 0.644 \tag{24}$$

Equation (24) expresses the equilibrium ratio of $\sigma_I(H_2O)/\sigma_I(\text{air})$ at exposure time $t = \infty$ and therefore is a minimum expected value where other factors of Eq. (14) remain invariant.

The critical Griffith stress σ_c for cohesive failure in HT424, with $\phi_c = 1 - \phi_I = 1.0$, may also be made for air and water immersion in similar fashion. In this case where the locus of failure is within phase 1, we make $\alpha_1 = \alpha_3 = H = 5.21$ $(dyn/cm)^{1/2}$, $\beta_1 = \beta_3 = K = 3.31$ $(dyn/cm)^{1/2}$ with $R_0 = 0$. The left view of Fig. 4-46 shows the argument in terms of the α vs. β coordinates. The right view shows the simple linear form for the Griffith function generated from Eq. (14) of Sec. 2.2.6 for $R_0 = 0$. For cohesive failure in HT424 the calculated ratio is as follows:

$$\frac{\sigma_c(H_2O)}{\sigma_c(\text{air})} = 0.63 \tag{25}$$

The close similarity between Eq. (24) and Eq. (25) indicates that a water environment should not preferentially promote interfacial failure and that a distinction cannot be made as to the expectation of interfacial vs. cohesive failure from the wettability analysis. However, a

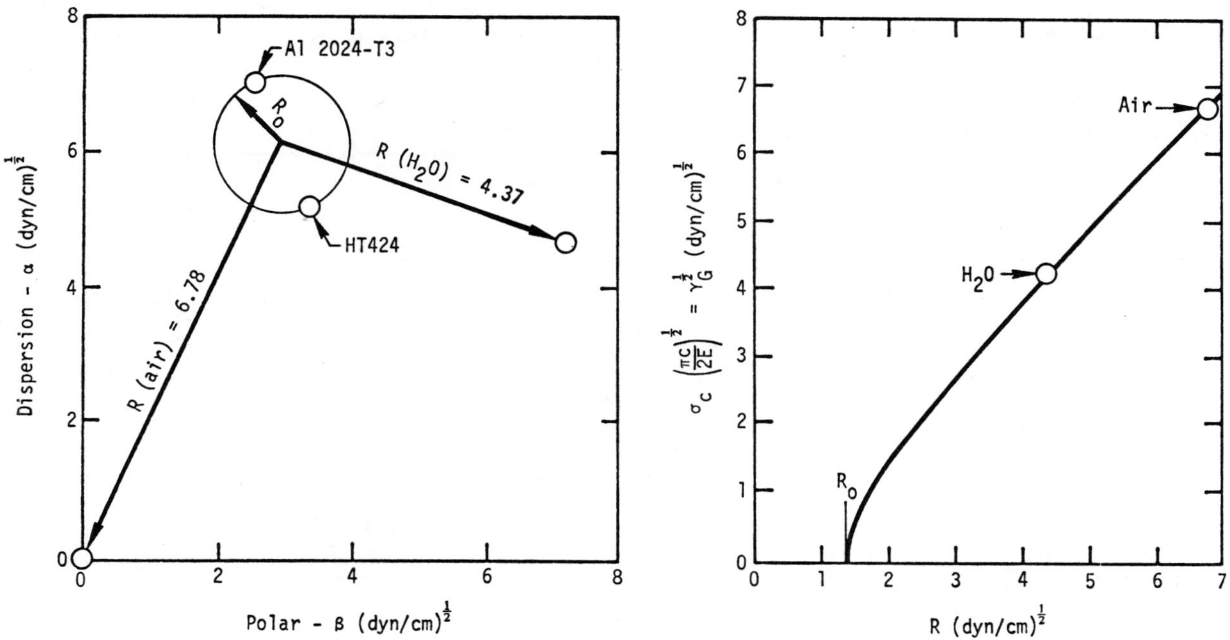

FIG. 4-45 Modified Griffith analysis of the effect of H_2O immersion in reducing critical failure stress σ_I for interfacial failure between HT424 and etched Al 2024-T3 ($\phi_I = 1 - \phi_c = 1.0$).

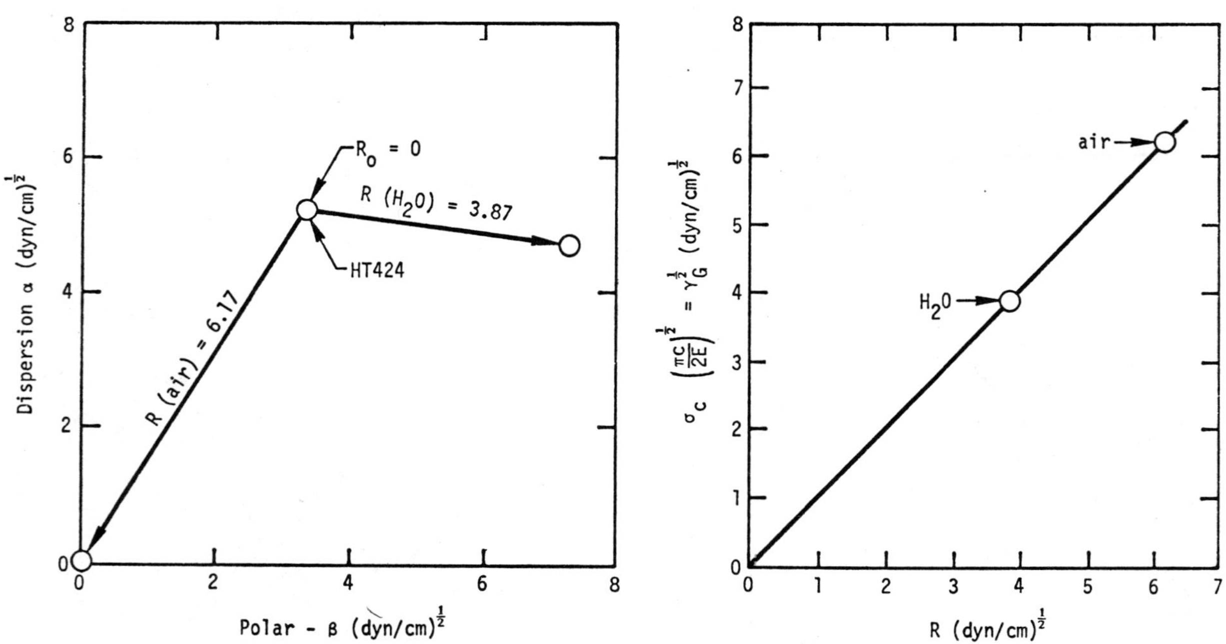

FIG. 4-46 Modified Griffith analysis of effect of H_2O immersion in reducing critical failure stress σ_c for cohesive failure in HT424 ($\phi_c = 1 - \phi_I = 1.0$).

comparison of predicted values of $\sigma_I(H_2O)/\sigma_I(\text{air})$ and $\sigma_c(H_2O)/\sigma_c(\text{air})$, calculated from the wettability analysis, can be compared with observed joint degradation in the next section.

3. 3. *Joint Aging (BET) Effects*

Al 2024-T3 samples were given the FPL etch and bonded at SET = 0 and SET = 20 hr. They were then fractured after various lengths of time in a humidity chamber at 95% RH and 54°C.

The average values of σ_b and ϕ_c are plotted in Fig. 4-47. The effect of bond exposure causes σ_b to decrease and then level off after about 200 hr. At about 700 to 800 hr σ_b decreases some more. The σ_b values for SET = 20 hr correspond to the lower values expected from Fig. 4-36. The fraction of the surface ϕ_c that failed cohesively remains almost constant with BET and also shows the expected decrease at SET = 20 hr (from Fig. 4-36). Since ϕ_c remains almost constant at about 0.8, it is concluded that degradation during BET is mainly due to weakening of the adhesive rather than degradation at the interface.

Finally, Fig. 4-48 shows a profile of the SET and BET response surface for lap shear bond strengths of 2024-T3 aluminum and HT424 adhesive. It should be noted that the initial and final values of BET, bond strengths $\sigma_b \approx 3200$ psi and $\sigma_b \approx 2200$ psi, correlate closely with the work of others (see Ref. 38, Table 15, p. 74) for the Al-HT424 system.

It is interesting to note that the predicted ratio $\sigma(H_2O)/\sigma(\text{air}) \approx 0.64$, for interfacial or cohesive degradation is in close agreement with the experimental ratio of σ_b at SET = 0 hr, BET = 0 hr, to that of σ_b at SET = 20 hr and BET = 1000 hr (Fig. 4-48):

$$\frac{\sigma_b(\text{20-hr SET, 1000-hr BET})}{\sigma_b(\text{0-hr SET, o-hr BET})} \approx \frac{2250 \text{ psi}}{3250 \text{ psi}} = 0.69 \qquad (26)$$

3. 4. *Summary of Results*

With respect to the Al 2024-T3—HT424, system, the following important observations have been made:

1. After surface preparation of Al 2024-T3 with the FPL etch, ellipsometry reveals that the oxide film varies in thickness and structure across the surface. The average thickness is of the order of 200 Å with refractive index of $n_f \approx 1.8$ close to that for α-alumina ($n_f \approx 1.7$) but with an effective absorption index much larger

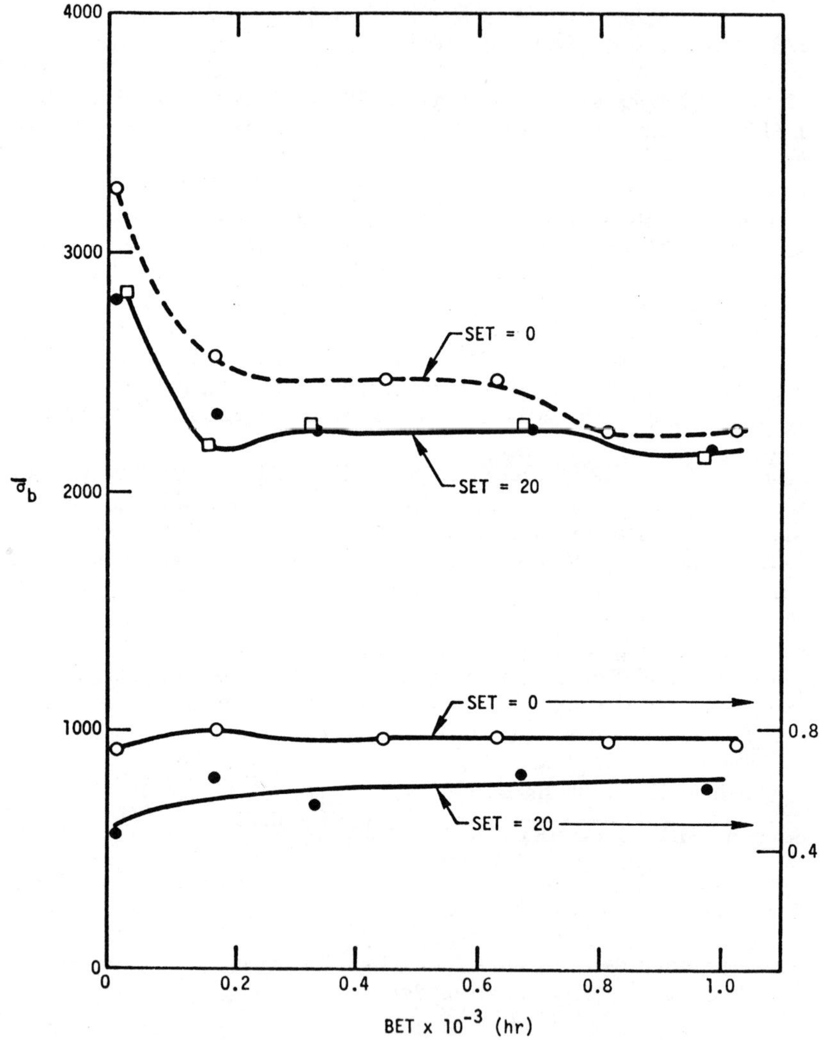

FIG. 4-47 Plot of σ_b and ϕ_c vs. BET at 54°C, 95% RH. SET = 0 and 20 hr.

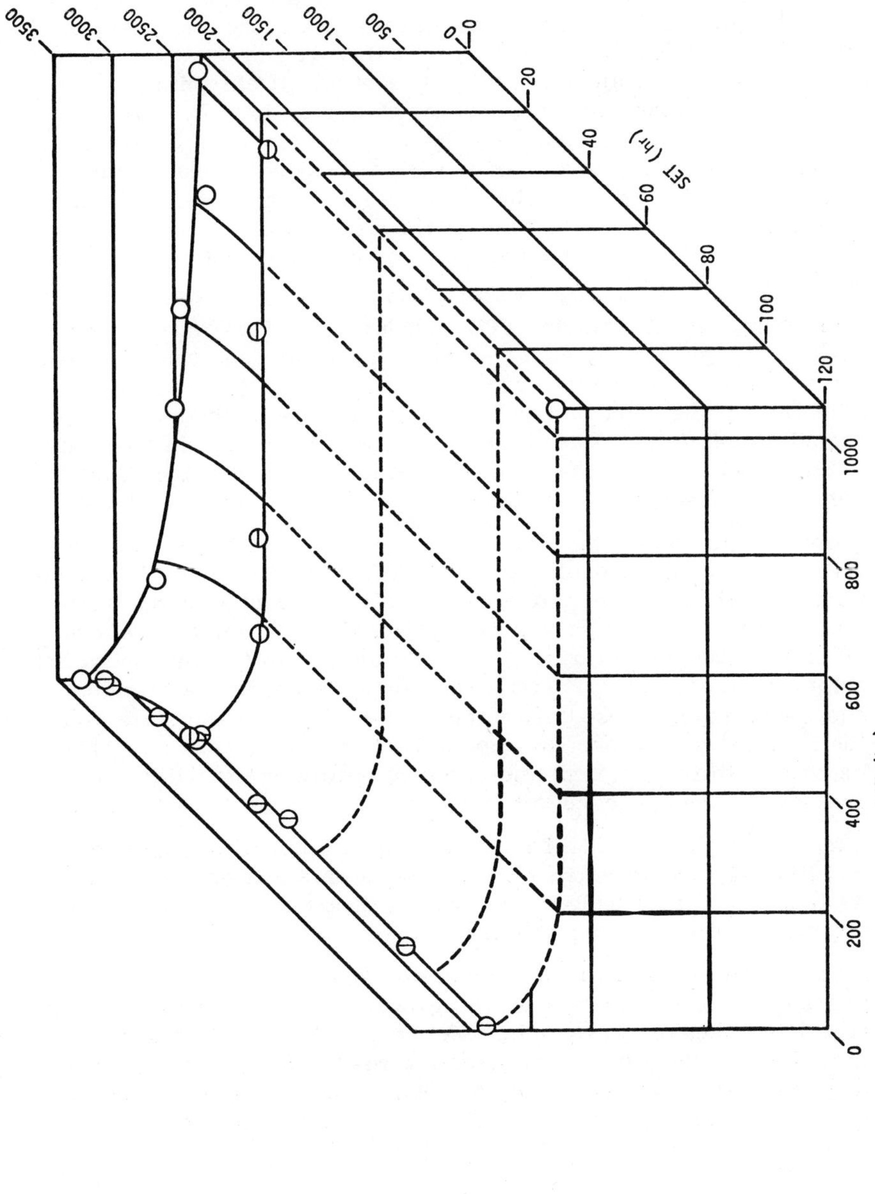

FIG. 4-48 SET and BET response surface for lap shear bond strength for Al 2024-T3—HT424.

($\kappa_f \approx 0.3$) than for α-alumina ($\kappa_f \approx 0$). The effective absorption index may be related to the surface roughness as well as the true absorption index. Ellipsometry of FPL-etch surfaces prior to and after bonding and fracture indicates that interfacial failure is in the oxide film for thicker films.

2. Auger spectroscopy and ion-sputter-etching give semiquantitative chemical profiles of the films formed during surface preparation. Other than the aluminum and oxygen, Cu, C, and S were detected on the 2024-T3 aluminum after the FPL etch. If chromium was present, it was removed by the vacuum system before analysis was begun. Iron segregates to the outer surface of the oxide. Copper segregates into the oxide with a maximum concentration near the central region of the oxide. Nickel segregates to higher concentration in the oxide than in the metal. Silicon is not observed in the FPL-etch surface but was markedly noticeable in an area of interfacial fracture.

3. Microscopy reveals that the FPL etch leaves an aluminum surface with very high density of pit holes with a large spectrum of dimensions. Primer flows into these holes and probably increases adhesion by the hook-and-latch effect. Failure is observed in the adhesive, in the glass carrier, in the primer, and in the primer-metal interface (shown to be in the oxide). Considerable natural porosity is observed in the adhesive.

4. Contact angle measurements provide data for wettability envelopes, which showed that freshly etched 2024-T3 aluminum alloy is wettable by the HT424 adhesive and primer. It is apparent that the wettability parameters, viz. the polar part of the solid surface tension γ^p and the dispersion part γ^d, relate directly to the thermodynamic ability of the surface to form strong or weak bonds. Although one would expect these parameters to be related to the state of the surface with respect to organic contamination, they may also be related to the physical and chemical nature of the oxide film. These relationships have yet to be explored.

5. Surface mapping of 2024-T3 aluminum with respect to ellipsometry and SPD have revealed that these instruments can be used as nondestructive testing tools to predict the probable loci of weak bonding prior to the bonding process.

6. Surface aging of 2024-T3 aluminum in humidity chambers causes degradation of the surface (oxide growth and contamination) with associated changes in the ellipsometry, SPD, and wettability parameters. The surface degradation results in bond strength degradation. The fraction of the total bond area (ϕ_I) increases and the associated bond strength σ_I decreases with SET.

7. Joint exposure in a humidity chamber causes degradation of the HT424 adhesive and therefore degradation of bonds for 2024-T3 aluminum HT424 joints. The Griffith critical stress analysis from the contact angle and wettability parameters faithfully predicted that joint exposure in a humidity chamber will cause degradation of the adhesive and interface for the aluminum system.

4. DISCUSSION AND CONCLUSIONS

4.1. Summary

This chapter described a study of aluminum 2024-T3 alloy adherend, adhesively bonded with an epoxy HT424 adhesive. The adherend surface properties, after standard surface preparation treatments, were characterized with a number of surface tools. Ellipsometry was used to estimate oxide film thickness and optical properties, surface potential difference (SPD) measurements were made to characterize the outer surface dielectric properties, Auger electron spectroscopy (AES) was used to establish chemical analysis of the outer surface and chemical profiles during ion-sputter-back-etching. Topography and morphology were established with electron, scanning electron, and optical microscopy. Wettability parameters were established by making contact angle measurements with a series of liquids with different surface tension. A study of bond strengths after adherend aging and bond aging in humidity chambers was correlated with surface character. Comparison of surface character with bond failure revealed probable failure mechanisms.

It was demonstrated that mapping of surfaces prior to bonding can reveal areas that will have a predisposition to fail interfacially and how the mapping technique can be used as an nondestructive testing (NDT) tool for monitoring surfaces.

4.2. Discussion

To improve adhesive joints, the investigator concentrates his efforts on the weakest link. If failure occurs primarily in the adhesive, emphasis is placed on a search for a stronger adhesive. If failure occurs primarily at a weak boundary layer due to contamination or in the oxide, emphasis is placed on cleanliness or surface preparations. Since new applications demand stronger and stronger adhesive joints, the tendency will be to continue improvement of the weakest link as it shifts from one locus to another. It is concluded, therefore, that the development of tools such as described in this chapter for identifying the weak

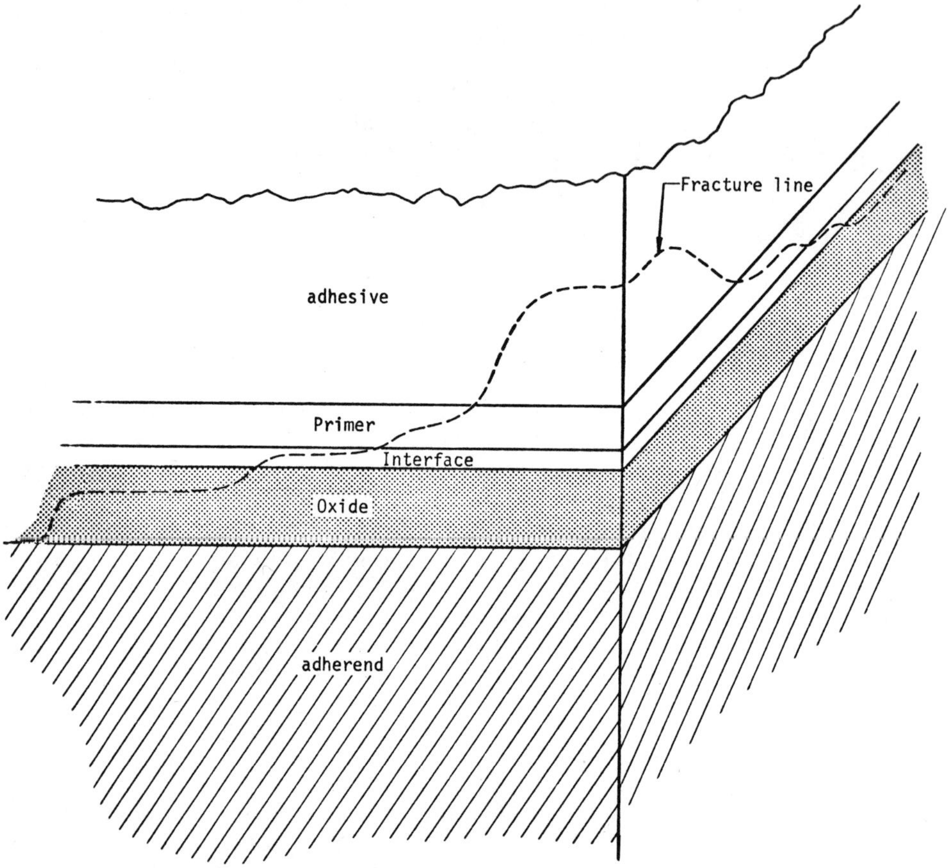

FIG. 4-49 Schematic representation of bond region with hypothetical fracture line.

regions is of importance now and will continue to be of prime importance in the development of stronger adhesive joints.

The most significant results of this research impacts on two important areas: one has to do with nondestructive testing (NDT) of surfaces prior to bonding, the other has to do with understanding failure mechanisms in the Al 2024-T3—HT424.

It has been demonstrated that surface regions that will result in a poor adhesive bond can be identified prior to bonding by ellipsometry and/or SPD. The potential of such NDT techniques is most significant for two reasons. First, supposing a completely satisfactory adhesive-bonding procedure is developed in the laboratory for a particular application, inadvertent problems can arise on a factory assembly line that may cause deviations from the necessary processing parameters. Examples of these problems include contaminated processing solutions, contamination of atmospheric ambient, malfunction of process equipment, etc. Even though it may be impractical to survey every part or every region of every part by the NDT equipment, spot checks could be valuable for taking corrective action on the assembly line before too many parts have been bonded. Second, to improve on a given adhesive system, it is necessary to identify those parameters that are causing weakness.

Comparison of surface properties for various surface treatments with bond strengths, as in Tables 4-3 and 4-17, does not reveal conclusive correlations. Proper surface preparations yield strong bonds with mostly cohesive failure as for the FPL-etch and plasma treatment in Table 4-3. The low bond strengths for the vapor-degrease process in Table 4-3 may relate to the large carbon contaminant revealed in the Auger analysis and the resultant low wettability revealed by the low solid surface tension γ_{SV}. For a properly prepared surface, (uncontaminated), the effect of surface treatment to bond strength is probably mechanical, by increasing surface area and/or producing a hook-and-latch property. However, this report demonstrates that there is a relationship between physical and chemical surface properties and bond strength for aged or contaminated surfaces.

Failure of an adhesive joint between Al 2024-T3 and HT424 can occur at the metal-oxide interface, in the oxide layer, in the primer or adhesive, at voids between primer and adhesive, or in the glass carrier. A schematic representation of half of a bond is given in Fig. 4-49. The area of any given bond can be subdivided into small regions or patches that have failed in one of the above-mentioned loci. The dashed line in Fig. 4-49 represents the line of fracture that might occur. In spite of the attempt to prepare surfaces with uniform character, the result was always a nonuniform surface with respect to surface properties and localized bond failure mode. Let the fraction of the surface that failed at the interface be ϕ_I and the corresponding average bond strength be σ_I, the fraction that failed in voids between primer and adhesive be ϕ_v ($\sigma_v = 0$), the fraction that failed in the adhesive be ϕ'_c with corresponding average bond strength σ'_c, and the fraction that failed in the glass be ϕ_g with σ_g. The total bond strength* σ_b can be expressed

$$\sigma_b \approx \phi_I\sigma_I + \phi'_c\sigma_c + \phi_g\sigma_g \tag{32}$$

where

*This equation is justified in reference 35.

$$\phi_I + \phi'_c + \phi_v + \phi_g = 1 \tag{33}$$

Then

$$\sigma_b \approx \phi'_c(\sigma'_c - \sigma_I) + \sigma_I(1 - \phi_v - \phi_g) + \phi_g\sigma_g \tag{34}$$

Except in a few instances (such as for Fig. 4-19) fracture in the glass was cohesive and since the glass matrix was always the same, ϕ_g and σ_g are constants, with ϕ_g estimated to be about 0.3. If for a set of experiments σ_I, σ'_c, ϕ_v, and ϕ_g are constants, then plots of σ_b vs. ϕ'_c will yield a straight line with $\sigma_b = \sigma_I(1 - \phi_v - \phi_g) + \phi_g\sigma_g$ at $\phi'_c = 0$ and $\sigma_b = \sigma'_c$ at $\phi'_c = 1$. If, however, any of the parameters σ_g, σ_I, σ'_c, ϕ_v and ϕ_g are not constant, a straight line will not result from a plot of σ_b vs. ϕ'_c. In our experiments we have made the approximation that $\sigma_g \approx \sigma'_c$ and $\phi_v \approx 0$ so that Eq. (34) reduces to a form of Eq. (19),

$$\sigma_b \approx \phi_c(\sigma_c - \sigma_I) + \sigma_I \tag{35}$$

where $\phi_c = \phi_g + \phi_i$.

Though considerable number of data cluster around a straight line as in Figs. 4-31 and 4-32, it is not surprising that there is large scatter such as in Fig. 4-33. If σ_c is constant but σ_I or ϕ_v is not constant for a set of experiments, then a plot of σ_b vs. ϕ_c will not fall on a straight line but will fan out from $\phi_c = 1$, $\sigma_c \approx 3800$ psi, as observed in Fig. 4-33. Values of σ_I can be calculated with Eq. (35) for each set of values of σ_b and ϕ_c, as seen in Fig. (36). It becomes obvious that results from the common practice of measuring σ_b as a function of experimental variables such as humidity, temperature, time, etc. is difficult to interpret in terms of physical characteristics and mechanisms unless the various parameters σ'_c, σ_g, σ_I, ϕ_I, ϕ_g, ϕ_c, and ϕ_v are separated. Each of these parameters may relate in a different way to a physical characteristic, such as oxide film thickness, oxide optical constants, work function (SPD), or chemical composition. Figures 4-16 and 4-17 also reveal that for lap shear joints the stress is not constant over the bond area. Failure tends to occur as shown in Fig. 4-19 from a paper by Bolger [33].

In spite of all of these unknowns, we have demonstrated that in certain circumstances (Figs. 4-28, 4-29, and 4-30) ϕ_I, ϕ_c, and σ_I are related to the oxide thickness and optical properties. As the oxide thickens the work function increases, as reflected in the SPD. The increase in interfacial failure with oxide thickness is shown to be failure in the oxide and is interpreted in the simple terms that the probability of the presence of a defect that can propagate under stress is directly proportional to the quantity of oxide (i.e., oxide thickness). The strength of the oxide σ_I is related to the oxide structure and is reflected in the optical properties. It follows that to discover more about the mechanism of bond failure for a particular adhesive system it is necessary to measure σ_I, ϕ_I, σ_c, ϕ_c, ϕ_v, σ_g, and ϕ_g in order to discover the relation between these variables and the physical and chemical properties within the bond. For example, chemical analysis by Auger spectroscopy has revealed that silicon contamination is in the oxide in the region of oxide failure but is not present in the oxide in the nonbonded region. This indicates that in addition to regions of the thicker oxide the surface that was contaminated with silicones or glass may also produce weakened areas.

Aging surfaces at various temperatures and relative humidity (SET) degrades the bond. The change in the physical and chemical properties associated with this degradation are not fully understood as yet, but it is clear that further research with ellipsometry, SPD, Auger spectroscopy, SEM, and electron diffraction is likely to reveal the details. Thus far it appears that during aging (SET) a structural transformation occurs in the oxide that is produced by the FPL etch. This is reflected in the optical properties. Simultaneous with oxide growth and structural transformation surface contamination may occur with resultant decrease in the polar nature (wettability) of the surface. Figures 4-21 and 4-28 show that at 95% RH and 54°C both σ_I and ϕ_c decrease during SET whereas σ_c remains essentially constant. On the other hand, during BET degradation is mostly restricted to the adhesive. During BET ϕ_c (and ϕ_I) remain essentially constant and σ_c decreases.

4. 3. Conclusions

The original purpose of this study has been fulfilled in that a better understanding of the physical and chemical properties of surfaces after various surface preparations has been obtained; the various fracture modes have been revealed and the important parameters of Eq. (34) are separated. It has been demonstrated in the case of FPL-etched aluminum that σ_I and ϕ_I are related to oxide thickness and structure (optical properties). Further study of the relationship between fracture parameters σ_I, ϕ_I, σ_c, ϕ_c, etc. and physical and chemical properties will be useful and hopefully will lead to an understanding of those properties that must be designed into surface preparations, primers, and adhesives to obtain high-strength and durable joints.

APPENDIX A

TABLE A-1 Details of Surface Preparations for Al 2024-T3

1. Vapor Degrease (VD)

   a. Vapor degrease for 1 min in a vapor degreaser containing trichloroethylene.

2. Sulfuric Acid-Dichromate (FPL Etch)

   a. Samples are cleaned ultrasonically for 15 min in trichloroethylene (TCE).
   b. Samples are cleaned ultrasonically again for 5 min in fresh trichloroethylene.
   c. The samples are alkaline-cleaned for 10 min at 66°C in a stirred solution 06 TURCO 4215 cleaner (31.2 g/liter in distilled water).
   d. The samples are rinsed with distilled water and then rinsed ultrasonically for 5 min in distilled water.
   e. Check for water break.
   f. The samples are immersed in a sulfuric acid-chromate solution for 13 min at 66°C. This solution is made as follows from reagent-grade chemicals: This solution has been reacted with Al metal previously to give a dark brown color to it:

Sodium dichromate	28.5 g
Sulfuric acid	285.0 g
Distilled water	to make 1 liter

   g. The samples are spray-rinsed with cold water (below 23°C).
   h. The samples are immersed in cold water and the spray rinse repeated.
   i. The samples are checked for water break and dried in a vented oven for 15 min at 40°C.

3. Glow Discharge (Plasma) Treatment of Al 2024-T3

   a. Vapor-Degreased (VD) Sample

      Wiped with TCE
      Cleaned ultrasonically 15 min in TCE
      Alkaline-cleaned 10 min at 66°C
      DI H_2O rinse (water break test)
      Air-dry

   b. Plasma Conditions:

      Gas = argon 150 μm

---

4. Glow Discharge (Plasma) Treatments of Al 2024-T3 for Wettability

   a. Clean 1 × 4 in. samples ultrasonically in trichloroethylene for 15 min.
   b. Place samples in plasma chamber and pump to 10^{-5} torr for 15 min.
   c. Adjust the plasma treatment variables to the following values and start plasma and maintain these settings:

Pressure	100 μm
Flow	2 liters/min argon
Power	100 W
Time	20 min

   d. Plasma is turned off and chamber is pumped to 10^{-5} torr.
   e. i. Open the chamber to a source of O_2 gas and adjust to atmospheric pressure for O_2 posttreatment for 1 hr.
      ii. Open the chamber to a source of H_2O vapor for 1 hr; H_2O posttreatment.
   f. Remove samples and perform contact angle measurements within 20 min.

5. For Plasma Treatment for Al 2024-T3 Bonding (Oxygen and Water Vapor Posttreatment)

   a. TCE wipe.
   b. Ultrasonic clean in TCE for 15 min.
   c. Place samples in plasma chamber and pump to 1 μm for 15 min.
   d. Purge chamber with argon for 10 min.
   e. Start plasma with following constants:

Pressure	300 μm
Flow	0.4 liter/min argon
Power	100 W
Time	20 min

   f. Turn off plasma and pump to 1 μm.
   g. Bleed in flowing oxygen at just below atmospheric pressure for 20 hr.
   h. Seal off chamber, place in glove box, pump to ~10 μm.
   i. Flood with argon to ambient pressure.

(continued)

(Table A-1 continued)

---

j. Primer coupons with HT424 primer parts A and B.
k. Remove coupons from glove box, cure primer in vacuum oven, and bond with HT424 adhesive.
l. Perform shear test in Instron.

6. For Plasma Treatment for Al 2024-T3 Bonding (No Posttreatment)

a. Wipe 12 Al 2024-T3 coupons with trichloroethylene-soaked cloth.
b. Alkaline-clean 10 min at 66°C in TURCO alkaline cleaner (4215).
c. Place samples in plasma chamber and pump down to 1 μm for 15 min.
d. Purge chamber with argon 10 min.
e. Start plasma with the following constants:

Pressure	300 μm
Flow	0.4 liter/min argon
Power	100 W
Time	20 min

f. Turn off plasma and seal chamber still at 300 μm
g. Place sealed chamber in glove box and pump down to ~10 μm.
h. Flood with argon to ambient pressure.
i. Primer coupons with HT424 primer parts A and B.
j. Remove coupons from glove box, dry primer in vacuum oven, and bond with HT424 adhesive.
k. Perform shear test in Instron.

TABLE A-2 Percentage Al (2024-T3) Alloy Composition

Element	%
Cu	3.8-4.9
Mg	1.2-1.8
Si	0.5
Fe	0.5
Zn	0.25
Cr	0.1
Mn	0.3-0.9
Ti	—

APPENDIX B

Determination of Adhesive/Cohesive Failure Areas

Although ϕ_I can be estimated by visual observation, a more quantitative estimate is needed. The most quantitative estimate thus far is made by photographing the fracture bonds and placing a fine grid (20 lines/in.) on cellophane over the bond area. The grid squares that cover the area that broke interfacially (metal color) were counted. The top lines of Table B-1 compare the visual estimate of observer (a) with the photographic technique (b). The mean percentage deviation of the visual values from the photo values is 55%. The second set of observations compares the phototechnique by observer (b) with the same technique by observer (c). The mean percentage deviation of the values obtained by (b) from those by (c) is 12%. A third set of observations by observer (b) was made visually and photographically, however, the visual observations were made by comparing the specimen with three samples that had been estimated by the photographic technique for comparison standards. By using the standards the mean percentage deviation of the visual observations from the photo observations is 4% in the region of 70 to 80% adhesive failure.

It would be advantageous if a technique could be found for measuring ϕ_I directly that would be rapid and accurate. A number of electrical methods were attempted. Measurement of the capacitance of the samples after putting lacquer on all areas except the bond area proved useless in that no reproducible result could be obtained; both sodium chloride aqueous solution and mercury were used. The last set of values in Table B-1 shows the estimates of ϕ_I by measuring the hydro-

TABLE B-1 Comparison of Estimates of ϕ_I by Various Techniques

		ϕ_I									%
a	Visual	0.16	0.18	0.18	0.24	0.24	0.25	0.28	0.39	0.45	55
b	Photo	0.15	0.25	0.10	0.20	0.40	0.35	0.60	0.65	0.70	
b	Photo	0.01	0.03	0.09	0.13	0.25	0.35	0.43	0.77	0.78	12
c	Photo	0.01	0.03	0.12	0.17	0.34	0.41	0.51	0.77	0.78	
b	Photo	0.72	0.76	0.74	0.75	0.76	0.81				
b	Visual	0.70	0.70	0.75	0.76	0.80	0.80				4
b	Photo	0.18	0.18	0.25	0.24	0.49	0.39	0.45	0.97	0.97	
d	Electro	0.11	0.20	0.16	0.37	0.33	0.52	0.21	0.76	1.3	34

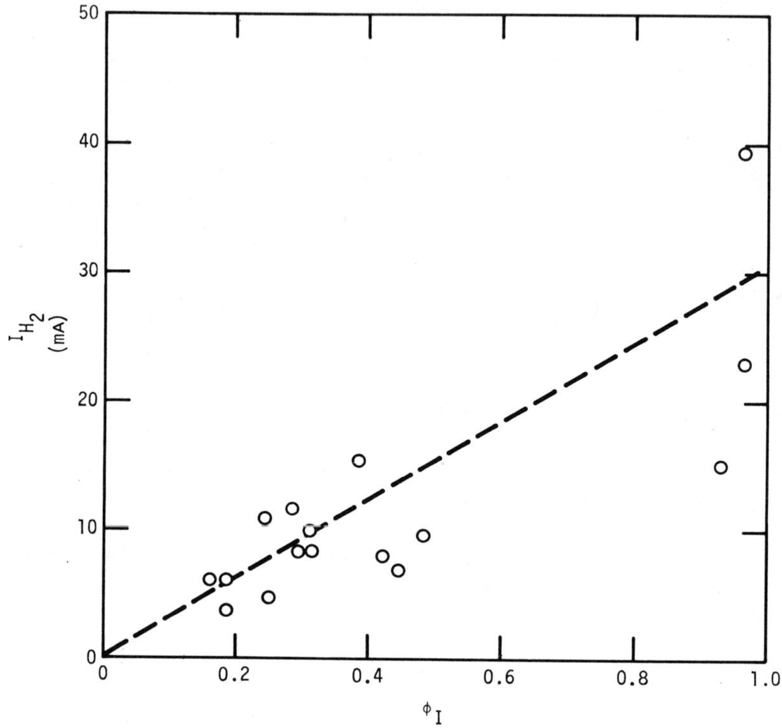

FIG. B-1 Plot of the fraction of the bond area that failed interfacially, ϕ_I vs. hydrogen current I_{H_2}.

gen evolution current, with a potentiostat, at a fixed time interval after placing a constant voltage between the sample and a reference electrode in a 3% sodium chloride solution. This technique does give a correlation between I_{H_c} and ϕ_I estimated by the phototechnique as seen in Fig. B-1, but the mean deviation of the electrochemical from the phototechnique is 34%. The fractional value of ϕ_I was obtained for the electrochemical measurement by dividing the current for a given specimen that had not been bonded but was masked except for the area where the bond would be (i.e., 0.5 in.^{2}). To see if a correlation exists between bond strength σ_b and cohesive failure areas as determined by the electrochemical technique, σ_b is plotted vs. ϕ_c in Fig. B-2. Since the scatter is worse than that determined by the phototechnique (see Fig. 4-32), the electrochemical technique was abandoned.

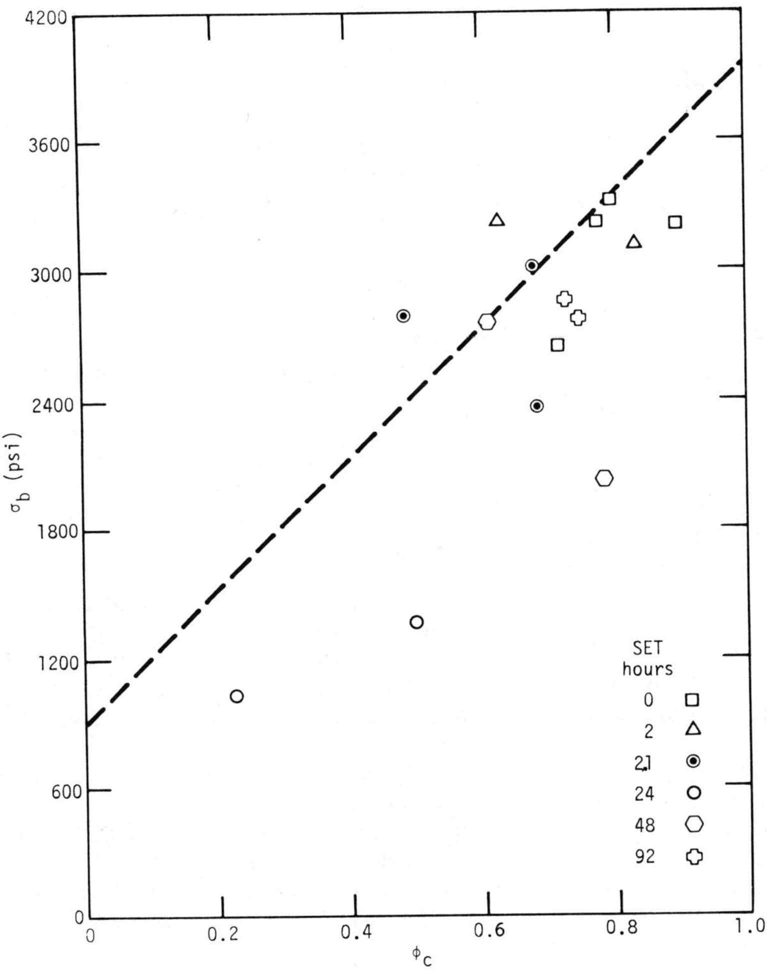

FIG. B-2 Plot of ϕ_b vs. ϕ_c, c determines with the electrochemical technique.

ACKNOWLEDGMENTS

The authors are pleased to acknowledge the help of P. J. Dynes, L. W. Crane, C. G. Rhodes, R. A. Spurling, and J. M. Harris for their help with the experimental work.

REFERENCES

1. O. Kubaschewski and B. E. Hopkins, *Oxidation of Metals and Alloys*, 2nd ed., Butterworths, London, 1962.
2. T. Smith, J. Electrochem. Soc. 112:560 (1965).
3. P. Kofstad, *High Temperature Oxidation of Metals*, Wiley, New York, 1966.
4. J. C. Bolger and A. S. Michaels, in *Interface Conversion for Polymer Coatings* (P. Weiss and G. Dale Cheever, eds.), American Elsevier, New York, 1969, Chap. 1.
5. F. Keller, M. S. Hunter, and D. L. Robinson, J. Electrochem. Soc. 101:335 (1954).
6. F. Liechti and W. D. Treadwell, Helv. Chim. Acta 30:1204 (1947).
7. R. A. Harrington and H. R. Nelson, Trans. Aime 137:62 (1940).
8. H. W. L. Phillips, Inst. Met. Monograph and Report Series No. 13, 237 (1952).
9. N. J. DeLollis, *Adhesives for Metals*, Industrial Press, New York, 1970, p. 35.
10. F. L. McCrackin, E. Passaglia, R. R. Stromberg, and H. L. Steinberg, J. Res. Nat. Bur. Std. A67:363 (1963).
11. F. L. McCrackin, NBS Technical Note 479, April 1969.
12. C. A. Fenstermaker and F. L. McCrackin, Surf. Sci. 16:85 (1969).
13. M. A. Barret and A. B. Winterbottom, 1st Int. Cong. Metallic Corrosion, London, 1961, Butterworths, 1962.
14. G. W. Simmons, J. Colloid Interfac. Sci. 34:343 (1970).
15. M. P. Seah, Surf. Sci. 32:703 (1972).
16. M. Klasson, J. Hedman, A. Berndtsson, R. Nilsson, C. Nordling, and P. Melnik, Physica Scripta 5:93 (1972).
17. M. L. Tarng and G. K. Wehner, J. Appl. Phys. 44:1534 (1973).
18. J. T. Grant, T. W. Hass, and J. E. Houston, accepted by Physics Letters A.
19. D. T. Quinto and W. D. Robertson, Surf. Sci. 27:645 (1971).
20. H. Shimizu, M. Ono, and K. Nakayama, Surf. Sci. 36:817 (1973).
21. J. R. MacDonald and C. A. Barlow, Jr., J. Chem. Phys. 39:412 (1963); 40:1535 (1964); 43:2575 (1965).
22. J. R. MacDonald and C. A. Barlow, Jr., Surf. Sci. 4:381 (1966).
23. R. V. Culver and F. C. Thompkins, Advan. Catalysis 11:74 (1959).
24. J. E. Boggio and R. C. Plumb, J. Chem. Phys. 44:1081 (1966).

25. W. A. Zisman, in *Adhesion and Cohesion* (P. Weiss, ed.), Elsevier, Amsterdam, 1962, pp. 176-208.
26. D. H. Kaelble, *Physical Chemistry of Adhesion*, Wiley, New York, 1971, Chaps. 5 and 12.
27. D. H. Kaelble, J. Macromol. Sci.—Reviews Macromol. Chem. C6(1): 85 (1971).
28. D. H. Kaelble, Adsorption and interdiffusion mechanisms of bonding in polymer composites, Proc. 23rd Int. Cong. Pure and Appl. Chem., Vol. 8, Butterworths, London, 1971, pp. 265-301.
29. American Cyanamide Co., Bloomingdale Department, Havre DeGrace, Maryland 21028.
30. S. Tajima, Adv. Corr. Sci. Tech. 1:234 (1970).
31. W. C. Hamilton and G. A. Lyerly, Tech. Report 4185, Picatinny Arsenal Contract DAAA-21-70C-0434, Gillette Research Institute, Rockville, Md., March 1971.
32. B. B. Bowen, Northrop Corporation (personal communication).
33. R. L. Patrick, *Treatise on Adhesion and Adhesives*, Marcel Dekker, 1973, p. 177.
34. D. H. Kaelble and E. H. Cirlin, J. Polymer Sci., Part C 35:79 (1971).
35. T. Smith and P. Smith, *Adhesion and Adsorption of Polymers*, Vol. 12A (L. H. Lee, ed.), Plenum, New York, 1980, pp. 123-139.

5

Mechanisms of Adhesion Failure between Polymers and Metallic Substrates: Titanium 6Al-4V with HT424 Adhesive*

Tennyson Smith and David H. Kaelble

Rockwell International
Thousand Oaks, California

1. INTRODUCTION

The general purpose, the problem, and the solution have been discussed in the previous chapter on aluminum (Chapter 4). The experimental approach for the Ti 6Al-4V HT424 system is essentially the same as for the aluminum.

The oxide film formed on titanium has been reported to be anatase (tetragonal TiO_2) by Burghers et al. [1] and Harrington and Nelson [2]. Other investigators [3] have found brookite (orthorhombic TiO_2) on stripped films. Rutile, anatase, and hydrated oxides have been identified [4] on anodized surfaces. On the other hand, Brusilavskii et al. [5] reported that no hydrates are formed, but that the oxides simply hold water by adsorption. Recent work has been reported by Hamilton and Lyerly [6] concerning the characterization of Ti and Al alloy surfaces during selected stages of prebond treatments. They suggest that bonding differences with variously treated titanium alloys were caused by variations in crystalline structure of the titanium oxide. In this report we find only amorphous oxide layers, as did Lively [7], for the phosphate-fluoride-treated Ti 6Al-4V alloy.

---

*This work was conducted under Contract No. F33615-72-C-1641 for the U.S. Air Force Non-Metallic Materials Division, Air Force Materials Laboratory, Wright-Patterson A.F.B., Ohio 45433.

2. EXPERIMENTAL SECTION

2.1. *Materials*

2.1.1. Adherends

The metal adherends are bare Ti 6Al-4V alloy cut from 0.063-in. sheet with dimensions slightly larger than 4 × 1 in. and then milled to precisely 4 × 1 in.

2.1.2. Adhesives

The adhesive used for most of the work reported here was an epoxy phenolic (HT424) from American Cyanamide. This adhesive has a glass fabric carrier and standard weight 0.135 ± 0.005 lb/ft^2. It was shipped in dry ice and stored at -10°C. Unfilled primer with parts A and B were used with the HT424 adhesive. The primer composition was 10 pbw HT424 part A, 10 pbw HT424 part B, and 30 pbw HT424 primer-thinner. Another adhesive was used in the initial stages of this study with Ti 6Al-4V alloy. The adhesive was FM-34B-32 with BR-34 primer.

2.1.3. Surface Preparations

As has been shown [8] and will be seen, every step of the surface preparation may be of significance with respect to bond strength. It is of the utmost importance that procedures be followed in every detail if experimental results are to be compared from laboratory to laboratory. Table A-1 (Appendix A) identifies the various surface preparations for Ti 6Al-4V. Most of the work on Ti 6Al-4V was done with the phosphate-fluoride-cleaning process (no. 1 of Table A-1). Approximate composition of the titanium alloy is given in Table A-2.

2.2. *Surface Characterization Techniques*

The same techniques were used for titanium as for aluminum: ellipsometry, Auger spectroscopy, chemical profiles, surface potential differences, microscopy, and contact angles. Details of these techniques are in the chapter on aluminum (Chapter 4).

2.3. *Adhesive Primer Procedures*

2.3.1. Applications and Curing Cycles

The priming procedure for Ti 6Al-4V with HT424 A and B primer is as follows: Freshly treated coupons are spray-primed to an approximate thickness of 0.001 in. using an atomizer operated on low-pressure, clean, dry (argon) gas. The primer has the following compositions:

10 pbw HT424 part A
10 pbw HT424 part B
30 pbw HT424 primer-thinner

The primed coupons are dried for 30 min under ambient laboratory conditions. The primed coupons are then oven dried for 60 min at 66°C. This priming procedure is recommended [9] by the manufacturer and was followed throughout the program.

The following is the BR-34 primer application for bonding Ti 6Al-4V:

1. Thin and spray on 2-mil dry coating.
2. 30 min air-dry.
3. 30 min oven-dry at 104°C (220°F)
4. 45 min oven-dry at 210°C (410°F)

The following is the BR-34 primer for contact angle analysis:

1. 1 part primer diluted with 5 parts BR34-32 thinner.
2. Coated on glass slide by dipping.
3. 30 min air-dry.
4. 30 min oven-dry at 104°C in dry N_2 atmosphere.
5. 60 min oven-dry at 210°C in dry N_2 atmosphere.
6. Uncured samples given 20 hr air dry and 5 hr vacuum dry

2.4. Bonding Procedures

For bonding fixture, see the chapter on aluminum (Chapter 4, Sec. 2.4.1).

The cure cycles for the HT424 system was that recommended by the manufacturer [9]:

1. Raise temperature to 171°C (340°F) in 60 min.
2. Cure for 60 min at 171°C (340°F).
3. Cool down.

For FM-34B-32 the cure cycle was:

1. Cut adhesive and store in desiccator prior to bonding.
2. Heat joint to 288°C (550°F).
3. Hold at 288°C (550°F) for 90 min at 40 psi.
4. Cool to room temperature at 40 psi.

See the chapter on aluminum (Chapter 4) for tensile lap shear testing (Sec. 2.5), surface-aging (SET) effects (Sec. 2.6), and bond joint-aging (BET) effects (Sec. 2.7).

3. EXPERIMENTAL RESULTS AND DISCUSSION

3.1. *Surface Character vs. Joint Strength*

3.1.1. Ellipsometry

Ellipsometry results for phosphate-fluoride films on Ti 6Al-4V are reported in Table 5-1. The values of n_f and d on the right of Table 5-1 are the refractive index and film thickness computed for a range of values of the absorption coefficient κ_f. For ion-sputtering experiments the film is removed in about 12.5 min (see Table 5-2). Sputter removal for other oxides have proved to be about 10 to 15 Å/min in our system. If the sputter rate is the same for the phosphate-fluoride film on Ti 6-4 the film is of the order of 200 Å thick. This value corresponds with the ellipsometer results if $n_f = 2.8$, $\kappa_f = 0.1$ (no solution was found for $\kappa = 0$). Since the refractive index of titanium dioxide is reported [10,11] to be about $n_f = 2.6$ to 2.9 and $\kappa_f = 0$ and since values of $n_f = 1.5$, $\kappa_f = 0.8$ for $d \approx 500$ Å seem to be unreasonable, a value of about 200 Å may be more correct. To check the ellipsometer measurements, the thickness of the film was measured by means of the reflection interference microscope. In this technique, the film was picked up on a glass slide and examined in reflection using a thallium light source ($\lambda = 5400$ Å). One beam is reflected off the glass slide while the other is reflected off the top of the film; the fringe displacement in the image of the combined beams is a measure of the film thickness. Figure 5-1 shows a piece of the phosphate-fluoride

TABLE 5-1 Ellipsometric Results for the Phosphate-Fluoride Film on Ti 6-4; $\lambda = 6328$ Å, Angle of Incidence = 70°, $\hat{n} = 3.09\,(1 - i\,1.11)$

		Computer solution		
Δ	ψ	κ	n	d (Å)
72.24	27.86	0	—	—
		0.1	2.85	200
		0.2	2.46	256
		0.3	2.16	310
		0.4	1.93	356
		0.5	1.75	397
		0.6	1.65	428
		0.7	1.58	464
		0.8	1.52	514
Interference microscope				540

TABLE 5-2 Ellipsometry and Sputter Time for Ti 6Al-4V with Various Surface Treatments (Angle of Incidence 70°)

Surface treatment	λ	Δ	ψ	n_f	κ_f	Film Thickness (Å)	Sputter time (min)	Sputter rate (Å/min)
Phosphate-fluoride	6328	72.86	27.86	2.84	0.1	200	12.5	16
Nitric acid-fluoride	5461	100.04	30.88	2.10	0.0	130	10	13
TURCO	5461	88.10	24.34	2.80	0.3	184	12.5	15

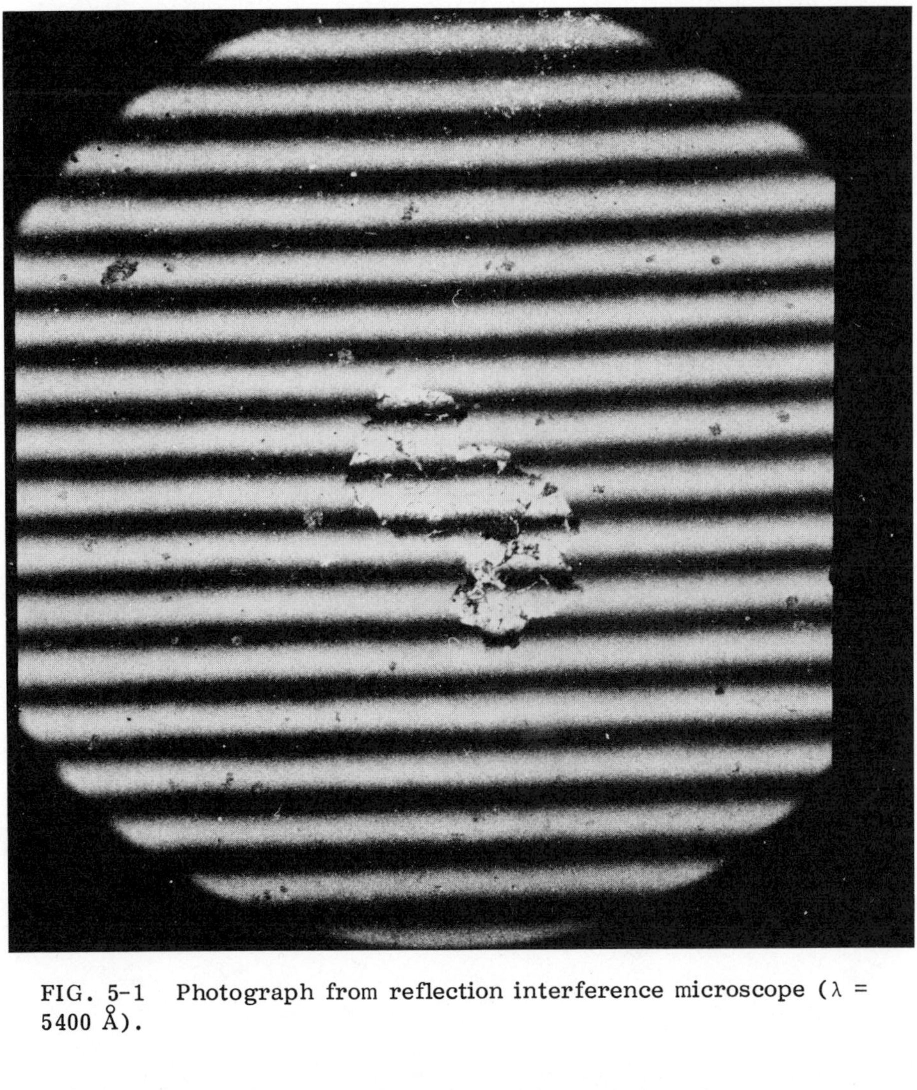

FIG. 5-1 Photograph from reflection interference microscope (λ = 5400 Å).

film on a glass slide. The displacement due to the film is about 0.2 of the fringe distance. Therefore, the film thickness is about 0.2 × 5400/2 = 540 Å. Since the interference microscope measurement is an absolute measurement, independent of the refractive index, it is concluded that the film removed from the Ti 6-4 and placed on a glass slide is indeed ~540 Å but that the removal process has increased its thickness. This appears even more likely when considering the observations made during the film removal.

The film that is removed from Ti 6-4 is yellow-brown with the oxide layer on top. When this film is washed with water the yellow-brown layer appears to dissolve, leaving the oxide layer on the microscope slide. This same observation was made by Hamilton and Lyerly [6] (personal communication). Two things could make the oxide films appear too thick in the interference microscope: an air gap between the film and the slide or some of the yellow-brown layer was not removed by the water. As for the aluminum surface after the FPL (Forest Products Laboratory) etch, the titanium surface after the phosphate-fluoride treatment is very rough. If the film is removed with the same topography as it has on the metal, the distance between a plane touching points along the bottom of the film and a plane touching points along the top would be much larger than the actual film thickness. Aside from effects of roughness of an order much less than the wavelength of the light used, ellipsometry does not suffer from this problem. Only light reflected from facets with the proper angle of incidence is accepted by the analyzer; light from all other facets of the rough surface are scattered away from the detector. It appears that the phosphate-fluoride films formed on Ti 6-4 by the Gillette researchers were quite different from ours. Their films were reported to be of the order of 3000 Å as compared to our 200-500; their films were crystalline—ours amorphous and their films contained fluorine—ours may not have.

It is believed that the differences may be due to the differences in phosphate-fluoride treatment used at Gillette as compared to that at the Science Center or to the oxide film removal process. Table B-1 of Appendix B shows the step-by-step treatment at Gillette [6] and General Dynamics [12] to compare with our Table A-1. Although each process is referred to as the phosphate-fluoride process, each has slightly different conditions. In a recent report [12] Alexander (at the Los Angeles Division of RI) points out the large effect that slight differences in the detail steps can have. He used the Battelle formulation and found that 80°F gave a completely different film (with respect to SEM) as compared to that found at 185°F. He also found that the rinse time was critical as to the film formed and the bonding characteristics. Both the Gillette workers and Alexander found fluorine in their films, whereas we found that if F was present it was masked by an Fe peak. This effect may be similar to our not observing chromium on FPL-etched aluminum. A 24-hr period in our vacuum system may have removed the fluorine.

3.1.2. Auger Spectroscopy

Figure 5-2 is an example of an Auger electron spectrogram for a Ti 6Al-4V sample. The spectrogram was performed after the phosphate-fluoride treatment but prior to ion-sputtering. The surface elements present are the alloy elements (see Table A-2) Ti, Al, V, Fe, C, N, and O as well as surface contamination P, Ni, S, and Cu. If F is present, it is masked by the Fe (650-eV) peak. Figure 5-3 shows the

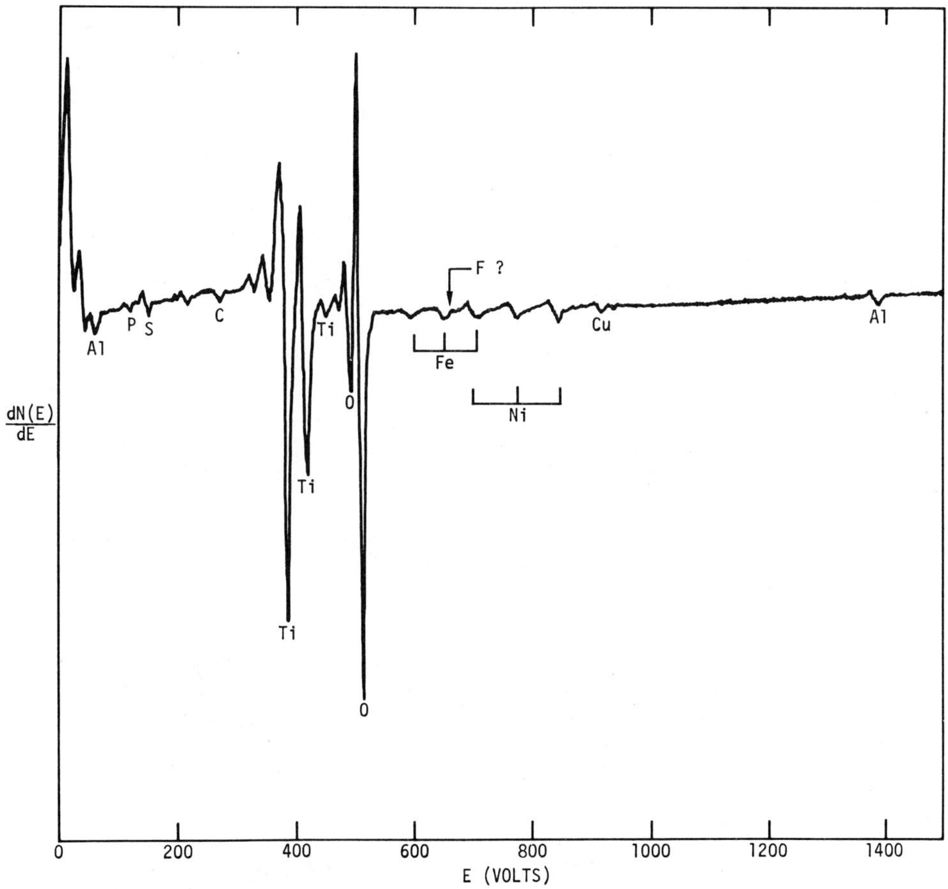

FIG. 5-2 AES for Ti 6Al-4V after the phosphate-fluoride treatment.

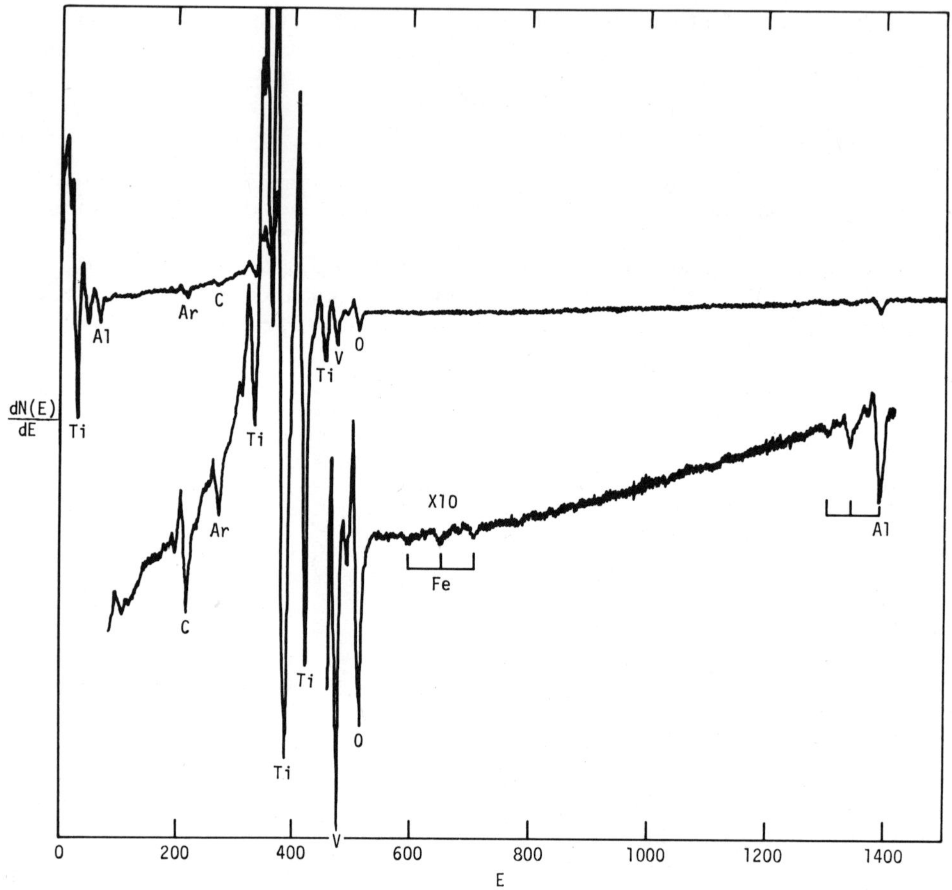

FIG. 5-3 Auger spectrogram for Ti 6Al-4V after phosphate-fluoride treatment and 40 min of Ar^+ bombardment.

AES after 40 min of Ar ion-sputtering. The surface contamination has been greatly reduced but of course the alloy elements are still present. The Auger peak-to-peak height (APPH) values prior to ion bombardment for surfaces prepared by the phosphate-fluoride, nitric acid-fluoride, and TURCO treatment are compared in Table 5-3.

3.1.3. Chemical Profiles

Figures 5-4, 5-5, and 5-6 are chemical profiles for films formed on Ti 6Al-4V by the phosphate-fluoride, the nitric acid-fluoride, and TURCO surface treatments, respectively. An interesting feature of

TABLE 5-3 Surface Characterization of Ti 6Al-4V Samples and Bond Strength for Various Surface Preparations: Adhesive HT 424

	Ellipsometry					Auger spectroscopy (APPH)											
Surface treatment	Film thick-ness (Å)	n_f	κ_f	SPD	γ_{SV}	Ti	Al	V	Cu	Ni	Fe	C	O	F	S	P	Bond strength (psi) (avg of 9 bonds)
Phosphate-fluoride	200	2.84	0.1	0.38	66	47	1.5	1.5	1	2	1.2	1	58	?	2	1	3400 ± 300
TURCO	160	2.80	0.3	0.04	66	47	5	2	1	5	0	5	47	0	4	0	3100 ± 400
Nitric acid fluoride	100	2.80	0.0	0.34	66	52	4.2	2	4.8	4.8	2	2.3	65	0	4.8	0	2400 ± 300

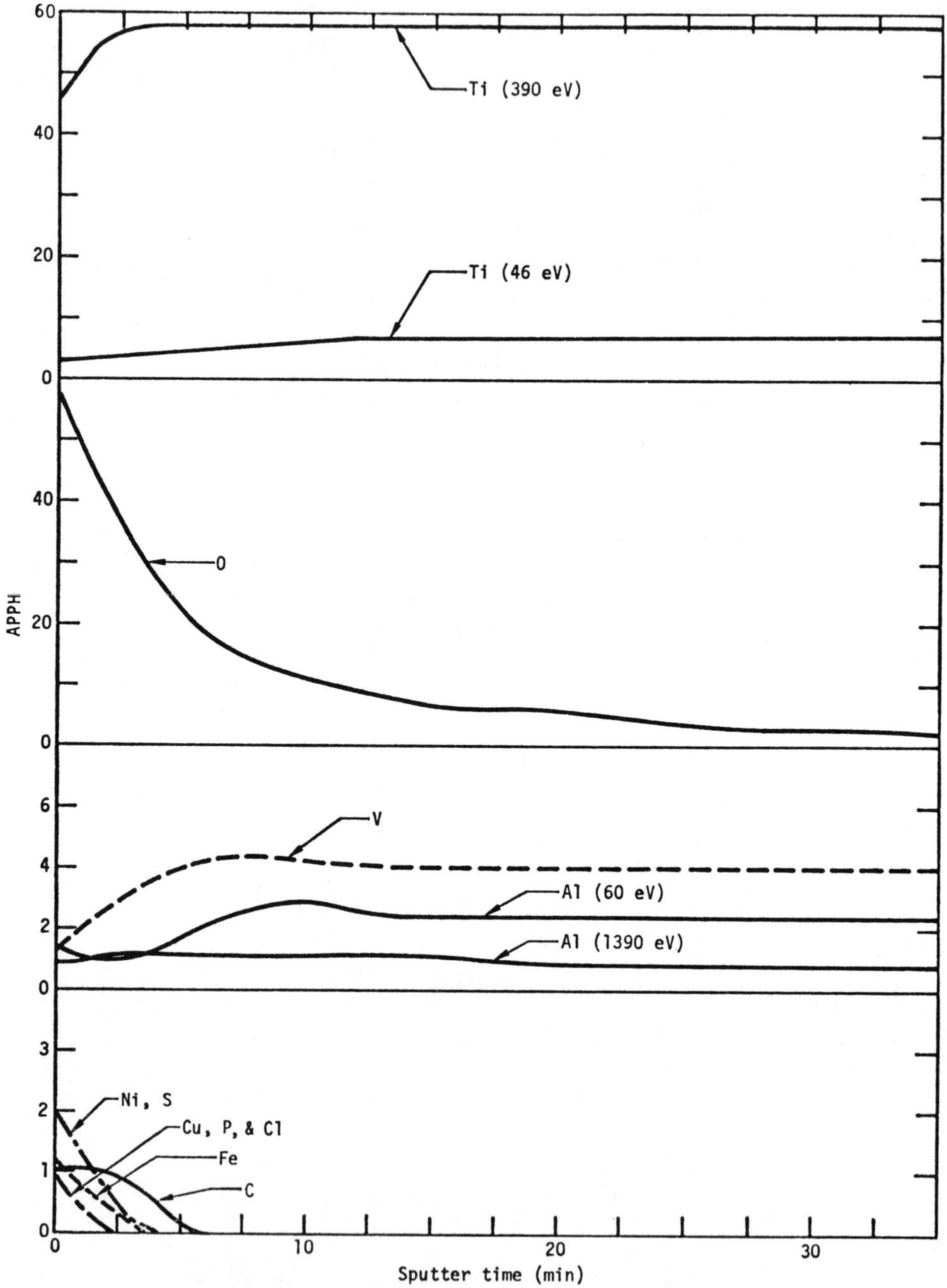

FIG. 5-4 Chemical profile of film on Ti-6Al-4V after phosphate-fluoride treatment.

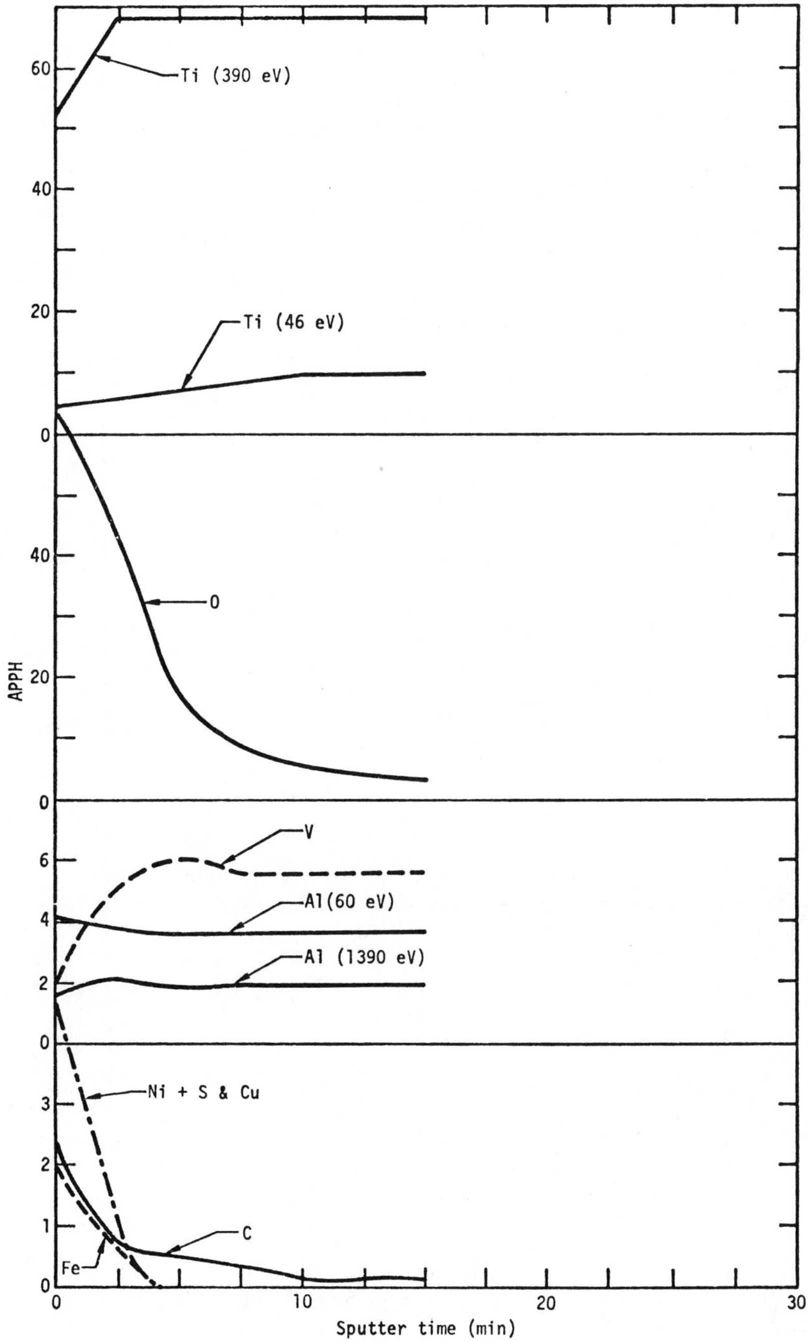

FIG. 5-5 Chemical profile of film on Ti 6Al-4V after the nitric acid-fluoride treatment.

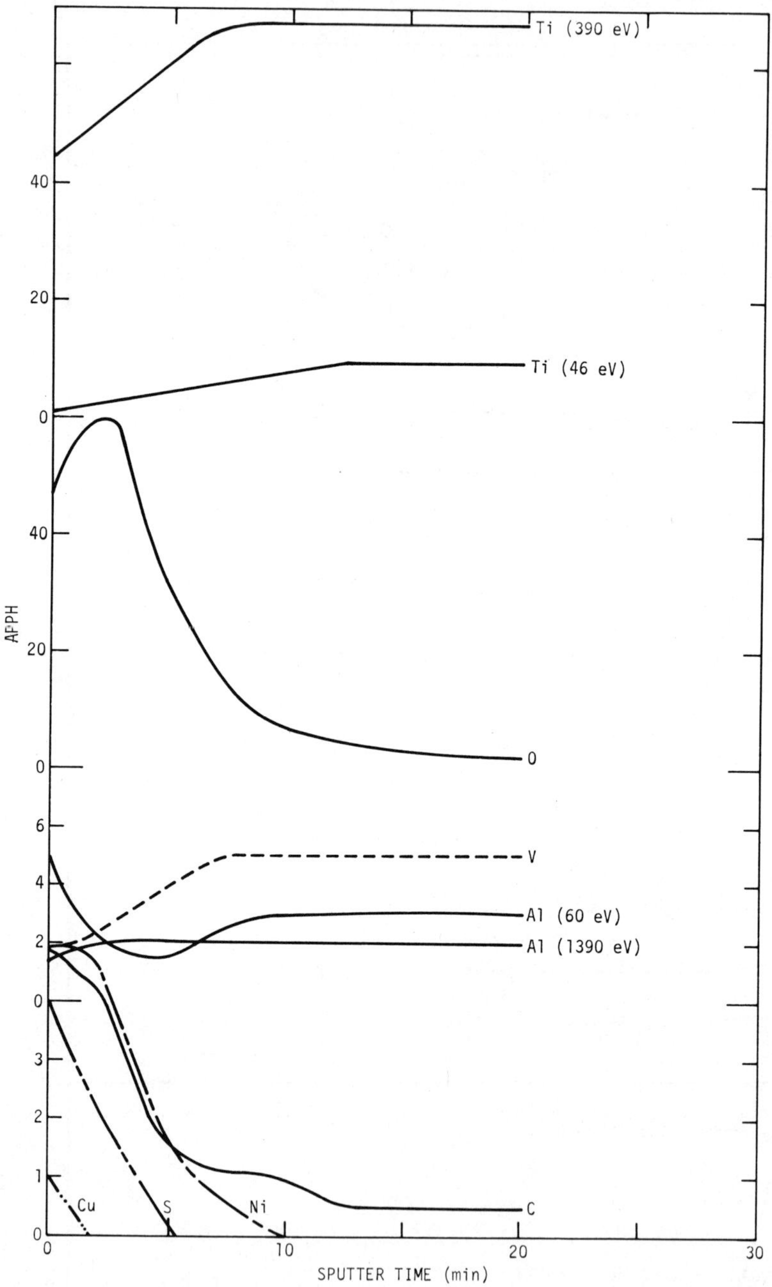

FIG. 5-6 Chemical profile on film on Ti 6Al-4V after the TURCO treatment.

Figs. 5-4, 5-5, and 5-6 and Table 5-3 is that the surface phosphate-fluoride treatment with the least surface contamination produced the strongest bonds. In Fig. 5-6 for TURCO-treated Ti 6Al-4V the oxygen APPH increases during the initial sputtering as the carbon contamination is removed. These chemical profiles give a qualitative analysis of the elements that are in the chemical films.

3.1.4 Surface Potential Difference

The surface potential difference (SPD) for three different surface treatments are recorded in Table 5-3. There seems to be no correlation between SPD and bond strength or any of the other surface parameters.

3.1.5. Microscopy

Figures 5-7 and 5-8 show scanning electron microscope (SEM) photographs at ×1000 and ×4000 for the phosphate-fluoride, TURCO, nitric acid-fluoride, and PASA JELL surface treatment of Ti 6Al-4V. The surfaces that result from the nitric acid-fluoride and the PASA JELL treatment (Fig. 8) look very much alike and are very similar to SEM pictures reported by Lively [7] at LTV Aerospace Corp. and Rauch and Reddy [13] at American Cyanimide. The SEM photographs for the TURCO treatment are quite different than for the nitric acid-fluoride or PASA JELL treatment and are similar to pictures reported by Alexander [12] at the Los Angeles Division of Rockwell International.

The most controversial SEM results are for the phosphate-fluoride treatment. Alexander demonstrated that what was considered to be minor changes in the phosphate-fluoride procedure could cause dramatic changes in the surface morphology. Although for this report we used the RI phosphate-fluoride treatment (per ST0610LA0004; see Table B-2) recommended by Alexander, our micrographs do not look like those reported by him [12] for the same treatment. The SEM pictures for the phosphate-fluoride treatment in Fig. 5-7a and b look more like those reported by Alexander (Ref. 12, Fig. 2) for the Battelle 185°F phosphate-fluoride treatment, but with less pronounced crystallite growth. There is some resemblance to the photographs reported by Hamilton and Lyerly [6] which show sharp grain boundaries but much less crystallite growth. A comparison of the details for the phosphate-fluoride treatment is given in Appendix B. It will be observed that slight differences in procedure do exist and may account for the differences in SEM results.

Microscopic examination of fractured bonds for Ti 6Al-4V and HT424 adhesive reveals the natural porosity of the adhesive as was found in the Al study. Figure 5-9b shows photographs of two halfs of a fractured bond that failed cohesively. About 50% of the bond area is void, although a layer of adhesive is attached to the metal

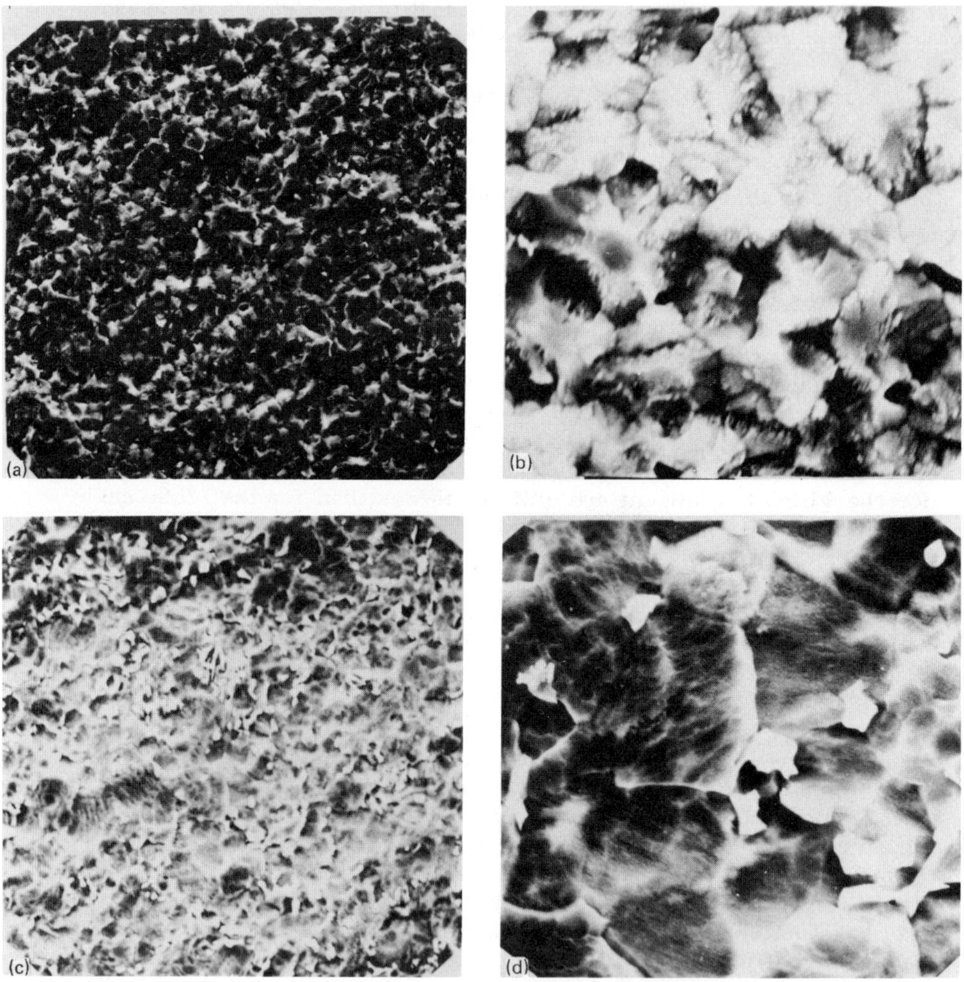

FIG. 5-7 SEM of a Ti 6Al-4V surface after the phosphoric-fluoride treatment (a—1000×, b—4000×) and after the TURCO treatment (c—1000×, d—4000×).

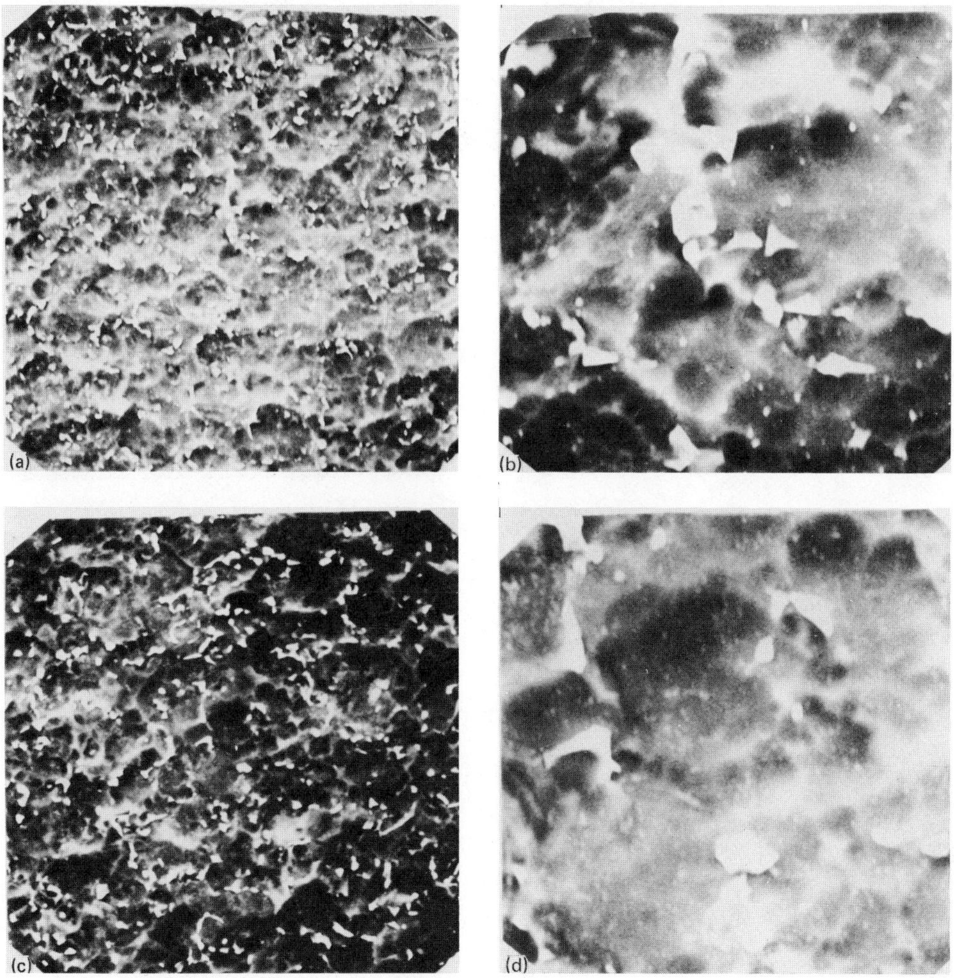

FIG. 5-8 SEM of a Ti 6Al-4V surface after the nitric acid-fluoride treatment (a—1000×, b—4000×) and after the PASA JELL treatment (c—1000×, d—4000×).

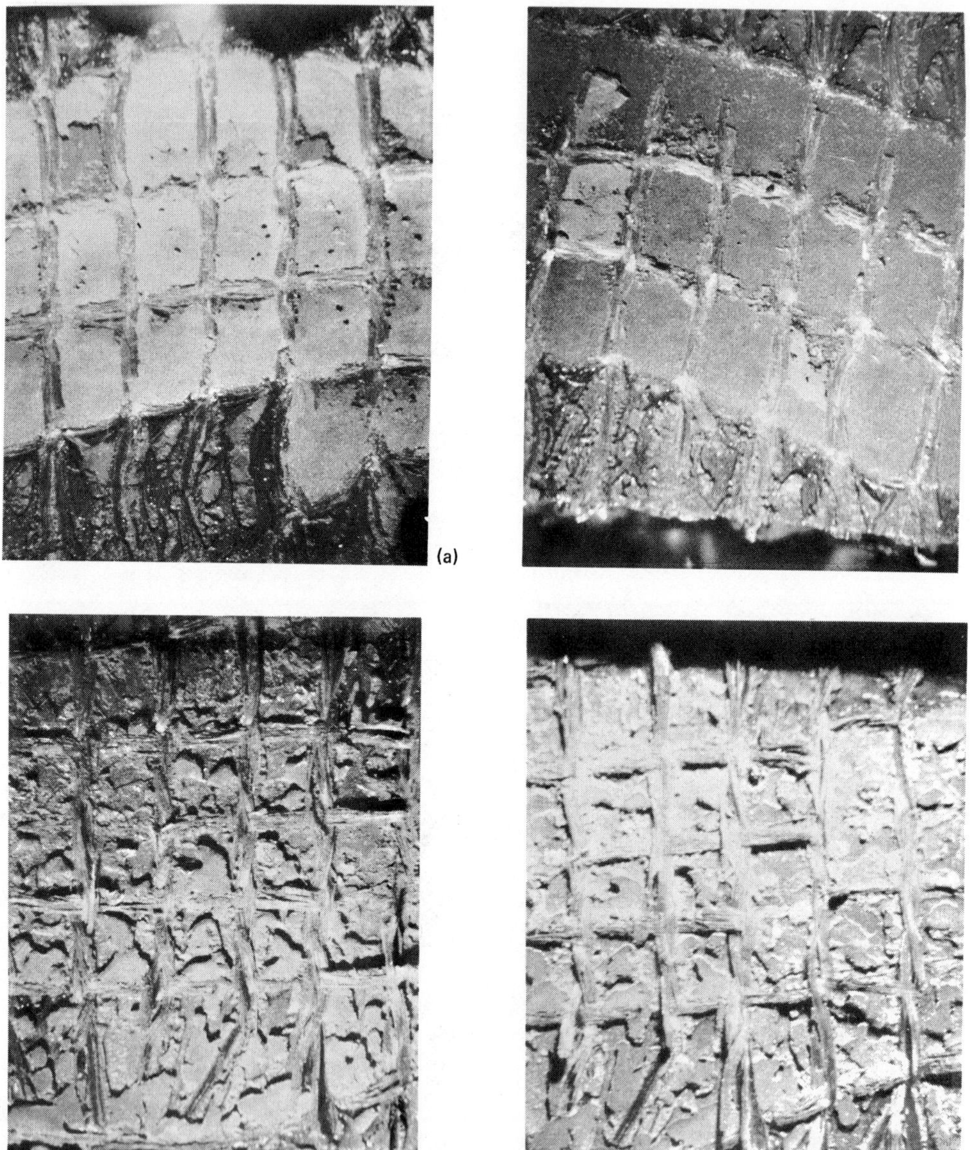

FIG. 5-9 Photographs of bond fracture areas for Ti 6Al-4V—HT424: (a) cohesive failure; (b) adhesive failure.

over the entire bond area. Figure 5-9a shows photographs of two matching halfs of a bond that failed interfacially in the center portion and cohesively at the top and bottom. The natural porosity is noticeable in the areas that failed cohesively. The picture on the right of Fig. 5-9a shows the oxide-primer interface between the glass fiber net. The picture on the left shows the matching adhesive surface. Contrary to photographs of Al 2024-T3—HT424 fracture surfaces, the bright areas are adhesive surface, the dull metal-oxide surface. Although the distinction is not easily determined from the photographs, there is no question when observing under the microscope which area is adhesive and which is metal.

The SEM photographs in Fig. 5-10 show fracture surfaces after a tensile test of a Ti 6Al-4V—HT424 lap shear joint. Figure 5-10a is a low-magnification picture (×20) showing the top of the adhesive (left side), glass fibers and metal-oxide interface (central). Considerable adhesive or primer adheres to the oxide, as seen in Fig. 5-10b (×200). Figure 5-10c is a high-magnification (×4000) of the oxide surface and Fig. 5-10d is the mating adhesive surface which replicates the oxide. The oxide surface after bonding and fracture (Fig. 5-10c) is similar to that prior to bonding (Fig. 5-7b) except the surfaces in each grain appear more rounded. There does not appear to be any oxide particles adhering to the adhesive surface in Fig. 5-10d, indicating failure to be at the primer-oxide interface in this region.

The film on a phosphate-fluoride-treated sample of Ti 6Al-4V was removed using the technique described by Hamilton and Lyerly [6] in Gillette Research Institute Technical Report 4185. One segment of the film was examined in the Philips EM-300 electron microscope and another segment was examined with a Zeiss interference microscope. The transmission electron micrograph in Fig. 5-11 reveals that the film roughly reproduces the underlying metal structure and is similar to that reported by Hamilton and Lyerly [6] (see their Fig. 8). There are regions 2-3 μm in diameter which likely are replications of primary alpha particles. In addition, there are many undulations in the film superimposed upon this structure; these undulations are apparently characteristic of the film and not related to the microstructure of the Ti 6-4.

A typical selected area electron diffraction pattern from the film which revealed very diffuse rings indicates that the film is amorphous. In addition to the diffuse rings there are discrete reflections appearing faintly. Dark field electron microscopy (Fig. 5-12) shows particles that give rise to the reflections in the diffraction pattern. The crystalline material (bright spots in Fig. 5-12) represents less than 1% of the surface layer. A crystal pattern of the minute amount of crystalline material was not indexable as any form of titanium oxide and is probably due to impurity particles. Although Hamilton and Lyerly report that

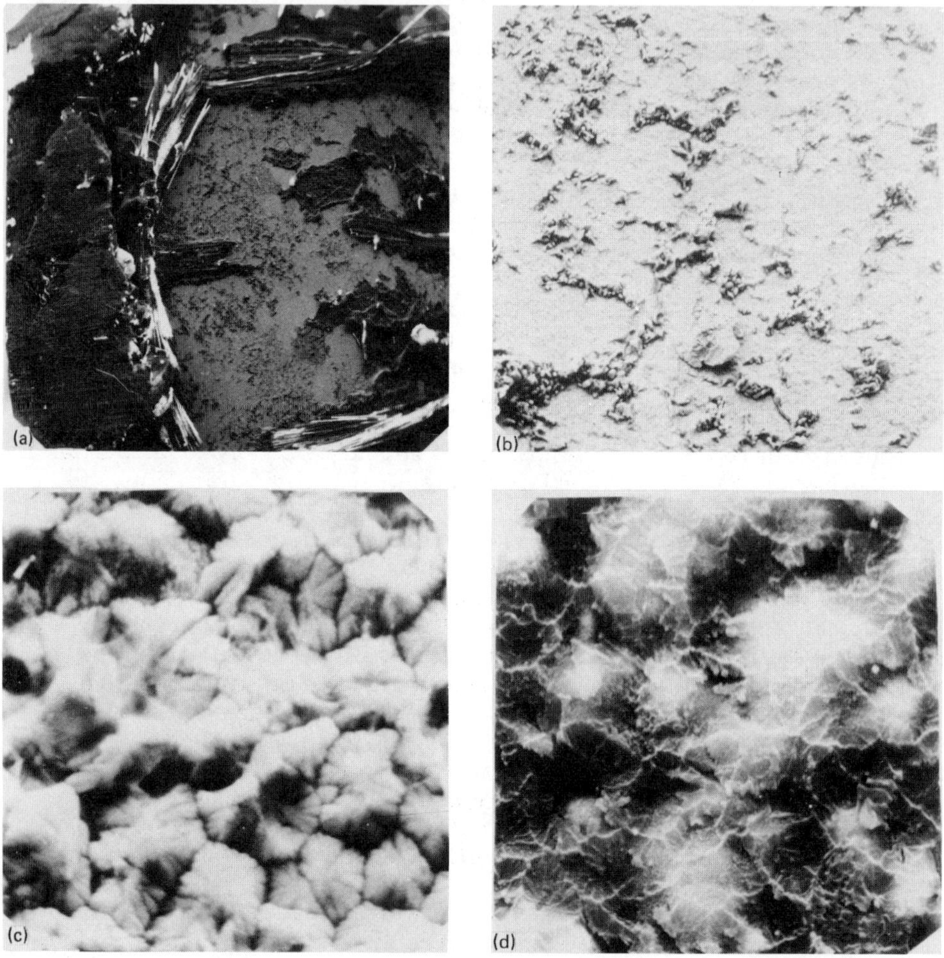

FIG. 5-10 SEM of Ti 6Al-4V surface after bonding with HT424 and fracture: (a) 20×; (b) 200×; (c) 4000×; (d) 4000×.

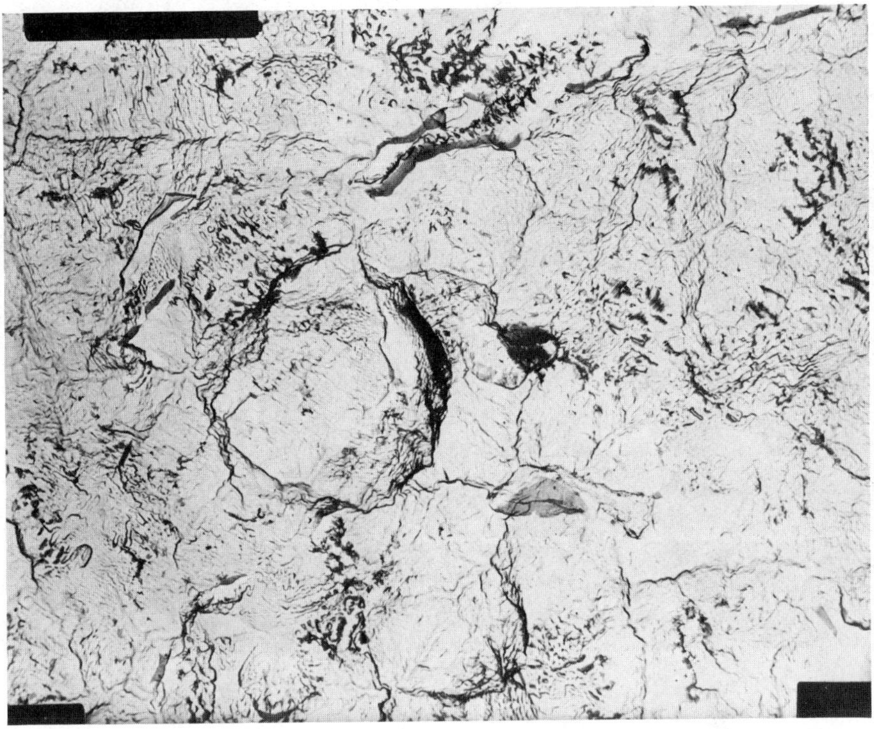

FIG. 5-11 Transmission electron micrograph of phosphate-fluoride film from Ti 6Al-4V.

their removed phosphate-fluoride films were crystalline anatase, Lively [7] found theirs to be amorphous as we report here.

3.1.6. Contact Angles

Contact angle measurements and surface energy analysis were conducted on cast films of the BR-34 primer and FM34B-32 adhesive (Bloomingdale Division, American Cyanamide). Both primer and adhesive films were evaluated in the uncured and cured states. The steps in film preparation are detailed in Table 5-4. The film preparation steps outlined in Table 5-4 provide smooth and homogeneous surfaces of the polymeric phase of the polyimide adhesive-bonding system for wettability study. Steps 4 and 5 of Table 5-4 utilize a dry N_2 blanket over the specimen in order to minimize hydrolytic or oxidative degradation at the film free surfaces.

FIG. 5-12 Typical electron diffraction pattern of phosphate-fluoride film from Ti 6Al-4V.

TABLE 5-4 Film Preparation for Wettability Analysis

BR-34 primer
1. 1 part primer diluted with 5 parts BR34-2 thinner
2. Coated on glass slide by dipping
3. 30-min air-dry
4. 30-min oven-dry at 104°C in dry N_2 atmosphere
5. 60-min oven-dry at 210°C in dry N_2 atmosphere
6. Uncured samples given 20-hr air-dry and 5-hr vacuum-dry

FM 34B-32 adhesive
1. Film dissolved in MEK
2. Coated on glass slides
3. Given 2-hr vacuum-dry
4. Cured for 90 min at 550°F in dry N_2 atmosphere
5. Uncured samples given only the vacuum-dry

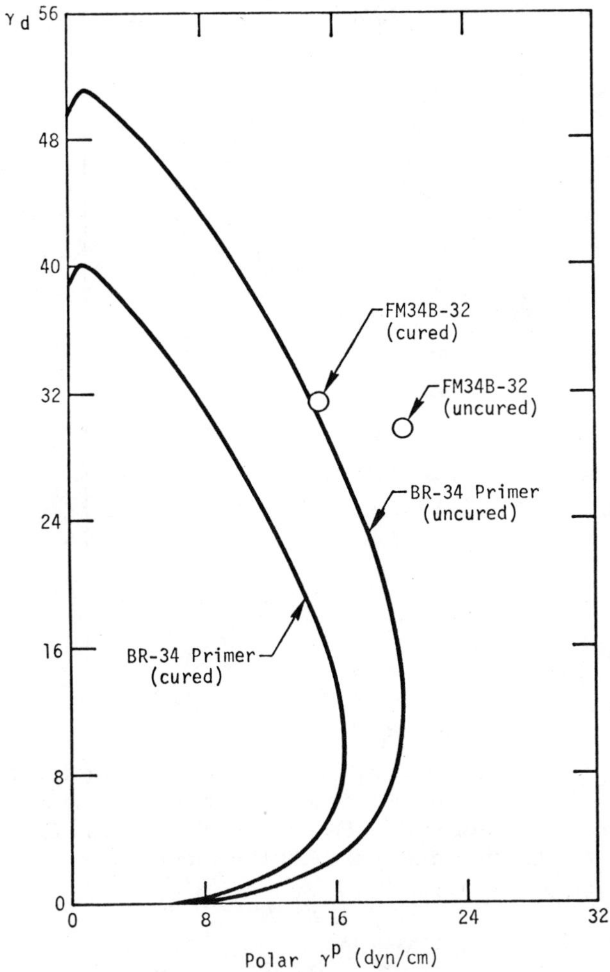

FIG. 5-13 Wettability envelopes for uncured and cured BR-34 primer.

The solid surface tension properties show that adhesive primer BR-34 displays significantly higher values γ_{SV}^{d} and lower values of γ_{SV}^{p} than the FM34B-32 adhesive in both the uncured and cured state. These results would imply that the BR-34 primer and FM34B-32 display differences in surface chemistry and also that the interface between primer and adhesive will display nonoptimum interfacial bonding. The wettability envelopes for BR-34 primer are shown in Fig. 5-13.

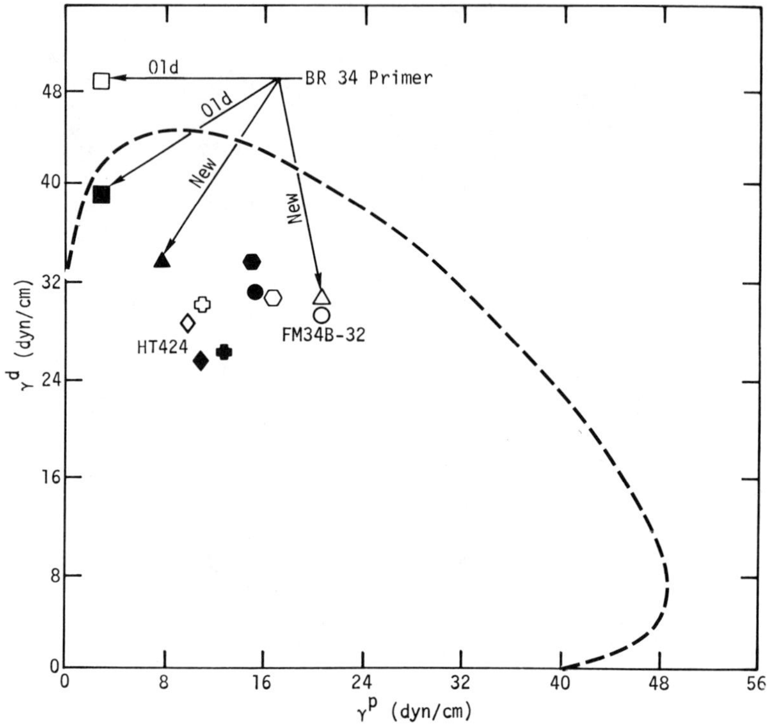

FIG. 5-14 Wettability envelopes for nitric acid-fluoride and phosphate-fluoride treatment of Ti 6Al-4V. BR-34 primer: old uncured □; old cured ■; new uncured △; new cured ▲. FM34B-32: old uncured ○; old cured ●; new uncured ⬡; new cured ⬢. HT424: uncured ◇; cured ◆. HT424 Primer: uncured ✣; cured ✚.

Wettability studies for Ti 6Al-4V alloy were initiated using the surface treatments recommended by the adhesive manufacturer for the FM34B-32 adhesive system. Table A-1 outlines the detailed steps for the metal surface treatments utilized in these experiments.

The prediction from Fig. 5-13 is that poor bonding will result between the BR-34 primer and the FM34B-32 adhesive. Figure 5-14 shows the wettability envelope for freshly prepared nitric acid-fluoride which is essentially identical to the phosphate-fluoride treated surfaces of Ti 6Al-4V. The prediction is poor bonding between BR-34 primer and either of the treated surfaces.

In order to test the predictions developed by the wettability envelope analysis of Figs. 5-13 and 5-14, a series of lap shear joints were prepared and tested, as indicated in Table 5-5. These single lap shear bonds all showed low bond strength and dominant interfacial failure as predicted from the wettability envelope analysis. The lower set in Table 5-5 shows that low bond strength also result when no primer is used.

To test whether extraneous stresses imposed by differential thermal expansion during cure influence bond strength, a Ti 6Al-4V bonding jig was constructed to replace the Al jig and applied in bonding studies for lap shear strength of Ti 6Al-4V adherends primed with BR34B-32 primer and bonded with FM34B-32 adhesive. In this experiment three sets of six adhesive joints were prepared using a different metal surface treatment for each set of six bonds. The results of these experiments are summarized in Table 5-6.

The low values of lap shear strength reported in Table 5-6 for several standard surface treatments for Ti 6Al-4V using the Ti 6Al-4V

TABLE 5-5 Results of Preliminary Study of Lap Shear Tests: Metal = Ti 6Al-4V, Primer = Br-34 (American Cyanimid), Adhesive = FM-34B-32 (American Cyanimid)

Bond no.	Surface treatment	σ_b Bond strength (psi)	Average σ_b (psi)	Type of failure
1		1052		
2		728		
3	1. Alkaline-cleaned	604	871 ± 211	~95% Interfacial
4	2. Br-34-primered	908		
5	(primer tended	652		
6	to bead up)	1288		
1	1. Alkaline-cleaned	870		
2	2. HNO_3-HF-etched	1070		
3	3. Br-34-primered	Broke in jig	970 ± 100	~95% Interfacial
4	(primer tended			
5	to bead up)			
6				
1	1. Alkaline-cleaned	930		
2	2. HNO_3-HF-etched	1030	807± 231	
3	3. PASA-Jell-107-	460		~50% Interfacial
	treated	Broke in jig		
4	4. No primer			

TABLE 5-6 Summary of Simple Lap Shear Bond Tests of BR-34B-32 Primer/FM-34B-32 Adhesive to Surface-Treated Ti 6Al-4V Adherends Using a Ti 6Al-4V Bonding Jig and Tested at 23 ± 1°C and 50 ± 5% RH

Bond no.	σ_b (psi)	Avg σ_b (psi)	Avg dev. (psi)	Type of failure
1. PASA-JELL treatment				
1	1220	1420	±557	~50-100% at primer-metal interface
2	1300			
3	1970			
4	2000			
5	70			
6	1960			
2. TURCO titanium treatment				
1	840	1278	±419	100% at primer-metal interface
2	460			
3	2020			
4	1540			
5	1300			
6	1510			
3. Phosphate-fluoride treatment				
1	1960	1965	±92	75% at primer-metal interface
2	1970			
3	1760			
4	2200			
5	1900			
6	2000			

alignment jig tend to reinforce the conclusion that poor wetting between BR34B-32 primer and adhesive or primer and Ti 6Al-4V described by the wettability envelopes of Figs. 5-13 and 5-14 is the primary cause of low joint strength.

Experimental observations show that the BR34-32 thinner supplied by American Cyanamide—Bloomingdale Division for use with BR34B-32 primer produced an autophobic reaction to surface-treated Ti 6Al-4V adherends. This phenomenon is characterized by initial wetting and spreading, followed by spontaneous retraction of the liquid film to form drops of finite liquid-solid contact angles. BR34-32 thinner shows autophobic effect with contact angle going from $\theta = 0$ to $\theta = 12°$ in 10 to 20 sec after contact with phosphate-fluoride-etched Ti 6Al-4V as monitored in the NRL goniometer at 23 ± 1°C. Figure 5-15a shows the

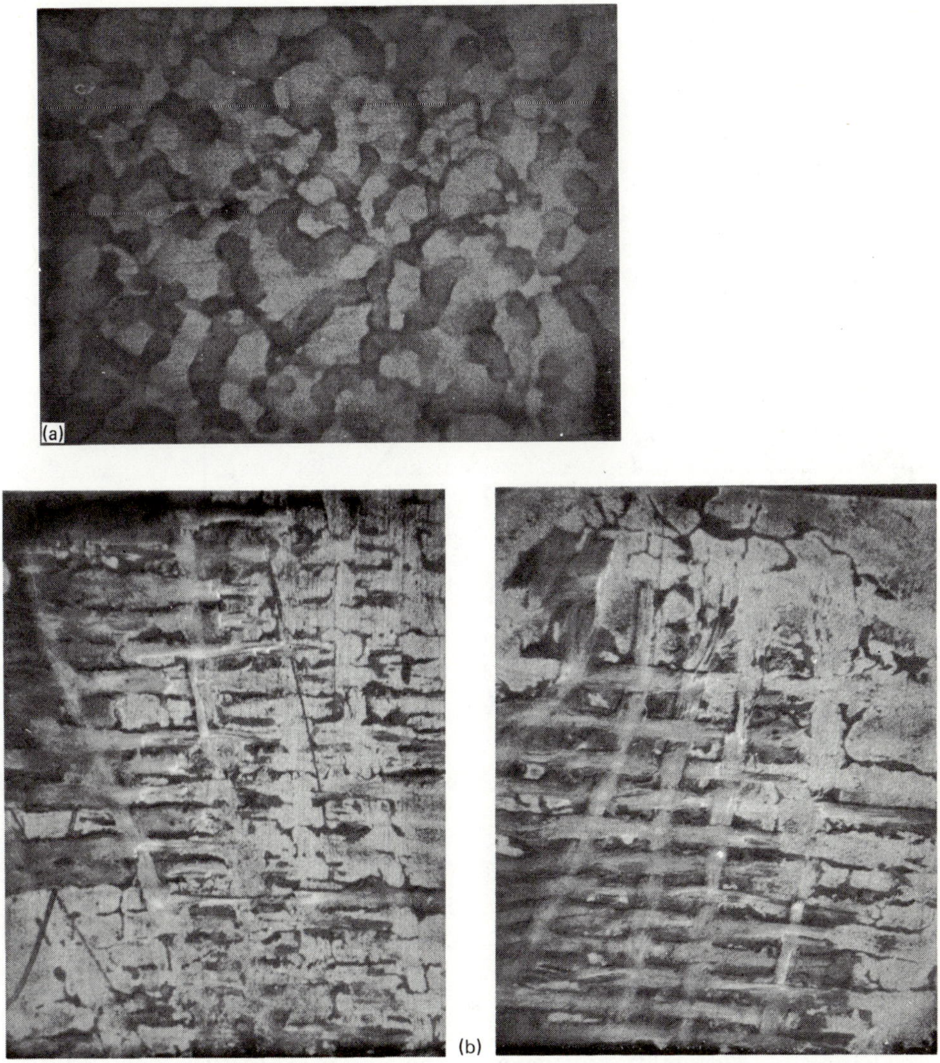

FIG. 5-15 (a) Rivulet pattern of FM34B-32 primer on HNO_3/HF-etched Ti 6Al-4V due to autophobic wetting-dewetting; (b) fracture surfaces of FM34B-32-primed Ti 6Al-4V bonded to FM34B-32 adhesive showing failure at the primer-adhesive interface.

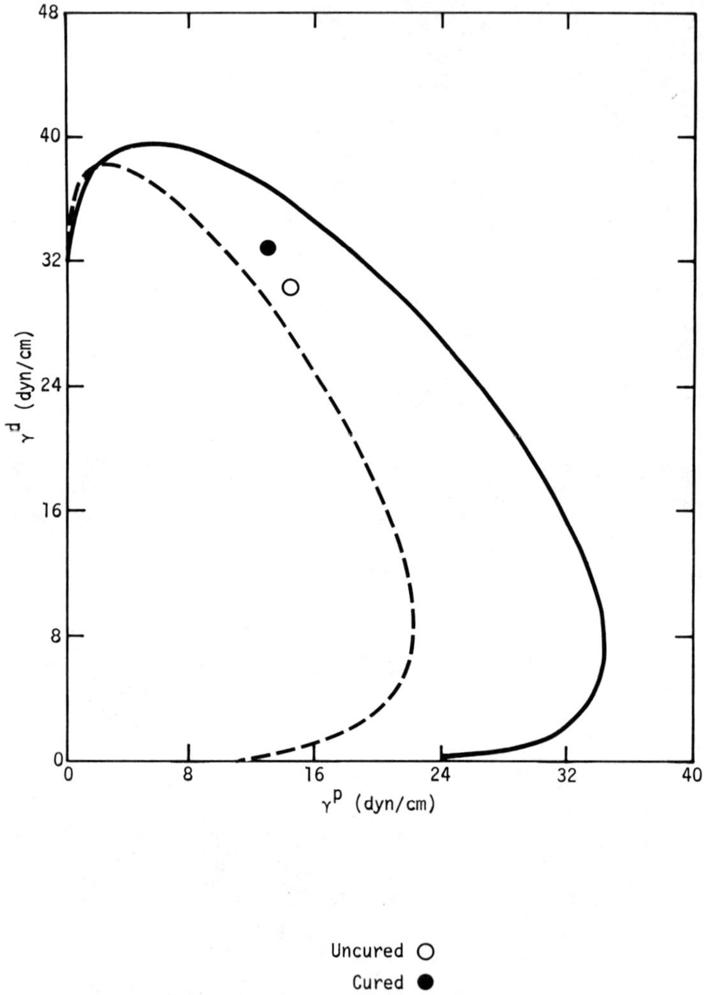

FIG. 5-16 Wettability envelopes for new BR34-32 primer, uncured (solid line) and cured (dashed line). New FM34B-32 adhesive: uncured ○ ; cured ● .

rivulet pattern produced on HNO_3/HF-etched Ti 6Al-4V by the dried film of BR34B-32 primer as a consequence of the autophobic wetting effect which leaves an uneven distribution of primer on the metal surface. Figure 5-15b shows the two fracture surfaces from the lap-shear joint prepared from these incompletely primed surfaces in which substantial failure is evidenced at the primer-adhesive interface as predicted by the wettability study. It was concluded that the BR-34B-32 primer was from a bad lot and a new supply of primer and adhesive were purchased. Figure 5-16 shows the wettability envelopes for the new BR34B-32 primer. The wettability parameters for the FM34B-32 adhesive falls within the envelope for the uncured primer. Figure 5-14 shows that the parameters for the new adhesive and primer fell within the envelopes for the metal surface. It is predicted that strong bonds will result between the new primer, adhesive, and Ti6Al-4V adherends.

The bond strengths for five lap shear joints with the new adhesive and primer are reported in Table 5-7. The bond strength values fall within the range reported by the manufacturer for this system and correlate well within the average value of 3121 psi reported by Lively [7]. The primer did not bead up and form a rivulet pattern as in Fig. 5-15a. It is concluded that the first batch of primer was bad and that the wettability analysis revealed the fault.

The surface tension properties of HT424 adhesive and primer fall well within the wettability envelope in Fig. 5-14. Strong bonds are expected from the wettability analysis. The bond strengths reported for Ti 6Al-4V-HT424 system in Table 5-8 average 3358 ± 352 psi and in Table 5-9 they average 3143 ± 533 psi for TURCO treatment, in accordance with the wettability analysis.

TABLE 5-7 Adhesive Bonding of Ti 6Al-4V (Phosphate Fluoride) with a New Batch of FM34B-32 Adhesive and BR-34B-32 Primer: Lap Shear Joint

Sample		Bond strength (psi)
1		3500
2		2440
3		2940
4		3080
5		3380
	Avg:	3068 ± 300

TABLE 5-8 Preliminary Examination Results of Phosphate-Fluoride Treatment of Ti 6Al-4V (single lap shear bond strength): Metal = Ti 6Al-4V, Primer = HT424 parts A and B (American Cyanimid), Adhesive = HT424 (American Cyanimid)

Bond no.	σ_b (psi)	Avg σ_b (psi)	Avg deviation (psi)
1	3550		
2	3830		
3	3040	3358	±352
4	2760		
5	3220		
6	3750		

3.2. *Surface Aging Effects*

3.2.1. Effect of Humidity and Time

Figure 5-17 shows σ_b vs. ϕ_c for various surface exposure time (SET) aging experiments. Although there is large scatter due to an unknown uncontrolled parameter, the trend is toward $\sigma_I \approx 0$ and $\sigma_b \approx 5000$ psi. These results are interpreted as failure of a weak boundary layer ($\sigma_I \approx 0$) at the oxide primer interface (as concluded from the SEM study) and failure in the adhesive of $\sigma_c \approx 5000$ psi. The average values of ϕ_c and σ_b ($\bar{\phi}_c$ and $\bar{\sigma}_b$) for six bonds at each of the various values of SET are (see Table 5-10) plotted in Fig. 5-18.

TABLE 5-9 Preliminary Examination Results of TURCO Treatment for Ti 6Al-4V (single lap shear bond strength): Metal = Ti 6Al-4V, Primer = HT424 parts A and B (American Cyanimid), Adhesive = HT424 (American Cyanimid)

Bond no.	σ_b (psi)	Avg σ_b (psi)	Avg deviation (psi)
1	3710		
2	2720		
3	3730	3143	±533
4	2320		
5	2790		
6	3590		

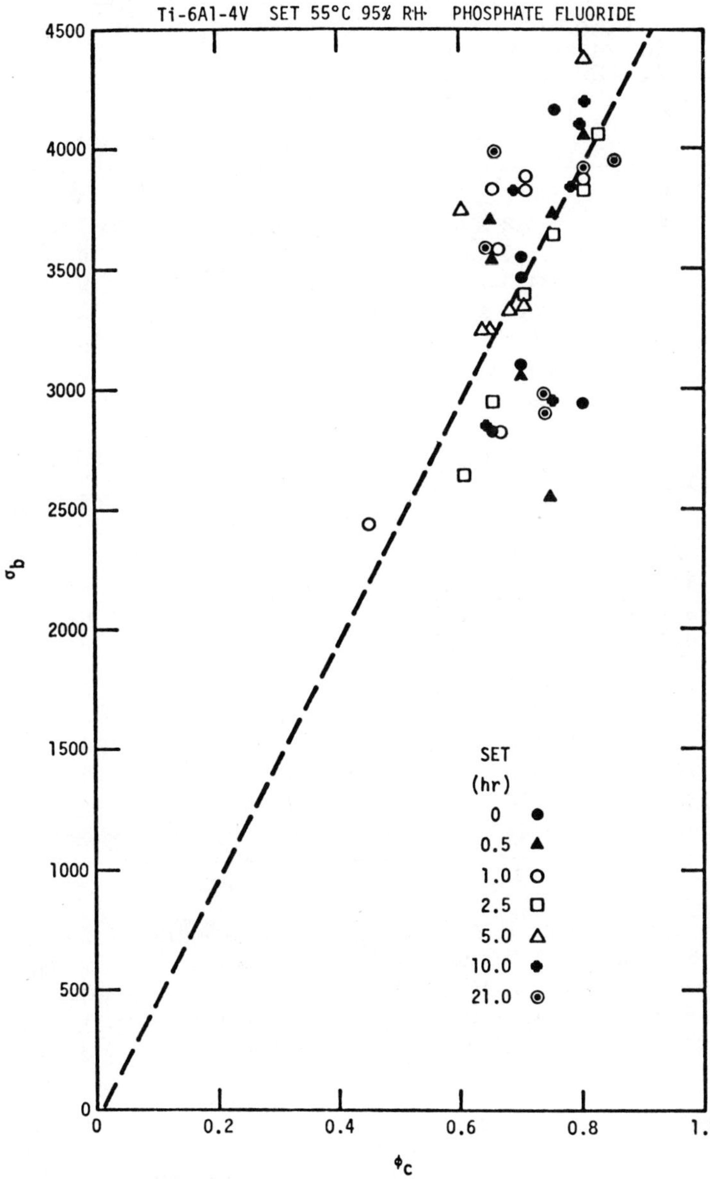

FIG. 5-17 Plot of σ_b vs. ϕ_c for SET, 54°C, 95% RH, phosphate-fluoride-treated Ti 6Al-4V—HT424.

TABLE 5-10 Brief Summary of Combined SET-BET Effects at 54°C and 95% RH upon Lap Shear Strength of HT424 Adhesive and Phosphate-Fluoride-etched Ti 6Al-4V

SET (hr)	BET (hr)	$\bar{\sigma}_b \pm \delta_b$	$\bar{\phi}_c$	Number of test specimen
0	0	3333 ± 387	0.70	6
0.5	0	3438 ± 432	0.70	6
1.0	0	3572 ± 384	0.60	6
2.5	0	3412 ± 425	0.75	6
5.0	0	3557 ± 332	0.70	6
10	0	3610 ± 493	0.75	6
21	0	3547 ± 414	0.70	6
0	0	3881 ± 412	0.50	6
0	6	3210 ± 163	0.70	6
0	24	3030 ± 367	0.75	6
0	47	2980 ± 380	0.70	6
0	99	2937 ± 371	0.70	6
0	194	2677 ± 508	0.60	6
0	374	2758 ± 449	0.70	6
0	670	2340 ± 393	0.67	6
0	1016	2490 ± 330	0.70	6
21	0	3777 ± 471	0.75	6
21	24	2880 ± 477	0.60	6
21	46	3243 ± 80	0.75	6
21	142	2435 ± 747	0.60	6
21	303	2653 ± 396	0.64	6
21	591	2922 ± 274	0.80	6

Within experimental scatter it appears that surface aging of the Ti 6Al-4V at 54°C and 95% RH has no effect upon either $\bar{\phi}_c$ or $\bar{\sigma}_b$. Ellipsometric measurement of unbonded side of the samples after fracture revealed that negligible changes had occurred during SET (Δ = 72.5 ± 0.8, ψ = 29.6 ± 0.8 after SET as compared to Δ = 72.9, ψ = 27.9 prior to SET).

3.2.2. Effect of SET on Contact Angles

Samples of Ti 6Al-4V sheet 0.063 × 1 × 2 in. were given the nitric acid-fluoride surface treatment. Wettability experiments were conducted on these etched surfaces subsequent to increasing surface exposure time (SET) at 23°C and 50 ± 5% RH. The results of one surface treat-

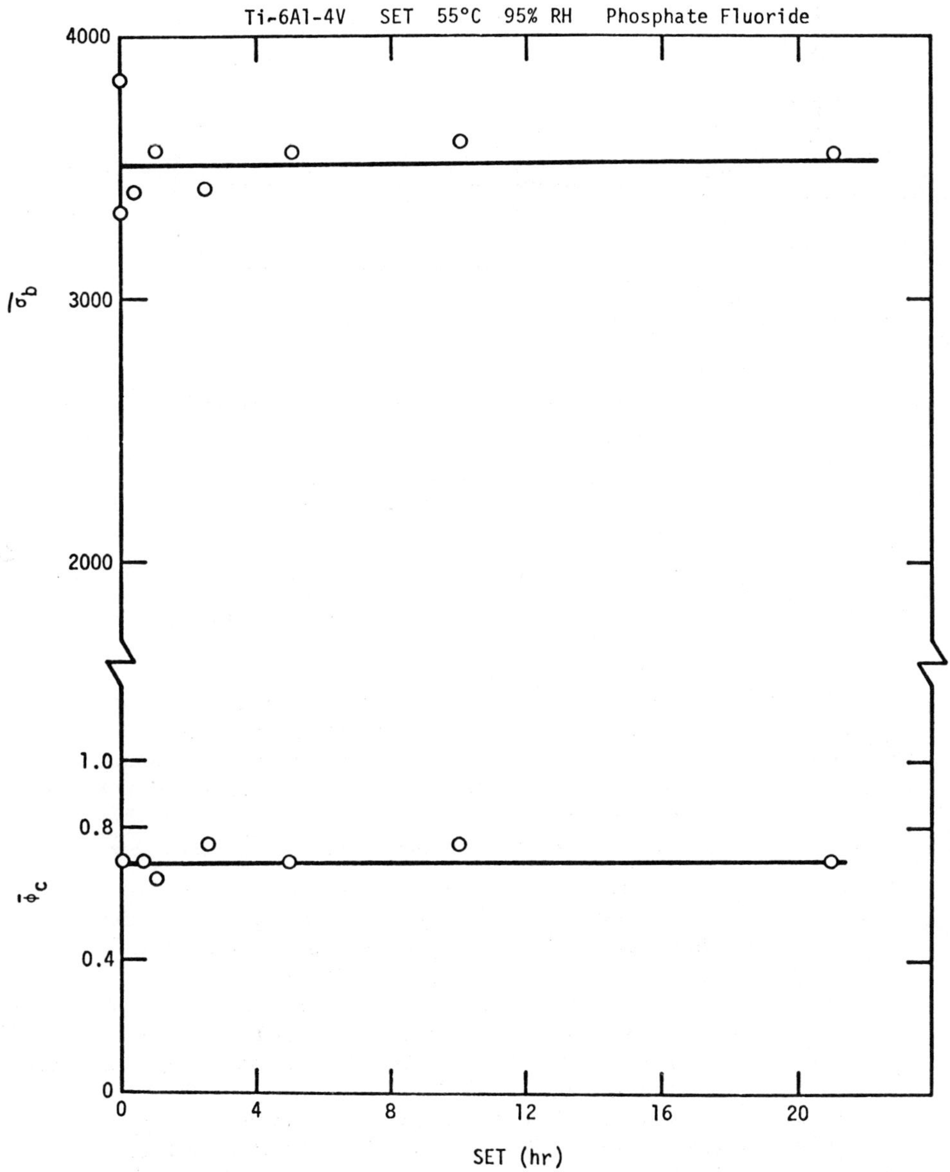

FIG. 5-18 Plot of σ_b and ϕ_c vs. SET at 54°C, 95% RH, for Ti 6Al-4V—HT424 with phosphate-fluoride treatment.

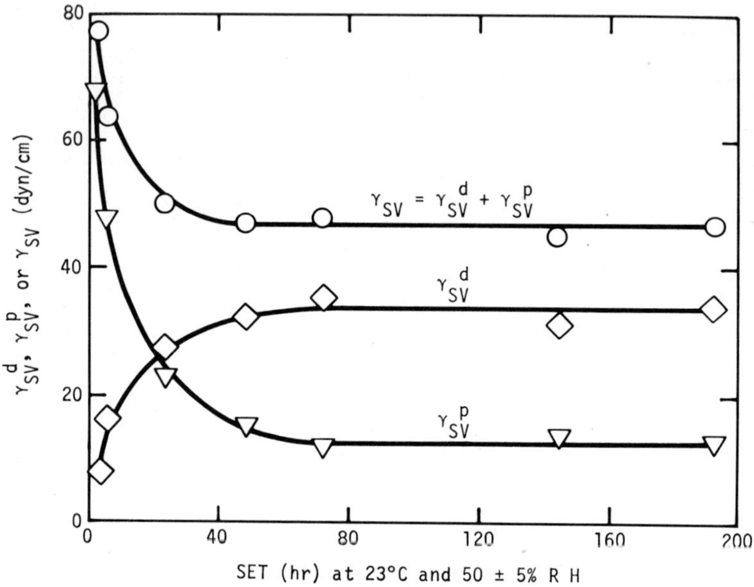

FIG. 5-19 Effect of ambient SET at 23°C, 50 ± 5% RH on calculated average values of γ_{SV}^d, γ_{SV}^p, and γ_{SV} for HNO_3/HF-etched Ti 6Al-4V.

ment of Ti 6Al-4V alloy display the same trends shown by earlier studies of Al 2024-T3 subsequent to the FPL-etch treatment. The initial exploratory results were reconfirmed in more detailed studies involving the same surface treatment followed by seven levels of ambient SET (23 ± 0.5°C, 50 ± 5% RH).

Figure 5-19 presents the mean values of γ_{SV}^d, γ_{SV}^p, and γ_{SV} plotted as functions of SET. The curves of Fig. 5-19 show that HNO_3/HF-etched Ti 6Al-4V alloy displays a regular time-dependent transition in surface properties with major changes during initial surface aging at SET = 0-24 hr at 23°C and 50 ± 5% RH.

The results shown in Fig. 5-19 for surface-treated Ti 6Al-4V may be compared with those previously reported for the FPL-etched Al 2024-T3 in Fig. 4-38 in Chapter 4. Whereas Al 2024-T3 showed approximately stable surface properties from SET = 0-50 hr at 23°C and 50 ± 5% RH, followed by changes to ~300 hr, it is evident that immediate changes occur with surface-treated Ti 6Al-4V alloy. This initial instability prevented meaningful wettability measurements on surface-treated Ti 6Al-4V at times where SET < 2.5 hr. The net effects of SET in reducing γ_{SV}^p and increasing γ_{SV}^d are similar for both Ti 6Al-4V and

Al 2024-T3 surfaces. The consistent shift of the values of $\alpha_S = (\gamma_{SV}^d)^{1/2}$ vs. $\beta_S = (\gamma_{SV}^p)^{1/2}$ properties of HNO_3/HF-treated Ti 6Al-4V are shown in Fig. 5-20 and these closely follow similar plots shown previously for the FPL-etched Al 2024-T3 (Fig. 4-43, Chapter 4).

The wettability envelope for the HNO_3/HF-etched Ti 6Al-4V surfaces at varied levels of ambient SET are displayed by the curves of Fig. 5-21. It may be recalled that if the surface tension properties of a simple liquid or adhesive material fall within the region between the origin $\gamma_{SV}^d = \gamma_{SV}^p = 0$ and the curve defining the wettability envelope, spontaneous spreading is predicted in the absence of rheological constraints. The data points in Fig. 5-21 represent the surface tension properties of the bad batch of polyimide primer BR-34 and adhesive FM-34B-32. It is seen in Fig. 5-21 that SET can only make a bad situation even worse as far as bondability is concerned.

The plots of γ_{SV}^d and γ_{SV}^p shown in Fig. 5-22 indicate that γ_{SV}^d appear to maximize at SET ≈ 16 hr while γ_{SV}^p minimizes at SET = 20-30

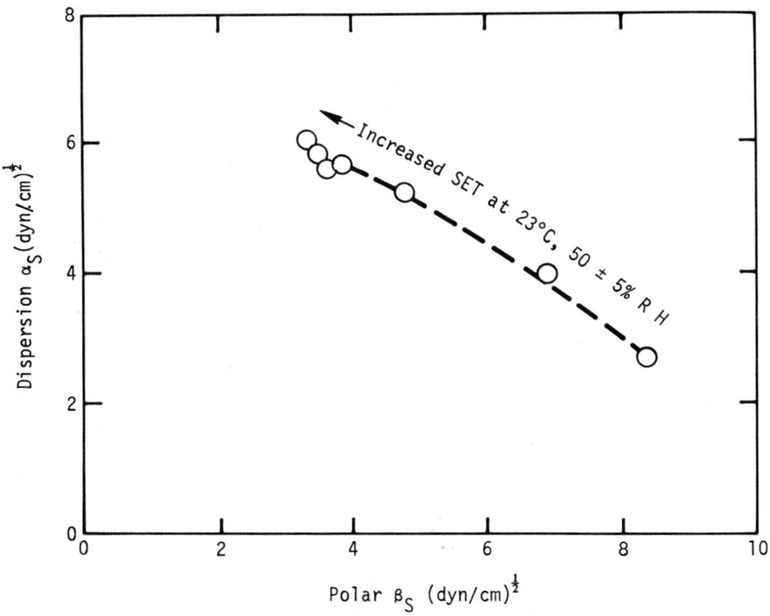

FIG. 5-20 Effects of increased ambient SET upon $\alpha_S = (\gamma_{SV}^d)^{1/2}$ vs. $\beta_S = (\gamma_{SV}^p)^{1/2}$ on HNO_3-HF-etched Ti 6Al-4V.

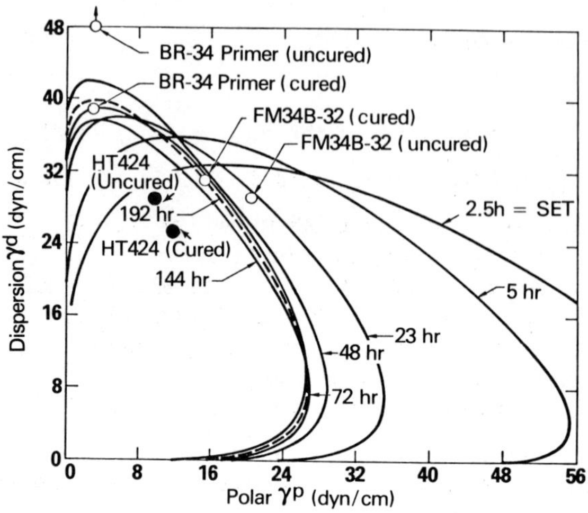

FIG. 5-21 Wettability envelopes for HNO_3/HF-etched Ti 6Al-4V at various SETs.

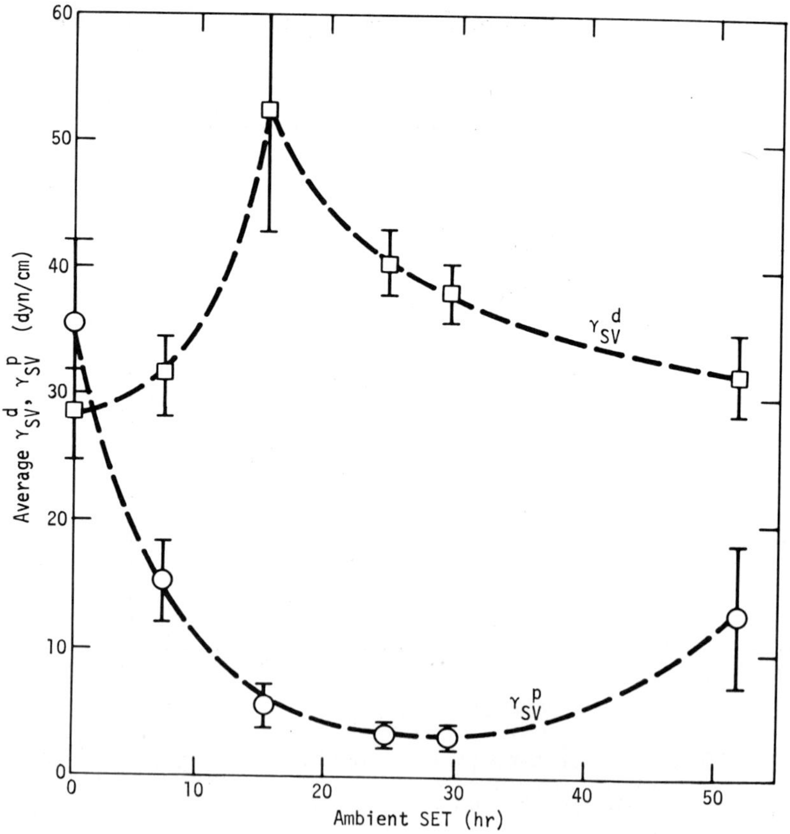

FIG. 5-22 Effect of ambient SET at 23°C, 50 ± 5% RH on calculated average values for γ_{SV}^{d} and γ_{SV}^{p} for phosphate-fluoride-treated Ti 6Al-4V.

hr. These results differ from γ^d_{SV} and γ^p_{SV} data for HNO_3/HF-etched Ti 6Al-4V shown previously in Fig. 5-19. These results of surface energy analysis of the phosphate-fluoride-treated Ti 6Al-4V panels do indicate that ambient SET effects at short times SET = 0-24 hr produce significant variations in wettability and should effect adherence to polymeric adhesives.

The wettability envelopes for phosphate-fluoride-treated Ti 6Al-4V developed from the γ^d_{SV} and γ^p_{SV} values are displayed in Fig. 5-23. The surface tension properties of uncured and cured HT424 adhesive are seen to fall well within the wettability envelopes for phosphate-fluoride-treated Ti 6Al-4V for ambient SET = $\leqslant$ 0.5 hr, 8 hr, and 16 hr and to fall near the boundary of the envelopes for SET = 25 hr and 30 hr. As seen in Fig. 5-18, the average bond strength remains high to SET = 21 hr in accord with the wettability analysis.

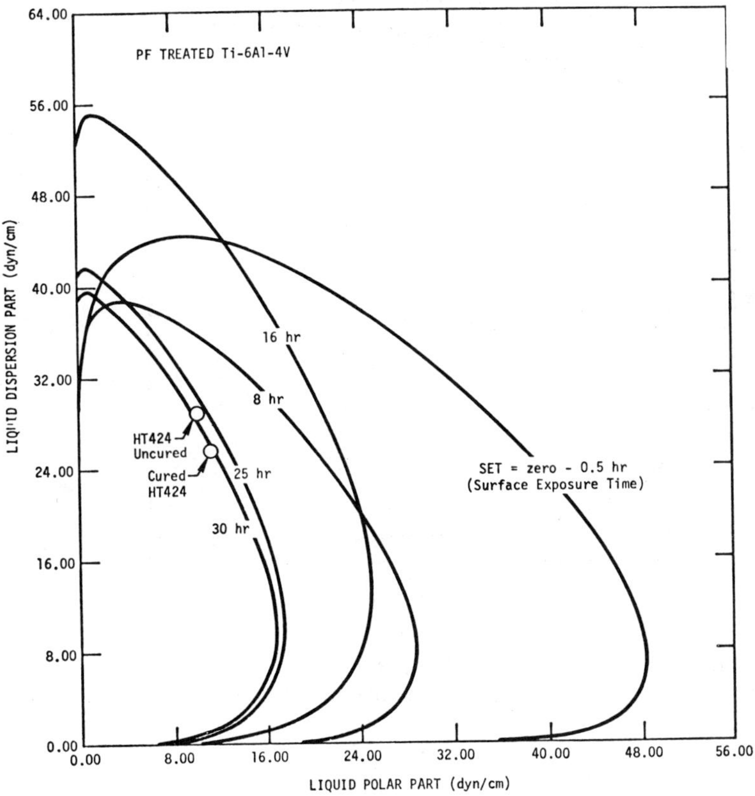

FIG. 5-23 Wettability envelopes for Ti 6Al-4V, phosphate-fluoride-treated and SET 23°C, 50 ± 5% RH ambient air.

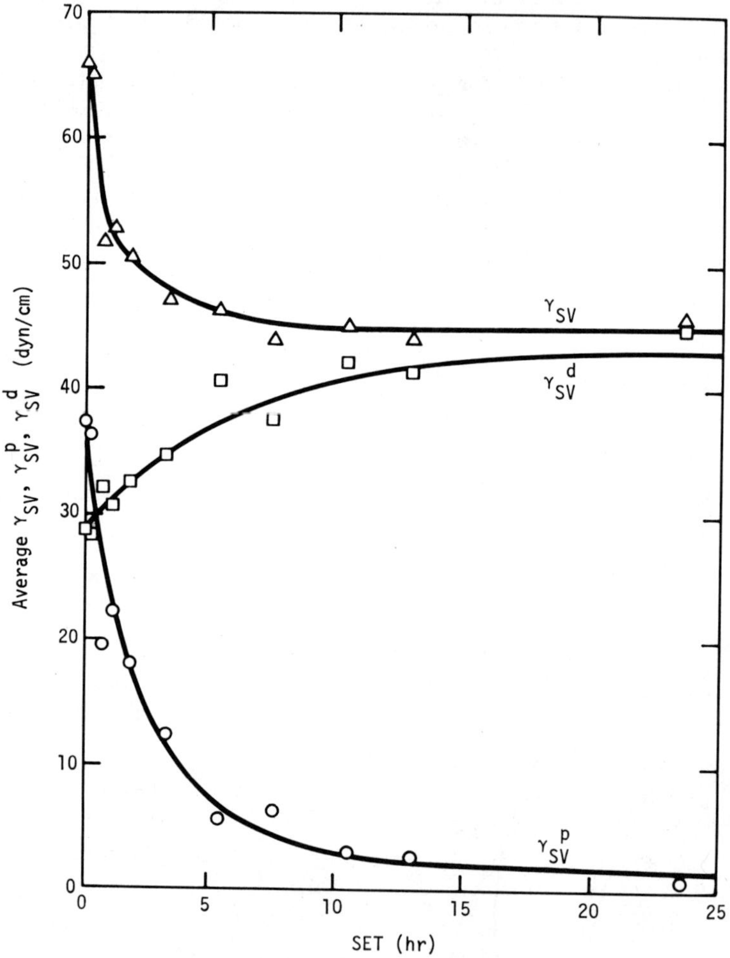

FIG. 5-24 Effect of SET at 54 ± 1°C, 95% RH on calculated average values for γ_{SV}^{d}, γ_{SV}^{p}, and γ_{SV} for phosphate-fluoride-treated Ti 6Al-4V.

Wettability measurements and surface energy analysis methods were applied to a detailed study of phosphate-fluoride-treated surfaces of Ti 6Al-4V under accelerated aging conditions of 54 ± 1°C and 95% RH surface exposure. Subsequent to surface treatment the surface exposure condition (SET) at 54 ± 1°C and 95% RH was applied for varied periods and the surfaces characterized by wettability measurements with the test liquids. The curves of Fig. 5-24 illustrate the rapid degradation of surface energy properties for phosphate-fluoride-treated Ti 6Al-4V due to SET effects. Figure 5-25 presents related information on surface energy degradation in terms of the rapidly diminishing wettability envelopes with increased SET. Inspection of Fig. 5-25 shows that the wettability envelopes for Ti 6Al-4V enclose the surface energy properties of uncured and cured HT424 epoxy phenolic adhesive for all SET conditions covering SET = 0-24 hr indicative of a wettable surface. The wettability analysis indicates that weak bonds should result for SET > 24 hr.

The averaged values of γ_{SV}^d, γ_{SV}^p, γ_{SV} for phosphate-fluoride-treated Ti 6Al-4V at SET (23°C, 50% RH) and SET (54°C, 95% RH) are combined on a single master curve in Fig. 5-26. A dimensionless time-humidity shift factor

$$a_{TC} = \frac{\text{SET}(23°\text{C, } 50\% \text{ RH})}{\text{SET}(54°\text{C, } 95\% \text{ RH})} \approx 3.5$$

is applied to the surface energy data to achieve the master curves of Fig. 5-26. The earlier discussion of SET effects on the surface properties of the FPL-etched Al 2024-T3 provides a substantially higher time shift factor:

$$a_{TC} = \frac{\text{SET}(23°\text{C, } 50\% \text{ RH})}{\text{SET}(55°\text{C, } 95\% \text{ RH})} \approx 58.5$$

The master curves of $\gamma_{SV} = \gamma_{SV}^d + \gamma_{SV}^p$ shown by Fig. 5-26 provide a compact means of combining the effects of time, temperature, and humidity effects.

The surface energy diagram of Fig. 5-27 illustrates the shift in dispersion (α) versus polar (β) components which are displayed by phosphate-fluoride-treated Ti 6Al-4V with surface exposure time SET. The averaged surface energy properties of HT424 ($\alpha_1 = 5.21$, $\beta_1 = 3.31$) are included in Fig. 5-27.

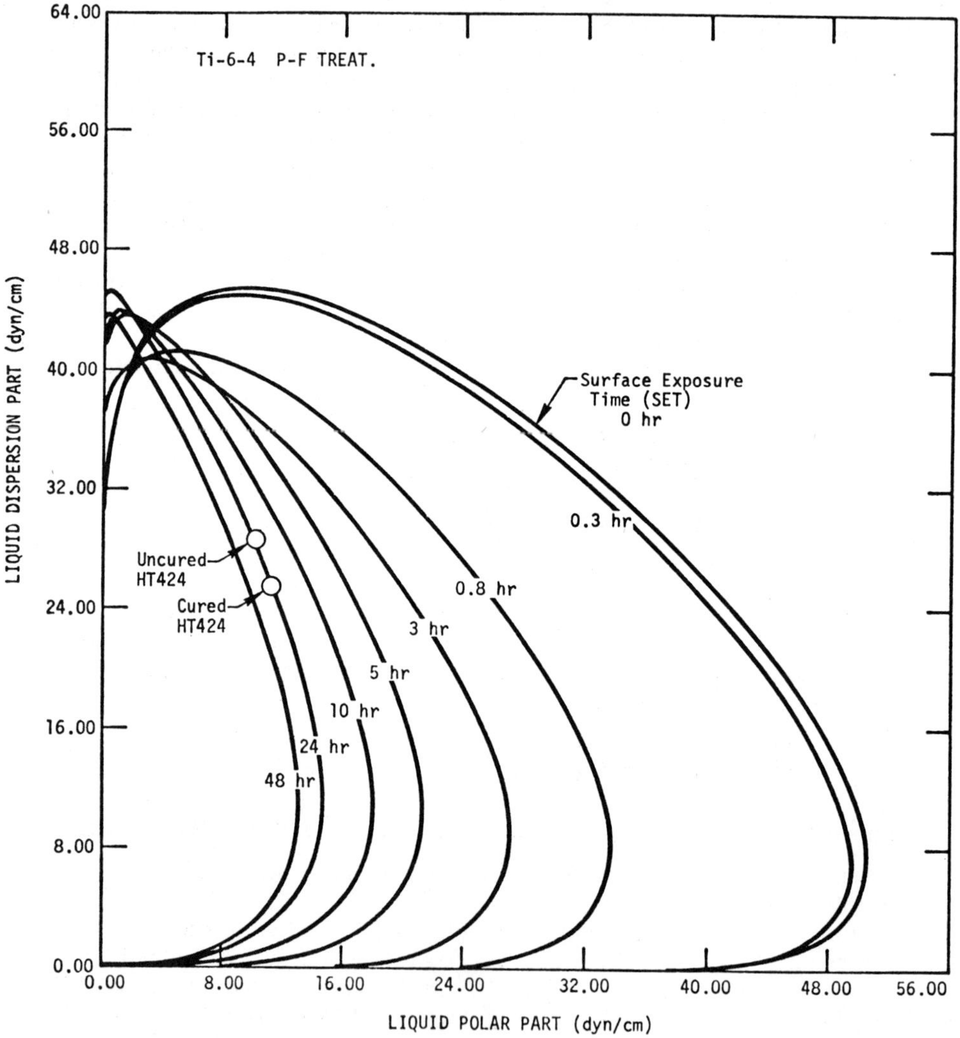

FIG. 5-25 Wettability envelopes for Ti 6Al-4V phosphate-fluoride, SET 54°C, 95% RH.

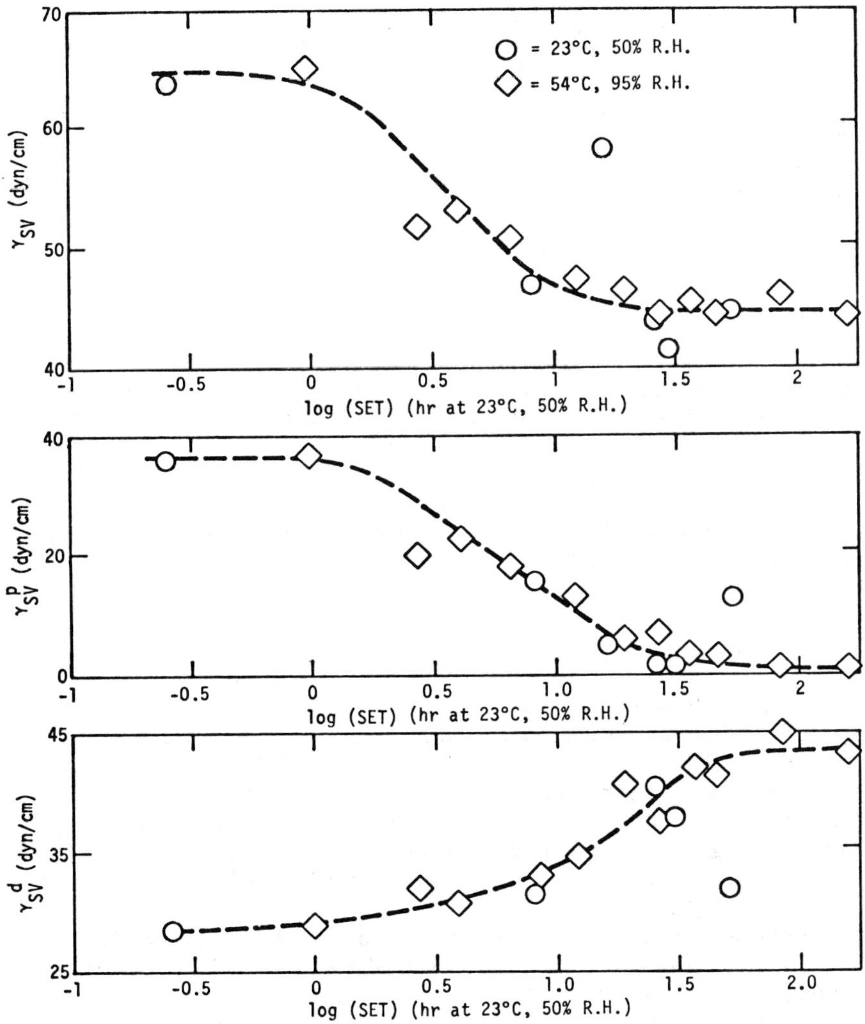

FIG. 5-26 Combined master curve for average values of γ^d_{SV}, γ^p_{SV}, and γ_{SV} for phosphate-fluoride-treated Ti 6Al-4V, SET 23°C, 50% RH and for 54°C, 95% RH.

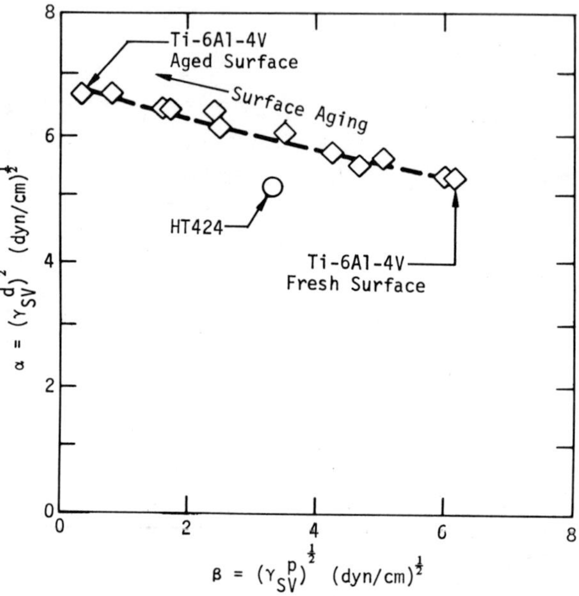

FIG. 5-27 Effects of SET on the shift of α_S vs. β_S for phosphate-fluoride-treated Ti 6Al-4V.

3.3. *Joint Aging Effects*

Figure 5-28 shows the bond strength σ_b versus ϕ_c for bond exposure times (BET) from 0 to 1016 hr at 54°C and 95% RH, Fig. 5-29 is for SET = 21 hr. The dashed lines in Figs. 5-28 and 5-29 have the same slope and intercept as for Fig. 5-25. The average values of ϕ_c and σ_b ($\bar{\phi}_c$ and $\bar{\sigma}_b$) at a particular value of SET and BET are plotted in Figs. 5-30 and 5-31. As for the SET data, BET has little effect on $\bar{\phi}_c$, but contrary to the SET data BET causes degradation of bond strength $\bar{\sigma}_b$.

Table 5-10 gives the average values of σ_b and ϕ_c for various BET and SET conditions and these data are plotted as a SET vs. BET response surface in Fig. 5-32.

The modified Griffith failure criteria defined by Eqs. (14) through (18) of Sec. 2.2.6 in Chapter 4 can be applied to evaluate water immersion effects on the critical failure stress σ_I of the HT424 to Ti 6Al-4V interfacial bond. The surface properties of HT424 are represented by $\alpha_1 = 5.21$ (dyn/cm)$^{1/2}$ and $\beta_1 = 3.31$ dyn/cm. The surface properties of surface-aged phosphate-fluoride-etched Ti 6Al-4V are represented by $\alpha_3 = 6.61$ dyn/cm and $\beta_3 = 0.60$ dyn/cm as described by

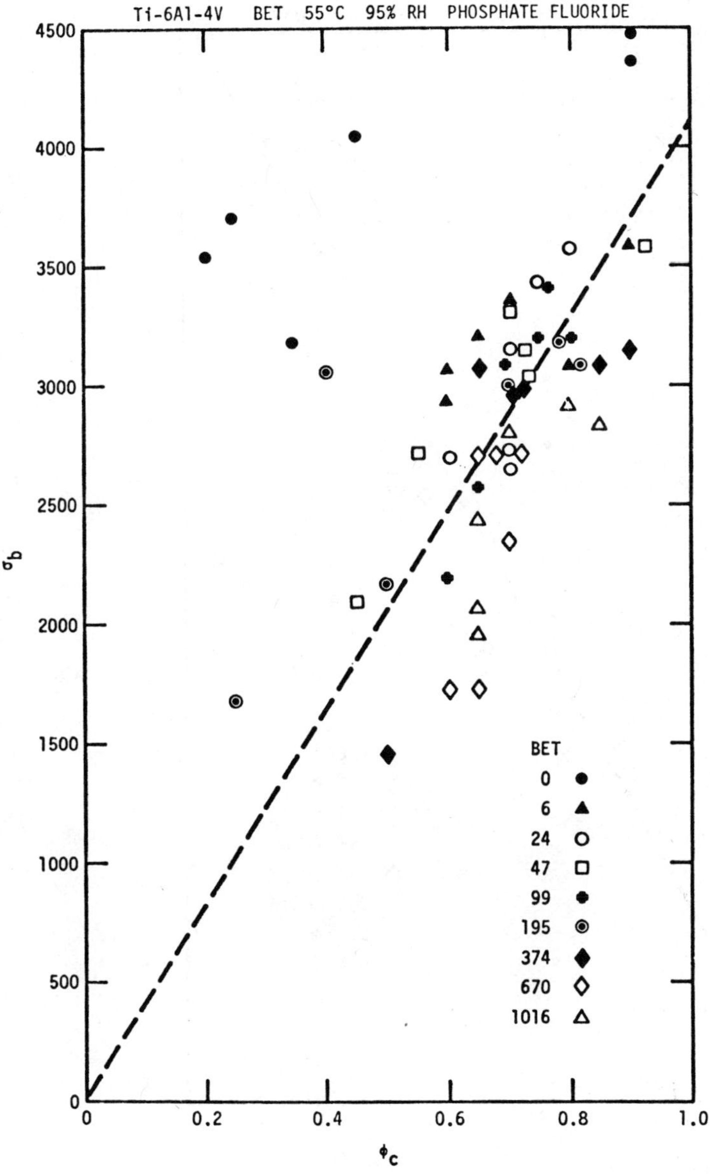

FIG. 5-28 Plot of σ_b vs. ϕ_c for BET 54°C, 95% RH, phosphate-fluoride-treated Ti 6Al-4V—HT424.

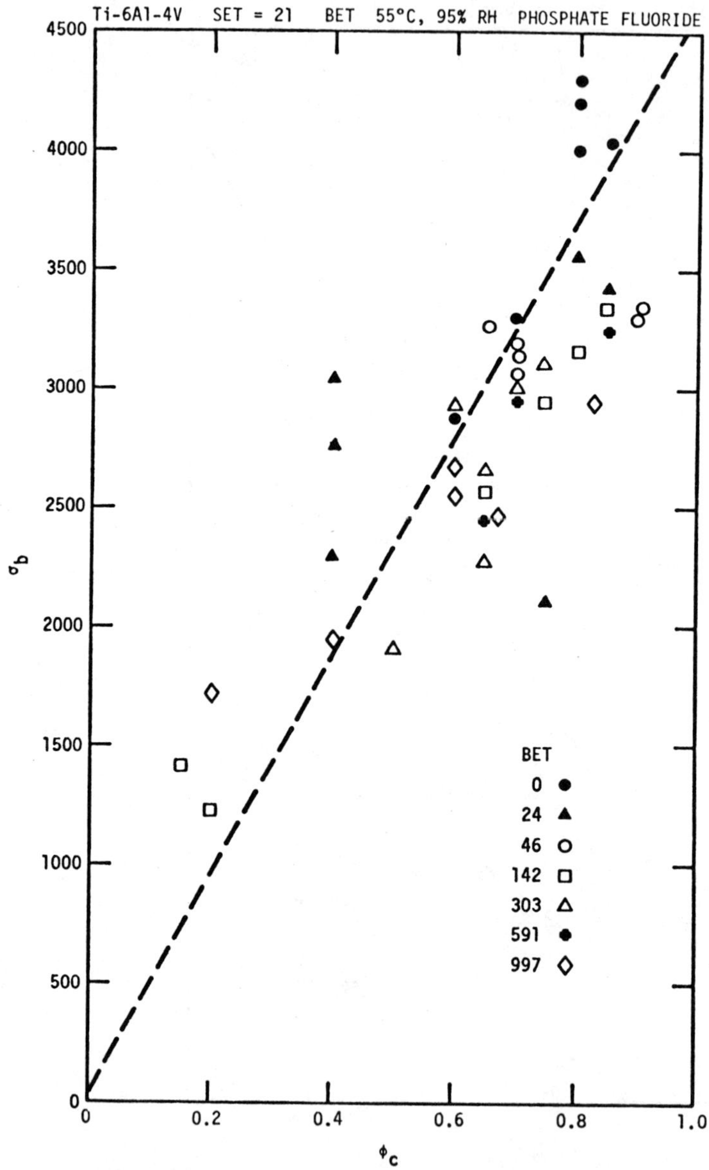

FIG. 5-29 Plot of σ_b vs. ϕ_c for SET = 21 hr and BET 55°C, 95% RH phosphate-fluoride-treated Ti 1Al-4V—HT424.

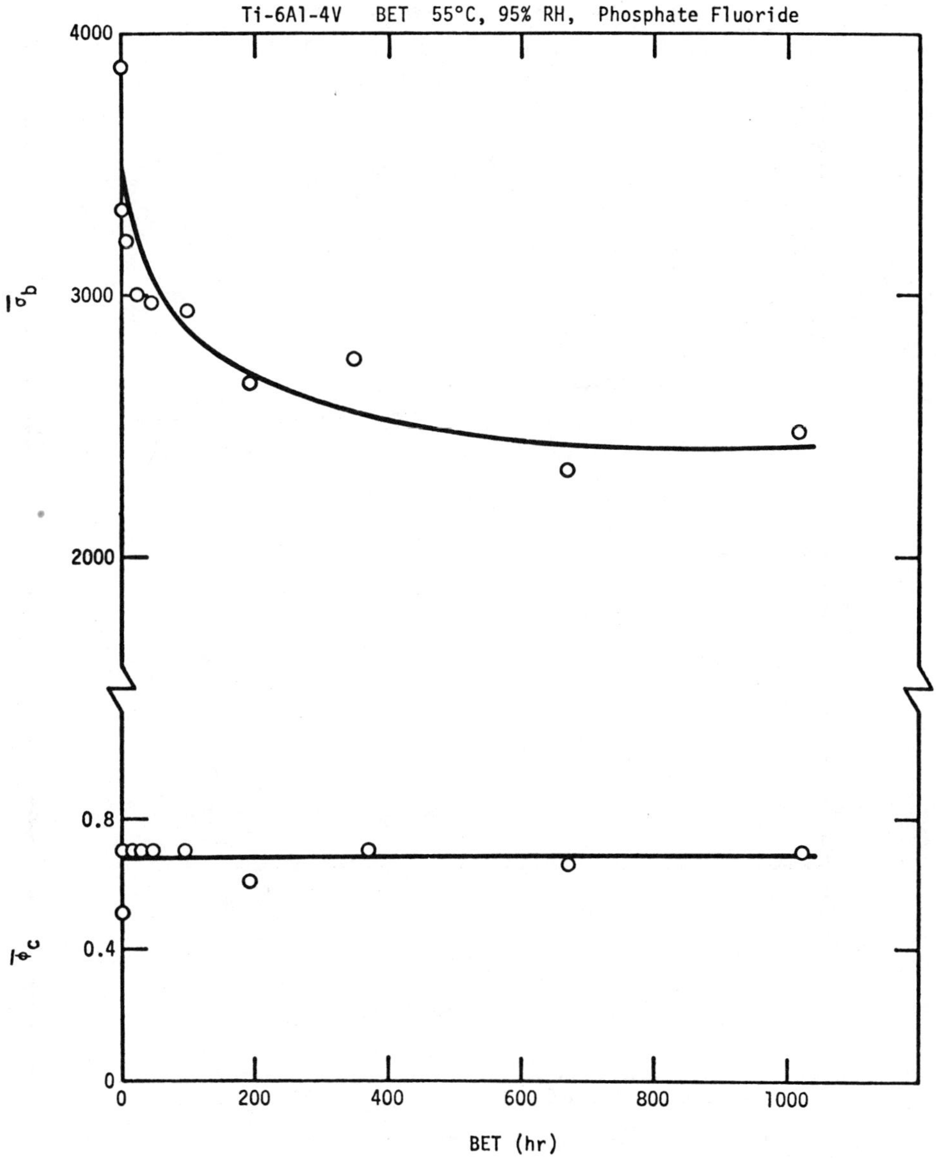

FIG. 5-30 Plot of $\bar{\sigma}_b$ and $\bar{\phi}_c$ vs. BET at 54°C, 95% RH for Ti 6Al-4V—HT424.

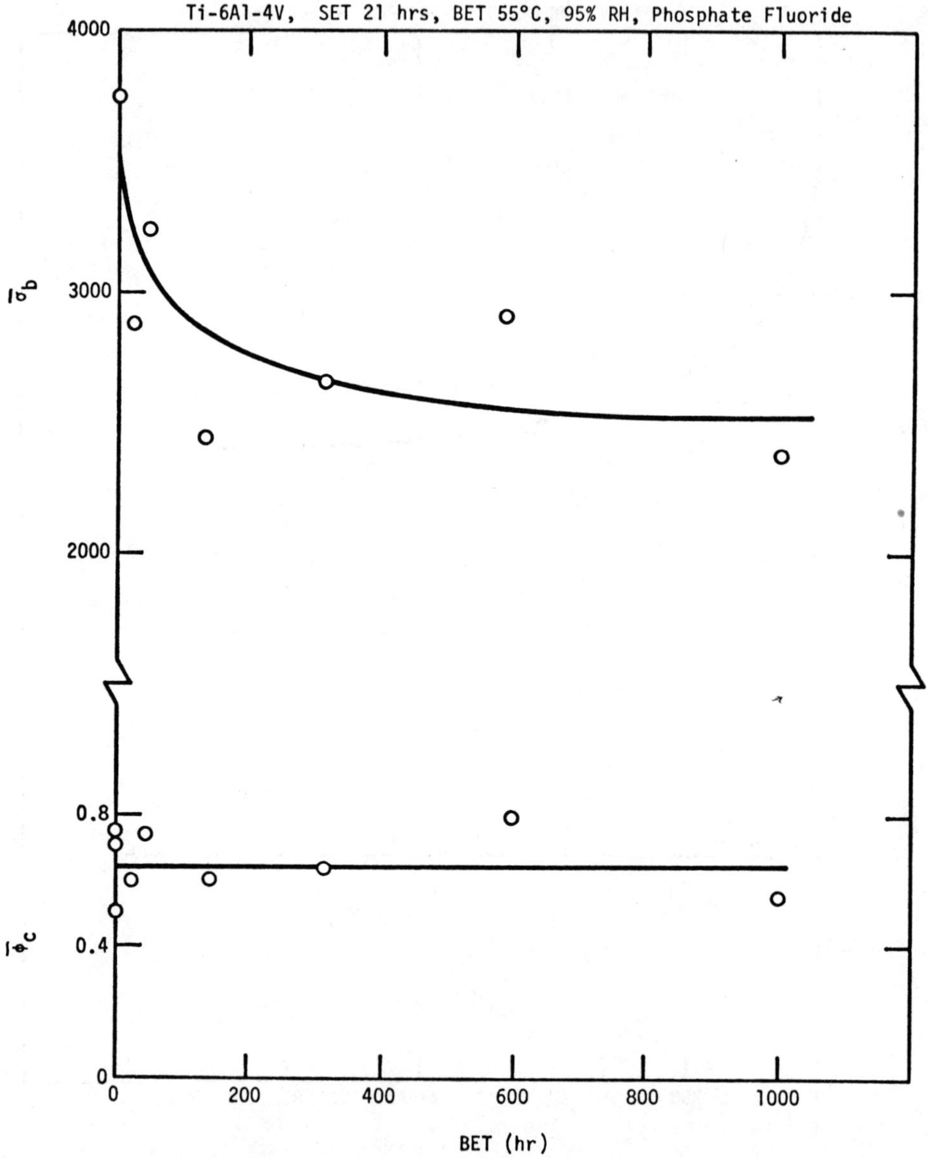

FIG. 5-31 Plot of $\bar{\sigma}_b$ and $\bar{\phi}_c$ vs. BET at 54°C, 95% RH for Ti-6Al-4V—HT424 with SET = 21 hr.

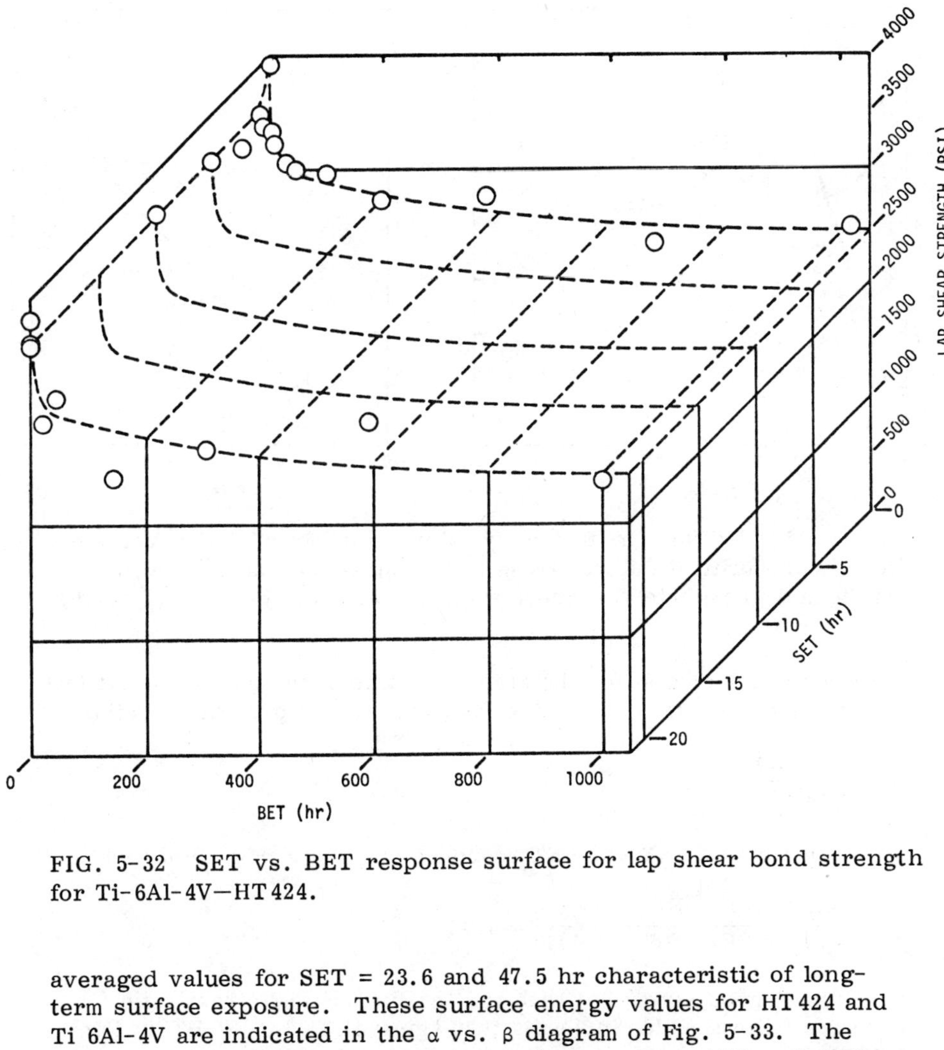

FIG. 5-32 SET vs. BET response surface for lap shear bond strength for Ti-6Al-4V—HT424.

averaged values for SET = 23.6 and 47.5 hr characteristic of long-term surface exposure. These surface energy values for HT424 and Ti 6Al-4V are indicated in the α vs. β diagram of Fig. 5-33. The point H = 5.91 and K = 1.96 dyn/cm defined by Eqs. (17) and (18) of Sec. 2.2.6 in Chapter 4 provides the locus for the vectors R_0, R(air), and R(H_2O), where $\alpha_2 = \beta_2 = 0$ for air and $\alpha_2 = 4.67$ dyn/cm and $\beta_2 = 7.14$ dyn/cm for H_2O. The right view of Fig. 5-33 presents the function of $\gamma_G(\propto \sigma_c)$ vs. R for interfacial failure, where $\phi_I = (1 - \phi_c) = 1.0$, between HT424 and Ti 6Al-4V. The prediction from the Griffith analysis is that

$$\frac{\sigma_I(H_2O)}{\sigma_I(\text{air})} = 0.84$$

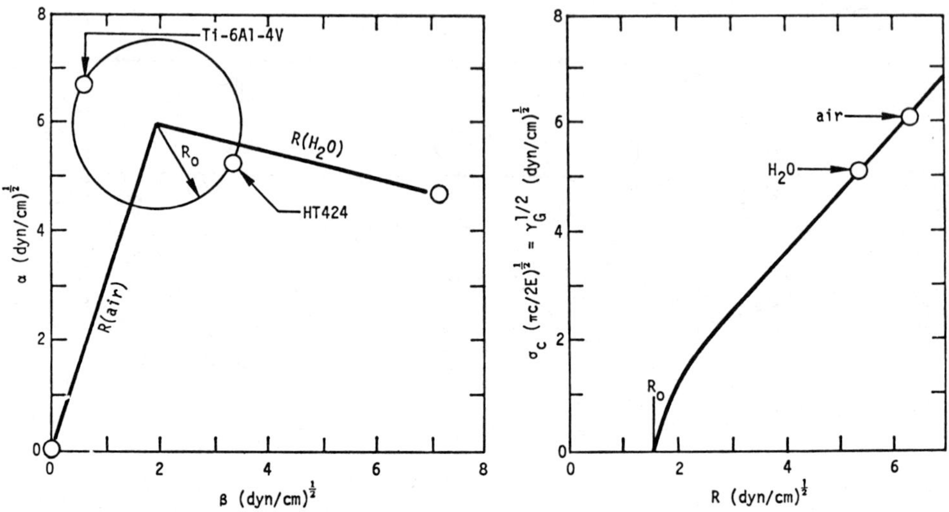

FIG. 5-33 Modified Griffith analysis of the effect of H_2O immersion in reducing critical failure stress σ_I for interfacial failure between HT424 and phosphate-fluoride-treated Ti-6Al-4V, ($\phi_I = 1 - \phi_c = 1.0$).

A previous calculation for the effect of water immersion on the critical Griffith stress ratio for cohesive failure of HT424 provided a ratio

$$\frac{\sigma_c(H_2O)}{\sigma_c(\text{air})} = 0.63$$

Inspection of the SET vs. BET response surface in Fig. 5-32 yields

$$\frac{\sigma_b(1000\text{ hr (BET)}}{\sigma_b(0\text{ hr SET, BET)}} \approx \frac{2500}{4000} = 0.63$$

which is in good agreement with the Griffith critical stress ratio for cohesive failure. This indicates that aging of Ti 6Al-4V-HT424 joints causes degradation mostly in the adhesive rather than at the interface.

3.4. *Surface Mapping*

We have prepared Ti 6-4 adhesive couples with the phosphate fluoride surface treatment, followed by deliberate contamination in certain areas. This was done to see if the contaminated areas could be detected by mapping with respect to surface potential difference

(SPD) and ellipsometry and the effect contamination would have on bond strength. It was demonstrated that failure areas on Al caused by humidity aging could be predicted in advance of bonding. It is demonstrated here that failure areas caused by contamination can be predicted in advance for Ti 6-4.

Figure 5-34 shows maps of ellipsometric parameters Δ and ψ, and surface potential difference (SPD) that were made after the phosphate-fluoride treatment but prior to bonding Ti 6-4 couples with HT424 adhesive. The map position along the abscissa corresponds with the position indicated on the schematic representation of the samples. After bonding and tests to failure on an Instron tensile tester the surfaces were examined under a low-power microscope to estimate the fraction ϕ_I of each map position square that failed at the metal-adhesive interface. Values of ϕ_I are also mapped in the figures.

Figure 5-34 is a control experiment for which no contamination was introduced. Values of Δ and SPD are fairly constant with respect to position indicating films of constant thickness and outer surface dielectric properties. The average value of Δ is 71.1 ± 1.6, $\psi = 28.3 \pm 0.9$, and SPD is 0.06 ± 0.04 V. Values of ψ and ϕ_I show unrelated patterns with respect to position. Variations of ψ probably reflect slight changes in substrate or film optical properties due to sample preparation. Interfacial failure ϕ_I proves to be essentially zero for positions 1, 2, 3, and 4 (the outer edge) but increases to about 0.3 for positions 5, 6, 7, and 8 (inner adhesive zone). The

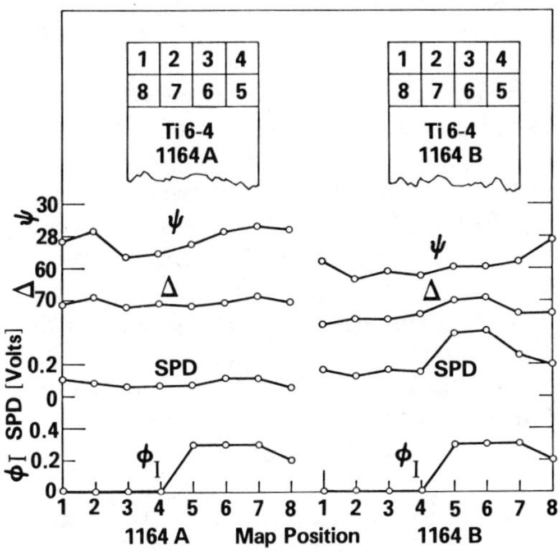

FIG. 5-34 Prebond surface character map, Δ, ψ, SPD as compared to postbond ϕ_I map for control test bond, 1164.

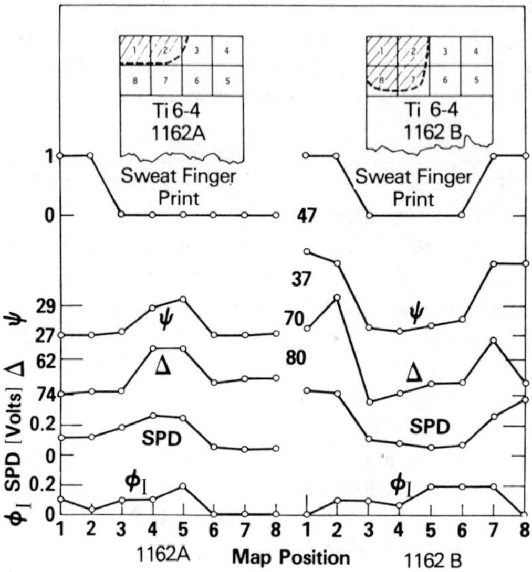

FIG. 5-35 Prebond surface character map, Δ, ψ, SPD as compared to postbond ϕ_I map for a Ti 6Al-4V contaminated sample 1162.

higher values of ϕ_I in zones 5-8 are not caused by variations in surface properties but are caused by higher stress levels in zones 5-8 as the lap shear joints tend to bend slightly in tension (see Fig. 4-18, Chapter 4).

Figure 5-35 is a map for specimens that were deliberately contaminated by a fingerprint, after wetting the finger with sweat from the forehead, in the areas indicated by the crosshatching (position 1, 2 for 1162A and 1, 2, 7, 8 for 1162B). A map of the contaminated areas is indicated at the top of the figure by giving a value of 1 for contamination and zero for no contamination. Values of ϕ_I for sample 1162 shows that interfacial failure was greater at the contaminated outer edge (position 1-4) and that in each case Δ, ψ, and SPD reflect the ϕ_I curve. The fingerprint map at the top of Fig. 5-35 for 1162B is identified by Δ, ψ, and SPD and shows that for 1162B contamination is much greater in the areas that were deliberately contaminated, but ϕ_I for both 1162A and B indicate that contamination has spread to the other areas.

Figure 5-36 shows the position of contaminants for 1161A and B and 1163A and B. Samples 1161A and B were contaminated by a dry finger. Samples 1163A and B were contaminated by smearing vacuum grease on a finger, rubbing the grease off, and then pressing the

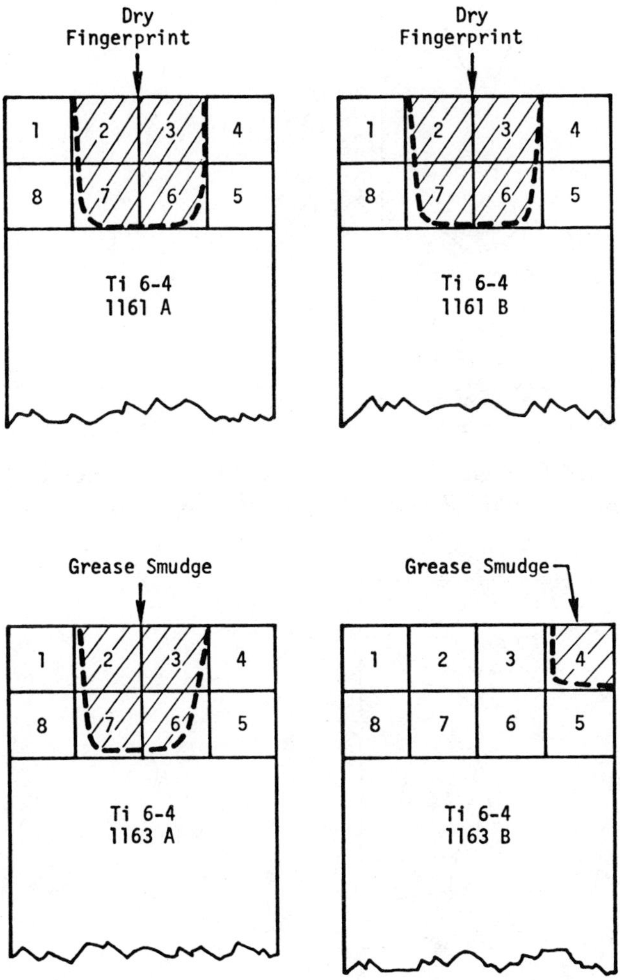

FIG. 5-36 Prebond surface character map, Δ, ψ, SPD as compared to postbond ϕ_I map for a Ti 6Al-4V contaminated sample.

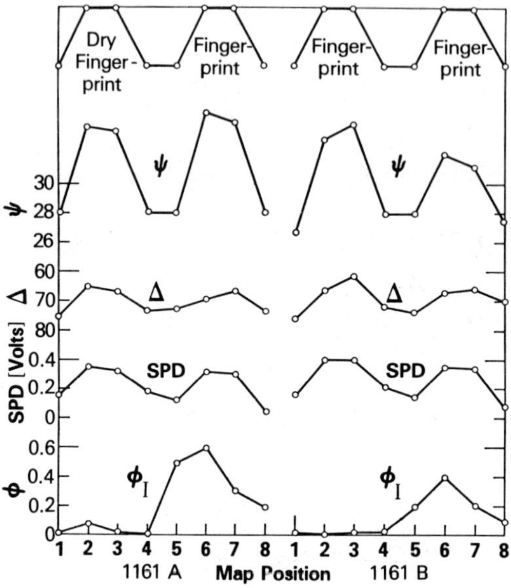

FIG. 5-37 Prebond surface character map, Δ, ψ, SPD as compared to postbond ϕ_I map for a Ti 6Al-4V contaminated sample 1161.

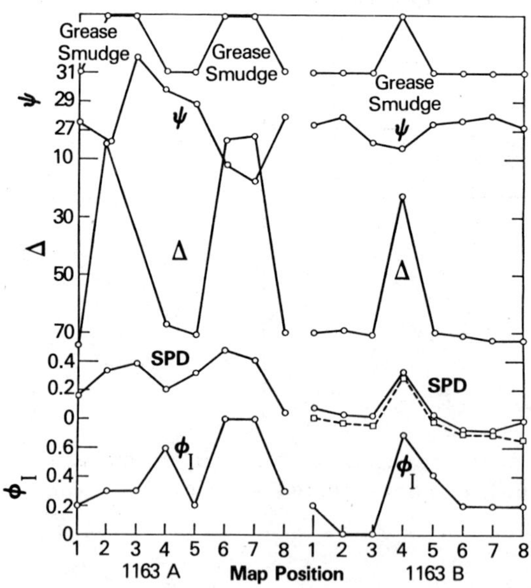

FIG. 5-38 Prebond surface character map, Δ, ψ, SPD as compared to postbond ϕ_I map for a Ti 6Al-4V contaminated sample 1163.

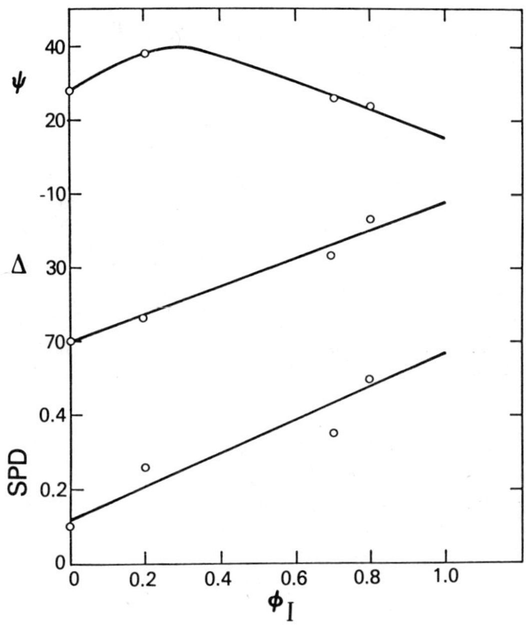

FIG. 5-39 Plot of SPD, Δ, and ψ vs. ϕ_I.

finger on the sample. Figure 5-37 shows that the dry fingerprint is clearly identified by Δ, ψ, and SPD and that ϕ_I increased above the average control values, except for positions 2 and 3 of 1161B. Figure 5-38 shows that the grease smudges are identified in the extreme by Δ, ψ, and SPD. Note that the pattern of ψ is inverted from that for other contamination (Figs. 5-35 to 5-37). This indicates that ellipsometry can be used to identify the type of contamination as well as its degree. The values of ϕ_I are much larger in the contaminated areas. The dashed SPD line for 1163B shows the reproducibility of the SPD scan. Average values of Δ, ψ, and SPD ($\overline{\Delta}$, $\overline{\psi}$, $\overline{SPD}$) corresponding to different average values of ϕ_I ($\overline{\phi}_I$) are plotted in Fig. 5-39. A correlation exists between $\overline{SPD}$, $\overline{\Delta}$, and $\overline{\phi}_I$. $\overline{\psi}$ does not show the same correlation because of the different types of contamination.

Table 5-11 is a table of bond strength σ_b and values of $\overline{\phi}_I$ for each bond couple. Comparison of average bond strength σ_b, $\overline{\phi}_I$, and $\overline{\phi}_c$ for different contamination reveals that the dry fingerprint bond is about the same as the control bonds even though the print could be detected by the surface tools. The grease smudge produced more interfacial failure and lower bond strength than the controls. Surprisingly, although the sweat fingerprint gave the lowest value of

TABLE 5-11 Bond Strength and Average $\bar{\phi}_I$ and $\bar{\phi}_c$ for Ti 6-4 (HT-424 Adhesive)

Sample couple	Contamination	σ_I (psi)	$\bar{\phi}_I$	$\bar{\phi}_c$
1166A-B	None	4040	0.13	0.87
1165A-B	None	3220	0.23	0.77
1164A-B	None	3800	0.14	0.86
1161A-B	Dry fingerprint	3800	0.16	0.84
1163A-B	Grease smudge	2780	0.34	0.66
1162A-B	Sweat fingerprint	2600	0.08	0.92

$\bar{\phi}_I$, it also gave the lowest bond strengths. It is concluded that either SPD or ellipsometry could be used to locate inadvertent contamination of Ti-6Al-4V surfaces after preparation but prior to bonding.

4. SUMMARY OF RESULTS

With respect to this system, the following important observations have been made:

1. After surface preparation of Ti 6Al-4V with the phosphate-fluoride treatment, the oxide is rather uniform across the surface. The average thickness is about 200 Å with $n_f \approx 2.8$ close to TiO_2 ($n_f \approx 2.85$) but with effective absorption index larger ($n_f \approx 0.1$) than the bulk oxide ($\kappa_f \approx 0$).
2. Auger analysis and profiles of Ti 6Al-4V with the phosphate-fluoride treatment revealed Ti, Al, V, Fe, C, N, and O as well as surface contamination P, Ni, S, and Cu. If F is present, it is masked by one of the Fe peaks. The concentration of O, Cu, Ni, Fe, S, and C is larger in the oxide than in the metal. The concentration of Ti, Al, and V are lower in the oxide than the metal.
3. Modes similar to failure in Al are observed for Ti 6Al-4V—HT424 with the exception that interfacial failure appears to occur at the oxide-primer interface. These observations indicate that the bond strength for a lap shear joint is the sum of the strengths in particular areas, i.e., the surface is not of uniform strength.

   Aluminum oxide film formed by the FPL etch could not successfully be removed, but titanium oxide film from the phosphate-fluoride treatment could. The oxide film from the phosphate-fluoride proved to be amorphous with maximum thickness ~540 Å. The larger thickness measurement for the removed oxide (measured by interferometry) with respect to the in situ film (measured by ellipsometry) is probably due to the very rough nature the film must have (revealed by SEM).

4. Contact angle measurements provide data for wettability envelopes, which showed that the phosphate-fluoride Ti 6Al-4V is wettable by the HT424 adhesive and primer. Aging these surfaces in humidity chambers reduces the wettability but not sufficient to cause weak bonds. One batch of FM34B-32 adhesive and BR-34 primer had wettability parameters that indicated nonwetting of the Ti 6Al-4V with varying surface preparations and therefore formed weak bonds. A new batch of FM34B-32 and primer proved to have proper wettability parameters and did form strong bonds.
5. Surface mapping of Ti 6Al-4V with respect to ellipsometry and SPD have revealed that these instruments can be used as non-destructive testing tools to predict the probable loci of weak bonding prior to the bonding process.
6. Surface aging of Ti 6Al-4V with the phosphate-fluoride treatment does not affect the oxide thickness, and negligible changes in ellipsometry parameters are noted. The fraction of the surface that fails at the interface remains essentially constant with SET, as does the bond strength. The wettability parameters do degrade with SET but not sufficiently to cause weak bonds.
7. Joint exposure in a humidity chamber causes degradation of the HT424 adhesive and therefore degradation of bonds for Ti 6Al-4V—HT424 joints. The Griffith critical stress analysis from the contact angle and wettability parameters faithfully predicted that joint exposure in a humidity chamber will cause degradation of the adhesive but not the interface for the titanium system.

5. DISCUSSION AND CONCLUSIONS

To improve adhesive joints, the investigator concentrates his efforts on the weakest link. If failure occurs primarily in the adhesive, emphasis is placed on a search for a stronger adhesive. If failure occurs primarily at a weak boundary layer due to contamination or in the oxide, emphasis is placed on cleanliness or surface preparations. Since new applications demand stronger and stronger adhesive joints, the tendency will be to continue improvement of the weakest link as it shifts from one locus to another. It is concluded, therefore, that the development of tools such as described in this chapter for identifying the weak regions is of importance now and will continue to be of prime importance in the development of stronger adhesive joints.

The most significant results of this research impacts on two important areas: one has to do with nondestructive testing (NDT) of surfaces prior to bonding; the other has to do with understanding failure mechanisms in the Ti 6Al-4V—HT424 system.

It has been demonstrated that surface regions that will result in a poor adhesive bond can be identified prior to bonding by ellipsometry and/or SPD. The potential of such NDT techniques is most significant for two reasons. First, supposing a completely satisfactory adhesive bonding procedure is developed in the laboratory for a particular application, inadvertent problems can arise on a factory assembly line that may cause deviations from the necessary processing parameters. Examples of these problems include contaminated processing solutions, contamination of atmospheric ambient, malfunction of process equipment, etc. Even though it may be impractical to survey every part or every region of every part by the NDT equipment, spot checks could be valuable for taking corrective action on the assembly line before too many parts have been bonded. Second, to improve on a given adhesive system, it is necessary to identify those parameters that are causing weakness.

Table 5-3 reveals that high bond strengths are obtained with the phosphate-fluoride and TURCO treatments but that low bond strengths result from the nitric acid-fluoride treatment. The only surface properties for the nitric acid-fluoride treatment that differ from both the TURCO and the phosphate fluoride treatments (see Table 5-3) are the larger concentration of copper within the oxide film and the smaller film thickness. In fact, there is almost a linear relation between film thickness and bond strength in Table 5-3. It is believed that this relationship is only coincidental or indirect in nature. The SEM pictures (Figs. 5-7 and 5-8) show that the surface roughness increases in the same order as the bond strength, i.e., nitric acid-fluoride, TURCO, and phosphate fluoride. For a properly prepared surface (uncontaminated) the effect of surface treatment to bond strength is probably mechanical, by increasing surface area and/or producing a hook-and-latch property. However, this chapter demonstrates that there is a relationship between physical and chemical surface properties and bond strength for aged or contaminated surfaces.

As for aluminum, titanium fails via the various modes illustrated in Fig. 4-49 (Chapter 4) for the lap shear joint. However, interfacial failure proves to be at the oxide-primer interface rather than in the oxide. The phosphate-fluoride-treated titanium does degrade with respect to wettability, probably due to surface contamination. Degradation of the adhesive, and thus bond strength, will occur when exposed to humidity chambers as for aluminum.

APPENDIX A SURFACE PREPARATIONS

TABLE A-1 Details of Surface Preparations for Ti 6Al-4V

1. Phosphate-Fluoride Surface Treatment
   a. Wipe metal with acetone-soaked cloth, followed by a 10-min ultrasonic cleaning in acetone.
   b. Emulsion-clean in TURCO 3878 for 15 min at 54-60°C.
   c. Water rinse spray and overflowing dip.
   d. Water break test.
   e. Pickle 1 min at 23°C in the following solution:
      70% HNO_3
      10 oz/gal NH_4HF_2
   f. Deionized water rinse for 2 min at 23°C.
   g. Dip in the following stirred solution for 1/2 min at 27°C:
      7.3 oz/gal $Na_3PO_4 \cdot 12\ H_2O$
      1.5 oz/gal NaF
      7.3% by volume HF (60% solution)
   h. Deionized water rinse—15 min.
   i. Deionized water blast to remove film.
   j. Dry samples under ambient air conditions.
2. Nitric Acid-Fluoride Treatment
   a. Solvent trichloroethylene (TCE) wipe metal with cloth.
   b. 10 min TCE—ultrasonic cleaning.
   c. 10 min cleaning in TURCO 4215 alkaline cleaner at 66°C.
   d. Rinse in distilled water.
   e. Pickle in the following distilled water solution at room temperature for 5 min:
      Nitric acid 15% by weight
      Hydrofluoric acid 3% by weight
   f. Rinse in distilled water at room temperature.
3. PASA-JELL Treatment
   a. through e: Same as no. 2, but add:
   f. Apply PASA-JELL 107 liquid to the titanium surface by dip or with an acid-resistant brush. Cover the entire surface by crossbrushing.
   g. Allow to dry for 20 min.
   h. Rinse thoroughly with water.
   i. Dry 15 min at 40°C in dry N_2.

(Table A-1 continued)

4. TURCO Treatment (LAD Modification)
   a. Wipe metal with acetone-soaked cloth, followed by a 10-min ultrasonic cleaning in acetone.
   b. Emulsion-clean in TURCO 3878 for 15 min at 54-60°C.
   c. Deionized water rinse.
   d. Etch in TURCO 5578 for 10 min at 66°C.
   e. Deionized water rinse.
   f. Dry in air.

TABLE A-2 Percentage Composition of Ti 6Al-4V Alloy Elements

Element	%
Al	5.5-6.7
V	3.5-4.5
Fe	0.3
C	0.08
N	0.05
H	0.015
O	0.2

APPENDIX B PROCESS DETAILS

1. Gillette: Bell process FW-4352, Rev. E, Phosphate-Fluoride Treatment

   a. Degrease.
   b. Alkaline-clean 5-15 min 140-180°F.
   c. Rinse.
   d. Pickle:
      2.0-3 oz/gal HF
      40-50 oz/gal HNO_3
   e. Rinse.
   f. Phosphate-fluoride 2 min max, RT:
      5.5-7.0 oz/gal Na_3PO_4
      2.5-3 oz/gal KF
      2.2-2.5 fluid oz/gal (70%) HF
   g. Rinse RT.
   h. Hot water soak 145-155°F 15 min.
   i. Final rinse (quick spray) RT—160°F 0.3-1 min.
   j. Dry air RT—160°F.

2. General Dynamics (Battelle formulation)
   a. Methylethylketone wipe.
   b. Trichloroethylene vapor degrease.
   c. Pickle in the following water solution at room temperature for 30 sec:
      Nitric acid 15% by volume of 70% HNO_3 solution
      Hydrofluoric acid 3% by volume of 50% HF solution
   d. Rinse in tapwater at room temperature.
   e. Immerse in the following water solution* at room temperature for 2 min:
      Trisodium phosphate 50 g/liter of solution
      Potassium fluoride 20 g/liter of solution
      Hydrofluoric acid (50% solution)—26 ml/liter of solution
   f. Rinse in tapwater at room temperature.
   g. Soak in 150°F tapwater for 15 min.
   h. Spray with distilled water and air-dry.

---

*This bath composition was developed at the Battelle Memorial Institute and is protected by U.S. Patent 2,864,732. Licensing arrangements can be obtained from the Battelle Development Corporation, Columbus, Ohio.

REFERENCES

1. W. G. Burghers, A. Claasen, and J. Zernike, Z. Phys. 74:593 (1932).
2. R. A. Harrington and H. R. Nelson, Trans. AIME, 137:62 (1940); H. W. L. Phillips, Inst. Met. Monograph and Report Series No. 13, 237 (1952).
3. S. Yamaguchi, J. Electrochem. Soc. 108:302 (1961).
4. B. Rivolta, Met. Ital. 50:173, 255 (1958).
5. A. M. Brusilavskii, B. E. Boguslauskaya, and E. G. Nasavich, Trudy Inst. Lakov i Krasok 2:23-35 (1939). C.A. 34:21427.
6. W. C. Hamilton and G. A. Lyerly, Tech. Report 4185, Picatinny Arsenal Contract DAAA-21-70C-0434, Gillette Research Institute, Rockville, Md., March 1971.
7. G. W. Lively, Tech. Report AFML-TR-73-270, Jan. 1974.
8. N. J. DeLollis, *Adhesives for Metals*, Industrial Press, New York, 1970, p. 35.
9. American Cyanamide Co., Bloomingdale Department, Havre DeGrace, Md. 21028.
10. H. H. Landholt and R. Börnstein, *Zahlenwerte und Funktionen*, Vol. II, Part 8, Springer, Berlin, 1923).
11. D. L. Johnson and L. C. Tao, Surf. Sci. 16:390 (1969).
12. E. L. Alexander, MA-71-545, July 1971.
13. F. C. Rauch and T. B. Reddy, Tech. Report AFML-TR-73-215, Nov. 1973.

6

Semantic Physics of Adhesion

Robert J. Good

Department of Chemical Engineering
State University of New York
Buffalo, New York

I. INTRODUCTION

The definitions of technical words are commonly taken for granted by scientists and engineers. Yet scientific breakthroughs have come about with the aid of definitions or redefinitions. For example, a major part of Darwin's great contribution, the theory of evolution and the origin of species, hinged on the meaning of the term "species." By the very nature of species, according to the pre-Darwinian view, every species that currently existed had always existed since the beginning of time. This semantic block was a great hindrance to the discovery that species could have "origins." Many other examples could be cited. In every case, after a satisfactory definition was achieved, the definition appeared to be totally obvious—by hindsight. Or (and this is extremely common) the era before the achievement of the definition was forgotten, or referred to as prescientific.

The science of adhesion is a field in which there has been no unanimity as to the meaning of the term that is most fundamental to the field—namely, "adhesion." More specifically, a definition with the appropriate physical content has long been needed. The problem is one in which semantics and physics are inextricably intertwined. This became evident to the present writer from observation of the way in which adhesion specialists who had been trained in different fields (e.g., chemistry, mechanical engineering, physics) appeared to be unable to communicate with each other in regard to important aspects of the field. The conclusion suggested itself that individuals were operating on the basis of different meanings of the word. Accordingly, an examination was made of the accessible definitions of the word "adhesion." The results of this survey can be summarized as follows:

The definition given in the 3rd edition of *Webster's International Dictionary* contains the following items that are pertinent to the technical use of the word. (The numbering is that in Webster; the definitions that are omitted pertain to usages such as in "adhesion to a religion or faith.")

2. The action or state of adhering, specifically, a sticking together of substances (as of glue to wood).
7a. A grip or sticking effect produced by friction.
7b. The force that must be developed to overcome this grip.
8. The molecular attraction that must be exerted between the surfaces of bodies in contact—distinguished from cohesion.

The *Oxford English Dictionary* gives the following:

> The action of sticking (to anything) by physical attraction, viscosity of surface, or firm grasping.

The ASTM gives the following [1]:

> *adhesion*, n: the state in which two surfaces are held together by interfacial forces which may consist of valence forces or interlocking action, or both. (See also *adhesion, mechanical* and *adhesion, specific*)
>
> *adhesion, mechanical:* adhesion between surfaces in which the adhesive holds the parts together by interlocking action. (See also *adhesion, specific*)
>
> *adhesion, specific:* adhesion between surfaces which are held together by valence forces of the same type as those which give rise to cohesion. (See also *adhesion, mechanical*)

The Russian authors Raevskii and Pritykin [2] have recently given a discussion of the various usages and definitions of adhesion. The following is an exerpt from their paper, together with their bibliography. The translation is by R. J. Mosley.

> The superiority of bonded joints over traditional has meant a considerable widening of their applications. Practically all polymers may be used as a basis for adhesive materials. Although attempts have been made to work out a classification and principles of codifying these materials [1] they are not satisfactory and have not been adopted in practice. The lack of a general unified classification meant that we cannot make a preliminary selection on the basis of the grade of bonding agent.
>
> A similar state of affairs applies to the theoretical conceptions of adhesion and bonding. The constant increase in the volume

of information naturally calls for clear systematization of the material accumulated using logically based characteristics. It is obviously necessary to create a single classification scheme, including principles of coding (nomenclature) of bonding materials.

Since the development of such a scientifically and logically incontrovertible scheme is impossible without clear definition of the fundamental concepts used in the theory and practice of adhesion, it is proper to begin the assessment precisely from these definitions.

Terminology. The actual concept of adhesion calls for fundamental clarification. Different definitions of this concept are given in the extensive literature [2-26]. From the basic viewpoint of meaning, they may be regarded as simply the translation of terms into Russian, i.e., without any physical meaning [3-8], or else defining adhesion as an index of the strength of an adhesion joint [2,9-27], as a phenomenon [18-24], or, finally, as a capacity, i.e., a potential property [7,25]. It is accordingly difficult to relate this concept with certainty to independent phenomena peculiar to the type of strength or to a potential property. At the same time it is evident that this formulation is necessary, since it decides the physical meaning of the concept, without which it is impossible to conduct the theoretical, and consequently also the practical, work.

In a simple translation of a term, it is difficult to express the physical meaning of a phenomenon. It is all the more difficult to show the limits of its applicability, and in addition such a definition is so general that we cannot draw any conclusions from it; yet it is precisely these three criteria that in our opinion every definition ought to meet. But even in encyclopedic publications, the translation of this concept is not accompanied by any kind of commentary [5,6]. Closely akin to this group of definitions is the definition of adhesion as a bond between bodies. It is possible to trace an interesting trend: for instance, adhesion has been defined as "the formation of a bond between bodies which have been brought into contact" [23] (a primary act), as a "molecular (?) bond" [19] (the presence of interaction), as the well-known mutual gripping of molecules on the surfaces of contacting bodies [20], of two surface layers [9], of two surfaces [8], or even of bodies as a whole [3,21], or, finally, as the strength of gripping (the result of the interaction) of materials [22]. It is thus evident that there is no unanimity of opinion as to whether adhesion should be regarded as a primary act of some particular interaction, the interaction itself, or the result of it.

It is indubitable that adhesion is, by virtue of the above survey of the meaning put into this term, not a property but a phenomenon. Then it is necessary to distinguish the actual phenomenon and its mechanism, i.e., the line of realization, and the final effect. When we start from these requirements we must define adhesion as a surface phenomenon, consisting of the occurrence of a physical or/and chemical interaction between the molecules (or ions) of the surfaces of condensed phases in molecular contact, which leads to gripping. The quoted formulation, in defining adhesion as a phenomenon, relates it to the class of surface phenomena; with the known mechanism of the phenomena the formulation characterizes it [not understood—RJM] by a whole spectrum of forces and indicates the conditions of realization (the presence of molecular contact) or definite objects (surfaces of condensed phases); speaking of the end result, it defines this last-named as the strength of the adhesion joint.

We must assess other terms as well, which characterize this phenomenon. In the first place these are objects of adhesion reaction substrates (or adherends) and adhesives (it is necessary to point out that in the definition of these terms too there is well-known confusion [7,25]). Moreover the difference of adhesive from adherend is not a formal one, since it is difficult to imagine a phase entering with equal success into the role of adherend or of adhesive. The latter may be defined as a phase taking part in adhesion reaction and having a greater potential capacity for its realization than an adherend.

Bibliography

1. D. A. Kardashov, Synthetic Bonding Agents, 1968, p. 69.
2. Ibid., p. 14.
3. N. I. Moskvitin, Bonding of Polymers, 1968, p. 198.
4. N. A. de Bruyne, in Sbornik, Structural Bonding Agents, Goslesbumizdat, 1959, p. 9.
5. A. D. McLaren and C. J. Seiler, J. Polym. Sci., 4, No. 1, 1949, p. 63.
6. Short Polytechnical Dictionary, Gostekhteorizdat, 1957, p. 30.
7. Encyclopedic Dictionary, Vol. 1, publ. Bol'sh. Sovetsk. Ents., 1953, p. 29.
8. P. A. Rebinder, Physical Encyclopedic Dictionary, Vol. 1, 1960, p. 19.
9. N. A. Krotova, On Bonding and Sticking, AN SSSR, 1956, p. 94.
10. G. Epshtein, Bonding of Metals, 1956, p. 71.
11. N. S. Enikolpyan and S. A. Vol'fson, Chemistry and Technology of Polyformaldehyde, 1968, p. 213.

12. V. P. Smigla, Thesis, 1961.
13. P. A. Rebinder, Short Chemical Encyclopedia, Vol. 1, 1961, p. 27.
14. I. Ya. Ben, Quick-setting Bonding Agents in the Footwear Industry, 1967, p. 7.
15. S. S. Voyutskii, Autohesion and Adhesion of High Polymers, 1960, transl. 1963.
16. N. A. Ivanov et al., Bonding Agents and Technology of Bonding Plastics in Furniture Production, 1962, p. 5.
17. G. M. Bartenev, Structural Materials, Vol. 1, publ. Sovetsk. Ents., 1963, p. 14.
18. R. C. Bowers and W. A. Zisman, in Sbornik, Structural Properties of Plastics (Physicochemical Basis of their Application), publ. Khimiya, 1967, p. 292.
19. W. A. Zisman, in Encyclopedia of Polymer Science and Technology, Vol. 1, 1964, p. 292.
20. V. G. Raevskii, *Trudy GIPK*, 1968, No. 1, p. 38.
21. V. L. Vakula, in Proceedings of Conference, Bonding Agents and Bonded Joints, MDNTP, 1967, p. 6.
22. ASTM D907-60.
23. W. Zeller and A. Franke, Das physikalische Rustzeug des Ingenieurs, 1966, p. 251.
24. A. S. Akhmatov, Molecular Physics of Boundary Friction, 1963, p. 277.
25. S. K. Zherebkov, Bonding of Vulcanized Rubber to Metals, 1967, p. 50.
26. A. D. Zimon, Adhesion of Powder and Particles, 1968, p. 9.
27. V. G. Raevskii et al., *Trudy GIPK*, 1968, No. 1, p. 54.

Discussions of the definition of "adhesion" have also been given by Salomon [3] and by Eley and Tabor [4]. These have been primarily in terms of intermolecular forces. Thus, Salomon gives the following:

> In physical chemistry, attraction between a solid surface and a second phase is called adhesion. . . . Adhesion leads to *sorption;* this may be *ad*sorption *on* a surface or *ab*sorption *into* a surface layer. . . . Adhesion can be due to electrostatic force, to van der Waals force, or to "chemical" valence forces; in the latter case, one speaks of *chemis*orption.

In addition to demonstrating the need for a satisfactory definition, the quotation from Raevskii and Pritykin points to certain characteristics which a valid definition of adhesion should have: It must embrace both the macroscopic phenomena and the microscopic, molecular mechanisms that are responsible for the phenomena; and it should not confuse the macroscopic with the microscopic. These characteristics are major components of what may be referred to as the semantic physics of adhesion.

Now, the macroscopic phenomena of adhesion are mechanical. The term "macroscopic" means, strictly, that the quantities of matter that are involved can be seen with the naked eye. However, there is no qualitative change in the mechanical phenomena of adhesion when the scale of observation is changed by introduction of light microscopy. (The qualitative change occurs when the scale of distances is entered, in which the discrete nature of atoms and molecules becomes important. See below.) So we can consider the macroscopic phenomena to extend to quantities of matter that can be observed directly with the light microscope.

The phenomena involve forces which can be directly applied or specified, and measured directly; and they involve energy in the same way that they involve force. Both energy and force can be treated by means of the stress and strain tensors. The macroscopic phenomena of adhesion also involve an interface, in a way that is different from the role of the interface in the mechanical processes of fastening by chain links, bolts, rivets, screws, nails, or hooks. This difference will be discussed in more detail below.

The macroscopic phenomena of adhesion are also treated, in the scientific literature, in terms of thermodynamics. We will examine the thermodynamic question later.

The microscopic phenomena that are pertinent to adhesion are the atomic and molecular contacts, and the interatomic attractions and repulsions that are present in systems where the macroscopic phenomena are observed. (The description in terms of the stress and strain tensors cannot be used on the microscopic scale. This is because these physical concepts are defined for continua, or, in the case of systems consisting of atoms and molecules, they are defined by averaging over regions of space that contain a large number of discrete particles.) The microscopic processes are, in most cases, known only by scientific inference; i.e. they cannot be observed directly *in the same way that the macroscopic phenomena are observed.* It will, however, be shown in Sec. 4 of this chapter that the macroscopic concepts employed in the mechanical definition, together with the principle that technical definitions must avoid circularity, lead logically to an absolute requirement that adhesion be given a definition in terms of microscopic phenomena also.

It is important that a definition of adhesion should not contain an implicit specification of the microscopic processes that are involved; a definition should be neutral as among inferred processes. This is because if a definition does imply any particular processes or forces etc., it restricts research to the study of those forces, processes, etc., and hence limits the kinds of answers that can come out of research. And anything that limits the kinds of answers that can be considered may foreclose research from providing correct answers.

There is a peculiarity that is commonly present in adhering systems, which complicates the problem of definition. In a great many such systems, one or more of the components is a polymer; and the mechanical behavior of polymeric materials involves cooperative processes. Polymer molecules can easily be large enough that, if linear and fully extended, they would stretch over dimensions that could be observed with the light microscope. Such molecules are mechanically anisotropic by a factor that is proportional to molecular weight; and the molecular anisotropy can contribute mechanical anisotropy to the system, which is vastly greater than that which can be achieved in nonpolymeric systems. It may well be that this macroscopic manifestation of cooperative, microscopic properties constitutes part of the reason why the definitions given heretofore have not distinguished between the macroscopic and microscopic aspects of adhesion.

To return to the macroscopic aspect of the problem, it should be noted that terms such as "gripping" or "sticking" are employed in the existing definitions in such a way as to indicate that the authors considered them self-explanatory. (If they are not self-explanatory, there is danger of circularity of definitions. See Sec. 5.) In the context of adhesion, there is a need to replace these undefined words with mechanical representations in terms of force and energy.

Finally, it may be noted that many definitions of "adhesion" (a noun) are expressed in terms of one or more verbs. Thus Raevskii and Pritykin use the phrases "mutual gripping" and "bodies that have been brought into contact." These suggest that it may be easier to define the verb associated with the phenomenon first, and then to define the noun.

2. DEFINITIONS

We will first give a set of definitions of the verb "to adhere"; and we will subdivide the definitions of the verb according to macroscopic and microscopic usages. A further subdivision will be made, here, according to whether the most general representation in mechanics is employed, this definition being of use only to those versed in tensor analysis. Since the vast majority of adhesion scientists and technologists have not had extensive preparation in tensors, we next provide some discussion of this definition which both amplifies it and makes it more specific. We then proceed to a set of macroscopic definitions which, though less elegant than the first, accomplish essentially the same result without explicit reference to the stress tensor.

2.1. *Macroscopic Definitions of "Adhere"*

2.1.1. Definition 1

Two solid bodies are said to adhere to each other when the following conditions are jointly fulfilled: (a) An interface of intimate contact exists between them, and persists under load which may be in an arbitrary direction. (b) Under an arbitrary distribution of stresses existing within the two bodies, the stress tensor is and remains continuous across the boundary. (c) When a load is applied to the system in an arbitrary manner, resulting in a noninfinitessimal change in any of the components of the stress tensor, the stress tensor remains continuous. (d) If the system is subjected to a tensile or shear loading, or a loading such that the resolved stress components in the interfacial region are predominantly tensile and/or shear relative to the interface, the tensile and shear components of stress are and remain continuous. The continuity of compressive stress components, or the transfer of work or force across the boundary by means of compressive stress, is irrelevant to adhesion.

Discussion

1. The system may consist simply of two solid bodies, or it may consist of two solid bodies and a third phase in the form of a thin layer, which is referred to as an *adhesive*. In the latter case, each of the solid bodies is said to *adhere* to the adhesive (or vice versa) and the composite comprises an *adhesive joint* [4]. See below.

2. The last part of item (d) in this definition is included in order to cover certain kinds of systems which clearly are not examples of adhesion, but which could be deemed to come under specifications (a) to (c). To lead into an example of such systems, consider Fig. 6-1.

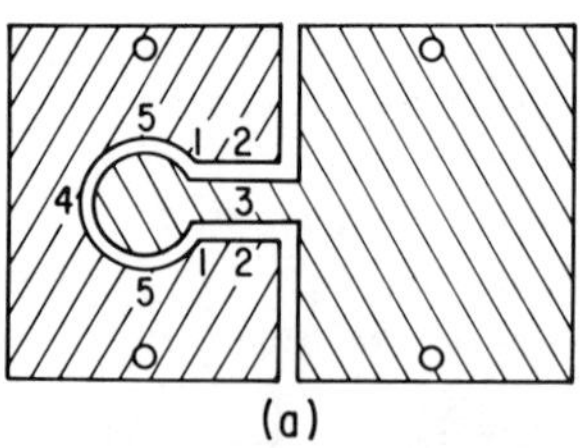

FIG. 6-1 Keying with loose fit.

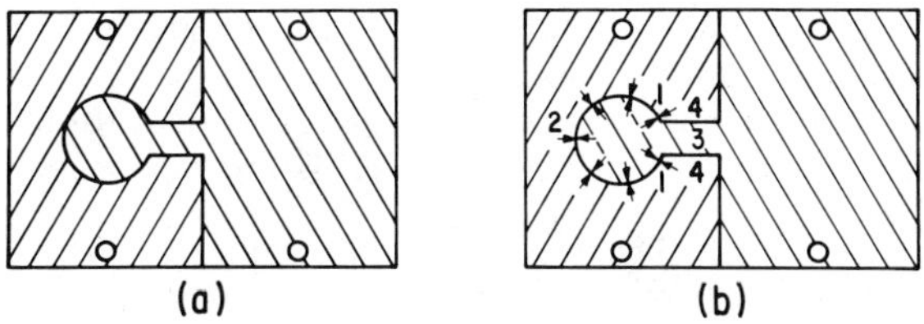

FIG. 6-2 Keying with close fit: (a) contact established; (b) slot has shrunk so that compressive stresses exist across the boundary.

Under tensile load, the cylindrical part of the key makes contact with the slot, in region 1, and tensile force is transmitted by compressive loading in that interfacial region and in region 2, and tensile loading in region 3. The stress tensor is discontinuous in regions 4 and 5, Fig. 6-1.

Now shrink the cylindrical hole continuously, until contact is made over the cylindrical part of the key. (See Fig. 6-2.) In the same sense that a stress has meaning in a body that consists of point particles, let the contact of the two bodies be "continuous." Let the shrinkage continue until there exists a compressive stress field normal to the boundary, as in Fig. 6-2b, which may reach a level higher than the *tensile* strength of the part of the key in region 3. Finally, reverse the process, so that the condition in Fig. 6-1a is restored. This process is not a mere thought experiment; it can be performed by starting with a slightly undersized slot in the left member, enlarging it by thermal expansion, and assembling the system. When the left member is hot, the condition in Fig. 6-1a exists; when cold, that in Fig. 6-2b.

If the system is in the configuration of Fig. 6-2b, and a horizontal, tensile load is applied, the actual load is transferred from one body to the other by virtue of the increased compressive stress across the interface in region 1, together with the compressive stress in region 4, and most particularly, tensile stress in region 3. Indeed, if the compression due to shrinkage is great enough, the limiting strength of the system is the tensile strength of region 3.

If the radial compression is less than the tensile strength of region 3, then before region 3 fails there will be a separation—and discontinuity of the stress tensor—in region 2. But if the system is one in which the two bodies adhere to each other, then instead of a gap opening up in region 2, a state of tensile stress will develop, and the stress tensor will remain continuous across the interface, under load.

Thus we can recognize that the formalism of the definition given above is satisfied.

This argument, and Figures 1 and 2, can easily be modified to treat shear loading; the same qualitative conclusions are reached.

3. A mechanical function that is accomplished by the existence of intimate contact, over the whole of an adhering region, is the prevention of stress concentrations at the boundaries of voids. Failure most commonly initiates at a region of stress concentration, where the local stresses exceed the local tensile or shear strength of the interface or the adjacent bulk solids. It is well known that this is one of the main physical reasons why the strength of an adhering system is often far below the strength of either bulk phase. In terms of the continuity of the stress tensor, the result of such stress concentration is the spatial propagation of the region of discontinuity.

4. In fluid mechanics, it is considered that an ordinary liquid always "adheres" to a solid, in the sense of there being continuity of the stress tensor at a liquid-solid interface where intimate contact exists. If gas bubbles are trapped at an interface, or if, due to the existence of a nonzero contact angle, the liquid does not penetrate into voids in the surface, then the extent of intimate contact is restricted.

2.1.2. Definition 2

Two elastic, solid bodies are said to adhere to each other when mechanical force or work can be transferred from one to the other, in tension or in shear, or any combination of tension and shear, across on an interface where intimate contact exists—without loss of that contact—the interface being perpendicular to the stress in the case of pure tension, or parallel to the stress direction in the case of pure shear [5].

2.1.3. Definition 3

For a system consisting of two bodies, one of which is plastic or viscoelastic and the other elastic, the bodies are said to adhere when mechanical force or work can be transferred from one to the other, as in definition 2, with the following further qualifications: Either, in the period of time under consideration, there is negligible plastic deformation; or plastic deformation occurs only in regions that are significantly removed from the interface. Interfacial (plastic) slip does not occur.

Comment

(a) The purpose of the specification about plastic deformation is the recognition of the fact that a system of two solids such as an ice skate on ice is not commonly considered to be an adhering system.
(b) If both phases are plastic or viscoelastic, the continuity condition, definition 1, still holds, of course. But the interface itself may deform quite appreciably, by expansion, contraction, flexing, or distortion, with maintenance of its topological integrity. The deformation considerations stated in definition 3 apply, but the second of the two qualifications is as follows: In the time period under consideration, there is not any greater shear deformation in material across to the interface than in the corresponding bulk material.

2.1.4. Definition 4

For a system consisting of two bodies, at least one of which is liquid, the bodies are said to adhere whenever (a) an interface of intimate contact between the two bodies exists and (b) tensile mechanical work or force can be transmitted across the interface between the two bodies, when suitable restraint against shear deformation exists. Also: (c) mechanical work or force in shear can be transmitted in a dynamic system, for example, in viscous drag.

Comment

Macroscopically, the adhesion between a liquid and a solid is quite different from that between two solids. Even though no static shear load can be supported by liquid-solid adhesion, and even though tensile loading can be supported only under special conditions of restraint, the same word is employed. Common usage is clear, and must be accommodated.

2.2. Microscopic Definitions of "Adhere"

2.2.1. Definition 1

An atom or molecule is said to adhere to a solid or liquid surface if, starting in a state of free translation or of migration, it collides with the surface inelastically, i.e., it is not specularly reflected.

2.2.2. Definition 2

An atom or molecule (regardless of its history) is said to adhere to a solid or liquid if it is present at the surface of the condensed phase, and remains (for a period of time that is long compared to the period of a molecular vibration) sufficiently close to the solid or liquid that there is an appreciable energy ε of interaction between it and the condensed phase. "Appreciable" means, in relation to the mean kinetic energy (3kT/2) of a gas molecule at the ambient temperature T, that

$$\varepsilon > kT \qquad (1)$$

Comment

The noun "adsorption" is usually employed to designate the *conditions* described by definitions 1 and 2; but the *verb "to adhere"* is used quite commonly enough that it must be considered explicitly here.

2.2.3. Definition 3

Two bodies in the form of condensed phases, i.e., two solids, a solid and a liquid, or two liquids, are said to adhere to each other if an interface of intimate contact exists over an appreciable period of time, and if the contact is close enough, over an appreciable region of the interface, that interatomic or intermolecular forces between the atoms or molecules in one phase and those in the other phase, when summed over all atoms or molecules, are appreciable. In this case, "appreciable" means that the total force or energy is commensurate with, or larger than, the loading to which the bodies are subjected.

Comments

1. Valence forces acting across an interface constitute a specific case, under definition 3.
2. The description of forces in terms of the Lifshitz theory of attractions [6] is a refinement on the theory of interaction of discrete atoms or molecules in separate phases. It includes cooperative effects in the electronic behavior of the atoms or molecules in a single phase, as manifested in the interaction across an interface or between bodies separated by a layer of a different phase. In terms of definition 3, the existence of forces of the type described by the Lifshitz theory is to be explicitly included.
3. The meaning of "appreciable," for the purposes of definition 3, is necessarily macroscopic. Thus, the microscopic definition of the verb "to adhere" necessarily requires the use of macroscopic concepts. There is no inconsistency in this requirement; and indeed, this mixing of the micro and macro levels of analysis constitutes the bridge between the two conceptual levels.
4. The nature of the adhesion that exists between two cells in a living tissue is currently a much-debated issue. There is evidence [7] that a "gap" exists between cells in certain kinds of tissues or artificial aggregates; this gap, which is some 50 to 100 Å wide, contains a variety of molecular species. It is held by some

that the most important components are polymeric molecules which have one end attached to the cell membrane of one cell or the other, and that it is these molecules that interact with each other to effect adhesion. Others hold that the long-range attractive forces, such as result from summed bimolecular interactions or as estimated by the Lifshitz theory, are large enough to account for the observed adhesion across the gap [8]. If the latter view prevails, it will be necessary to include in the molecular definition of "adhesion" the concept of molecular forces acting across a thin, fluid-containing region that separates two macroscopic bodies.

2. 3. Definitions of "Adhesion"

The noun "adhesion" can now be defined as follows.

1. *Macroscopic definition:* The condition in which two macroscopic bodies adhere to each other, or one to the other, in accordance with, definitions 1 to 4 in Sec. 2.1.
2. *Microscopic definition:* The condition in which two phases adhere to each other, in the microscopic sense specified in definition 3 of Sec. 2.2.

Comment

We do not include the microscopic senses of "to adhere" specified in definitions 1 and 2 of Sec. 2.2 in the definition of the noun "adhesion" for two reasons. First, as already noted, even though a gas atom or molecule which strikes a solid and is not specularly reflected, but remains close to the surface, may legitimately be said "to adhere," its condition (while in close proximity to the solid) is always referred to as a state of adsorption, not a state of adhesion. Second, we are concerned in this examination of the semantic problem with the interaction between quantities of matter which are large enough to qualify as phases. There is no *semantic* problem in the interaction of a single atom or molecule with a solid.

3. DEFINITIONS OF ASSOCIATED TERMS

Much of the confusion that is apparent in the multiplicity of definitions cited in Sec. 1 can be resolved by formally assigning a number of concepts to other words, in accordance with common usage. Thus it is appropriate to separate out the concept of an adhesive joint (cf. above).

An *adhesive joint* is a system in which two effectively parallel interfaces exist: It consists of two solids of macroscopic dimensions, which may or may not be of the same material, and a layer of adhesive, which may be of microscopic thickness, but in any case must be thin relative to the dimensions of the two solids that are joined. The composition of the adhesive is different from that of either of the two bodies.

The *condition* of adhesion exists with respect to the first body and the adhesive, and with respect to the second body and the adhesive. The two bodies do not adhere to each other, but to the adhesive.

Strength of adhesion and *adhesive joint strength* are to be distinguished by the specification that, with respect to the former term, the system consists of two adhering bodies with an interface between them; and for the latter term, it consists of two macroscopic bodies with a thin layer of another material in between them. *Adhesion* between the adhesive and the bodies that are to be joined is a necessary, but not a sufficient, condition for the existence of a *strong adhesive joint*. The latter depends, in addition, on the geometry of the system and on the bulk mechanical properties of the three phases.

Is adhesion a material property? The macroscopic and microscopic phenomena of adhesion differ intrinsically, in two important ways. The first is that adhesion in the microscopic sense is not observed directly in the same way that the transmission of force or work may be observed. It is inferred (and its existence is accepted as scientific fact) on the basis of indirect evidence.

The second way that the phenomena differ is that on the microscopic level, adhesion is independent of interfacial geometry and loading. It depends only on the molecular forces and on the existence of an interface across which the forces act. The macroscopic phenomenon is observed to depend strongly on the geometry of the system, on its history, and on the direction, rate, and duration of loading.

Hence the question of adhesion as a "material property" has the following answer: On the microscopic level, adhesion is a two-material property: it can be specified on the basis of the physical properties of the two materials that are in contact at the specified interface. On the macroscopic level, adhesion is not a material property. In this, "adhesion" has a semantic status somewhat similar to that of "brittleness": Brittle behavior, as is well known, depends strongly on specimen geometry, e.g., notching, and on the direction and rate of loading. It cannot generally, in practice, be specified on the basis of composition and of molecular forces alone. Unlike adhesion, brittleness has no microscopic definition.

Quite recently, Baszkin and Ter Minassian-Saraga [9] in France have introduced the word "adhesivity." By this they mean a material property of a single substance or body. It may be a property intrinsically associated with a substance, as is "tack" or "tackiness." (Tack is a property that depends on the intrinsic viscoelastic behavior of a material, together with its propensity for wetting other materials. It is relatively insensitive to the properties of the other body which is brought in contact with the tacky body.) Adhesivity is also a property that may be developed by surface treatment—for example, surface oxidation—or grafting of a wettable or reactive polymer onto a polyolefin. Thus, it is difficult to find adhesives or coatings that will adhere to an untreated polyolefin with a high strength of adhesion; but after the treatments just noted, it is found that a considerable number of coating or adhesive materials will adhere fairly strongly. So it makes sense to define a property which qualitatively measures the probability that strong adhesion will exist in the combination with an arbitrary second body or material.

Recently, cell biologists have started using the term "adhesiveness" to describe the behavior of cells in aggregates—whether living tissues, or clumps that form spontaneously when cells are suspended in an appropriate medium [10]. A hierarchy of adhesiveness has been established for different kinds of cells. We will not here go into the biological details, but it seems that this biological property is closely related to, or identical with, that described above, under "adhesivity."

The thermodynamic property *work of adhesion* is defined by the Dupré equation [11]:

$$W_{12}^{a} \equiv -\Delta G_{12}^{a} = \gamma_1 + \gamma_2 - \gamma_{12} \qquad (2)$$

where ΔG^a is the free energy change when 1 cm^2 of interface between phases 1 and 2 is created out of 1 cm^2 of free surface of phase 1 and 1 cm^2 of free surface of phase 2; γ_1 is the specific surface free energy of phase 1, γ_2 is that of phase 2, and γ_{12} is that of the interface between phases 1 and 2. If (but only if) a phase is a liquid, the surface free energy is identical with surface tension. The corresponding *work of cohesion* of phase i is

$$W_i^c \equiv -\Delta G_i^c = 2\gamma_i \qquad (3)$$

The thermodynamic work of adhesion is not directly related to the mechanical strength of adhesion or to the strength of an adhesive joint. It is found empirically that estimates of strength [12,13] based on Eqs. (2) and (3) are in disagreement with observations by several orders of magnitude. This is also true of estimates made using the Young-Dupré equation

$$W^a = \gamma_{\ell v}(1 + \cos \theta) \qquad (4)$$

The Young-Dupré equation is derived [11] by combining Eq. (2) with the Young equation:

$$\gamma_{SV} - \gamma_{S\ell} = \gamma_{\ell V} \cos \theta \tag{5}$$

Here θ is the equilibrium contact angle formed by the liquid surface on the solid. The qualitative trend of W^a with $\cos \theta$ is in accord with Eq. (4), but the magnitude of W^a predicted is in error by several orders of magnitude [14].

The thermodynamic work of adhesion is directly related only to the adhesion as defined above (Sec. 2.2.3), the *microscopic* definition of adhesion between two phases. Its relation to the practical adhesion of macroscopic systems is only through the process of wetting: It is well known that if one phase is liquid when the phases are brought into contact, wetting with zero contact angle is very beneficial to the forming of intimate molecular contact and to the development of good mechanical strength in the system. (See below, Sec. 4.)

It has been suggested to the author that a definition be included, for the term "thermodynamic adhesion" [4]. There does not seem to be a strong requirement for such a definition. If given, it would pertain only to the adhesion between liquids and to the quantities computed by the Young and Dupré equations. With respect to the force of adhesion of solids, it would be relevant only to the theoretical strength of an ideal adhering system in a very special geometry: plug separation, without any lateral contraction of the facing bodies before separation, and with alignment of the facing surfaces during separation to better than 10^{-10} cm/cm width [13].

The thermodynamic analysis of adhesion is adequately comprehended in the term "work of adhesion." There is no need for defining an additional "kind of adhesion," which would have no other meaning than that the work associated with it can be computed with the aid of the Young and Dupré equations. Thermodynamics does not provide the desired bridge between the microscopic and macroscopic aspects of adhesion, even when statistical thermodynamics is employed [13].

Cling is an important adhesion phenomenon which merits a separate definition, as: a special example of "to adhere." If one body is in the form of a sheet, film, or foil, the stiffness of which is low, but the strength of which (under tension parallel to its surfaces) is appreciable, then such a body is said to cling to another body if, when they are brought into contact, the body that is in sheet form conforms to the shape of the other body, either spontaneously or under external manipulation, and persistent contact is achieved over an appreciable part of the surface area of either body, and if appreciable energy or force can be transferred from one body to the other in shear.

4. DISCUSSION OF "MECHANICAL ADHESION" AND "HOOKING"

The ASTM definition of "mechanical adhesion," quoted in Sec. 1, refers to one kind of geometry that may be present in an adhering system. In comment 2 of Sec. 2.1.1.1, we demonstrated that this geometry might exist, and a fastening function be performed, in a system which nobody would consider as an example of adhesion. The physical relation between the geometry of such a system and its mechanical strength has been misinterpreted in the past. The nature of this misinterpretation will be described below.

The actual role of the adhesive in the phenomenon referred to in the ASTM definition of "mechanical adhesion" is to eliminate reentrant voids at the interface. When this elimination is complete, there is total continuity of the stress tensor, i.e., there are no longer any stress concentrations at the edges of voids. This contributes a great improvement in the mechanical strength of the system (see comment 3 on definition 1 in Sec. 2.1.1.1). One important mechanical property which is involved here is dilational strength of the adhesive. The boundaries of the void that is filled are essentially fixed; so shear deformation is constrained, and dilational strength is converted to tensile strength. Thus, a region of extreme local weakness due to stress concentration is replaced by a region of higher local strength. The contribution of the adhesive is far greater than what it could contribute by means of its practical bulk shear or tensile strength, or stiffness.

The mechanical strength of an interlocking or dovetailed system arises from geometric relations (and for chain loops, topological relations) in combination with the existence of certain bulk properties: high shear modulus, tensile strength, and stiffness. (See above, Fig. 6-2b, and the role played by tensile strength in region 3.) As opposed to this, an adhesive, when tested in bulk, is very often found to have relatively low stiffness and a low shear modulus, or low tensile strength even after cure.

As noted above, in true mechanical hooking or interlocking, an immediate result of a reversal of stress is the loss of contact in those regions where contact existed. (See the discussion of Figs. 6-1 and 6-2.) This is followed by establishment of contact in other regions of the surfaces of the two phases. This sequence of events needs to be distinguished from that in which, on load reversal, intimate contact is maintained. Such a distinction can most easily and logically be made by restricting the term "adhesion" to the latter situation, i.e., the condition in which truly intimate contact at the boundary exists *and persists under load*.

The words "by an interlocking action" employed by ASTM D907-70 imply that the contribution of the adhesive is analogous to that of the interlocking in a jigsaw puzzle, or the action of a hook, a dovetail joint, a rivet, or a chain link. The same implication is present in the discussion by Bikerman [15], under the heading of "hooking" adhesion. As we have just shown, this analogy is not a valid one.

The ASTM definition was probably intended by its authors with respect to adhesive joints; the critique just given was written with joint strength in mind. The same argument holds in every way with respect to strictly two-phase systems, i.e., the adhesion of one bulk solid to another, or (in a joint) the interaction between one solid body and the adhesive, considered without respect to the interaction of the second body with the adhesive.

In some adhering systems, the component which started in, or passed through, a fluid state during processing may reach a final condition of considerable brittleness. In such cases, there may occur a spontaneous fracturing near or at the interface as a result of the contraction of one phase or the differential contraction of the two phases after the attainment of rigidity. After this fracturing, there still may remain a considerable strength of fastening of one body to the other, on account of an interlocking action. But the mechanical state is no longer that of a system with a continuous stress tensor. The fact that an interlocking geometry may *also* be present, in an *adhering* system, does not vitiate the argument just given: The interlocking action is independent of adhesion, and so should not be defined as a form of adhesion.

The term or concept of "mechanical adhesion" has not turned out to be a particularly useful one, in the study or application of adhesion. The author, in three decades of reading the current literature in this field, has not found any instance where it has been usefully employed. Numerous workers in the field, in private conversation with the author or in symposium discussions, have indicated their inability to use the concept with any advantage.

The critique just given of the ASTM definition shows why the term has been so useless: It is based on a misinterpretation of the mechanics of adhering systems. This misinterpretation can be corrected by means of explicit discussion of the stress tensor. It can also be corrected with the aid of explicit consideration of the void-filling function of a practical adhesive, and of the need (or rather the lack of consistent need) for high stiffness, shear modulus, and bulk tensile strength in an adhesive.

The primary purpose of an ASTM definition is to aid in the design and fabrication of practical structures. This purpose is not served by a definition that gives the same name (even when qualified by adjectives such as "mechanical" vs. "specific") to unrelated phenomena. Since this irrelevance of the term is recognized implicitly by the engineering and scientific community, it is recommended that the term "mechanical adhesion" be eliminated.

5. GENERAL DISCUSSION

A primary requirement which must be observed particularly carefully with technical definitions is freedom from circularity. A notable example of circularity is to define "adhesion" as a "sticking together." When "sticking" is looked up in the same source, it is found that it is defined as "adhesion." The mechanical definitions given above eliminate this logical defect: They are given in terms that are independent of the phenomenon.

There is a physical-semantic need for a microscopic definition of "adhesion" in addition to the macroscopic, besides the reasons already discussed. This may be seen from the fact that by the nature of an interface, there is a very steep gradient of mechanical properties over a microscopic region. If there were no examination of the molecular interactions, it would be necessary to treat this steep gradient as a discontinuity (infinite gradient) of properties. If there were a real discontinuity of mechanical properties, then it would be impossible for the stress tensor to be continuous across the interface. If, then, "adhesion" were defined as continuity of the stress tensor under deformation, that definition of "adhesion" would be circular.

A final and very important comment must be made on dualism: Do the macroscopic and microscopic definitions correspond to two distinct phenomena? This can be answered by analogy to the modern theory of light and of the electron. An electron (and correspondingly, a photon) behaves like a particle in some experiments and like a wave in others. Yet there are not two distinct entities. In the one case, it is useful to think of the electron as analogous to a macroscopic particle, e.g., a baseball with an electric charge; and in the other case, one may usefully employ the analogy of a wave on water. For adhesion, there are two different kinds of analogies that are (in fact, though not explicitly) employed by workers in the field. With respect to the microscopic definition, the analogy is to a chemical reaction involving two atoms or molecules. (See the ASTM definition of "specific adhesion.") With respect to the macroscopic definition, the analogy is to a *one*-phase solid which has appreciable tensile and/or shear strength. A full understanding of adhesion requires a comprehension of both the microscopic and the macroscopic aspects, just as a full understanding of the electron requires a comprehension of both its wave and its particle character.

Lexicography is the assemblage of definitions of words according to usage. Semantics is the deep examination of the meanings of words; and by means of semantic studies, circularities of definitions may be revealed, as well as incompatibilities such as the simultaneous use of two inconsistent definitions. In the case of "adhesion," semantics allows us to establish the fact that, *physically*, adhesion involves both chemical, microscopic interactions and macroscopic bulk interactions, and hence that our definitions must be such that we can think of both kinds of interactions simultaneously. Otherwise, we are very likely to think in terms of only one kind of interaction to the exclusion of the other.

A severely abridged form of the material in this chapter was published in the *Journal of Adhesion* in 1976 [16].

ACKNOWLEDGMENT

The author thanks Dr. Herbert Reismann for discussion of the manuscript, and particularly for his contribution to the discussion of mechanics.

REFERENCES

1. ASTM D907-70.
2. V. G. Raevskii and L. M. Pritykin, RAPRA - 9(12)4.
3. G. Salomon, in *Adhesion and Adhesives*, Vol. 1, 2nd ed. (R. Houwink and G. Salomon, eds.), Elsevier, Amsterdam, 1965, p. 3.
4. D. D. Eley and D. Tabor, in *Adhesion* (D. D. Eley, ed.), Oxford University Press, London, 1961, p. 2.
5. R. J. Good, in *Aspects of Adhesion*, Vol. 6 (D. J. Alner, ed.), University of London Press, London, 1971, p. 29.
6. V. A. Parsegian and B. W. Ninham, Biophys. J. 10:664 (1970).
7. D. J. Robertson, Progr. Biophys. Biophysical Chem. 11:344 (1960).
8. R. J. Good, J. Theoret. Biol. 37:413 (1972).
9. A. Baszkin and L. Ter Minassian-Saraga, Comp. Rend. 268C:315 (1969); J. Polym. Sci., Ser C 34:243 (1971); J. Coll. Interfac. Sci. 43:191, 473, 478 (1973); A. Baszkin, personal communication.
10. M. S. Steinberg, Science 137:726 (1962); J. Exp. Zool. 173:395 (1970).
11. N. K. Adam, *Physics and Chemistry of Surfaces*, 3rd ed., Oxford University Press, London, 1941, p. 8.
12. W. D. Harkins, *Physical Chemistry of Surface Films*, Reinhold, New York, 1952.
13. R. J. Good, *Treatise on Adhesion and Adhesives*, Vol. 1 (R. L. Patrick, ed.), Marcel Dekker, New York, 1967, p. 50.
14. R. J. Good, Proc. Macromol. Secretariat, 169th Nat. Meeting Amer. Chem. Soc., Philadelphia, 1975; *Adhesion Science and Technology*, Vol. 9A, Plenum, New York, 197.
15. J. J. Bikerman, *The Science of Adhesive Joints*, 2nd ed., Academic Press, New York, 1968.
16. R. J. Good, J. Adhesion 8, p. 1 (1976).

7

Permanence and Endurance of Structural Adhesive Joints

Armand F. Lewis and R. T. Natarajan Gounder*

Lord Corporation
Erie, Pennsylvania

1. INTRODUCTION

Among the most important considerations in designing engineering structures employing adhesive joints are the permanence, endurance, and reliability of the joints with regard to the service performance of the structure. Much literature exists on the subject of failure criteria of materials in general [1] and adhesive joints in particular [2,3]. Many factors affect the permanence, endurance, and reliability of adhesive joints. Such factors as temperature, mechanical stress, and environmental aging effects are the subject of study in many adhesive materials and engineering research and development laboratories. In this chapter, a systematic overview of the subject of performance and endurance of adhesive joints will be presented. This should serve the purpose of reviewing the progress being made in the field and perhaps help to guide future work.

In most load-bearing structures of devices employing adhesive joints, the adhesive materials used are thermosetting systems. By nature, these adhesives are chemically irreversible. Such joints, therefore, are designed to function as permanent entities in the structure. By *permanence* of the joint is meant that it will continue to function or endure in the same state, place, or the like without any marked change. To the extent of this definition, the joint is an integral part of the structure and should not alter its behavior or deteriorate in its function, whether the structure or device is in its operating mode or not. From this standpoint, then, one must refer to the overall permanence of an adhesive joint in terms of its role of existing in its total, intended service-life environment. *Endurance*, on the other

*Present affiliation: RCA Laboratories, Princeton, New Jersey

hand, refers to the ability of the adhesive joint to perform its intended joining action and withstand its limit of operational stresses without behavioral deviation or failure in a functioning environment. Here, the joint is being acted upon by the external "forces" imposed upon it by the service conditions, e.g., mechanical stresses (static and/or dynamic), thermal fluctuations, and thermooxidative environments, corrosive atmospheres, etc. From this the *endurance limit* of the joint can therefore be defined as the critical limit in service conditions' intensity variable (e.g., highest temperature, highest mechanical stress, etc.) below which the joint will not suffer any detrimental changes that will lead to eventual failure of the adhesive joint. Furthermore, *reliability* refers to the probability that the adhesive joint will succeed (without failure, deviation, or alteration) in performing or enduring in its intended service environment.

A systematic overview of adhesive joint permanence can be initiated by examining the problem from first principles. Hence, the basic problem of the materials developer in designing a structural adhesive for engineering performance is identified as being the prediction, by laboratory tests and theoretical and/or mathematical modeling, of the performance behavior of an adhesive joint in a real service-life environment. Of course, in all instances service-life information can be obtained by long-term performance in actual service. However, this is time-consuming. What is needed are realistic accelerated, predictive test programs that are carried out in the laboratory under intensified conditions. However, this in itself is paradoxical. The only way to truly prove out a predictive test is to compare its predictions with real service experience on the real system or structure. If, after some time, the adhesive performs well and shows no signs of deterioration, the credibility of the developed accelerated predictive test is established. In effect, the completion of this circle of understanding, substantiation, and credibility of any particular test program is a long-term detailed task. Such a task, along with the development of useful engineering design parameters, is the goal of many present-day researchers in the field of structural adhesives.

With this philosophy as background, a materials engineering systems approach can be taken to describe the parameters and variables necessary in defining and stimulating the adhesive joint in its use environment. In this systems analysis, direct and indirect actions and interactions of all the various parameters may be listed and factored. A block diagram illustrating this approach is presented in Fig. 7-1. In interpreting this diagram, it can be assumed that from specific knowledge, understanding, and proper identification of the service parameters, a perception of the real conditions can be estimated using either the mathematical modeling or a simulation (laboratory test) technique. As shown, for exactness, the real service-life situation can

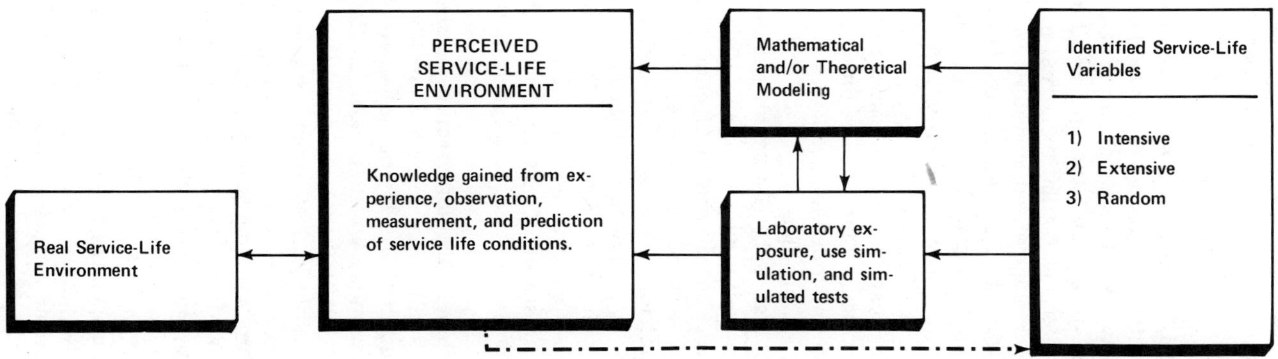

FIG. 7-1 Systems approach to studies of permanence, endurance, and reliability of adhesive joints and materials (see Table 7-1 for details).

be perceived only from present and prior knowledge and understanding. One can never be completely certain that he has identified all the service variables, especially when ultralong service-life conditions are being considered.

Basically, Fig. 7-1 illustrates the elements of systematic scientific thought—namely, theoretical modeling of a problem being carried out concurrently with laboratory experimentation. One approach serves as a cross-check on the other, and thought reinforcement results; a clearer view of the problem of simulating a real situation is formulated. Approaches that consider theoretical modeling and experimentation individually have limitations. Here all the parameters necessary to define a problem may not be identifiable or tractable using one approach by itself. Using the mathematical modeling/laboratory simulation systems approach, the impact of identified variables influencing the behavior of the real system can be factored and their relative importance considered in greater detail. This can result in a more complete understanding of the problem and perhaps lead to a more rapid optimization of the approach to solve a problem.

As a supplement to Fig. 7-1, the list of service-life variables in Table 7-1 shows that adhesive joints employed in engineering applications are subjected to a great diversity of conditions. As structural entities, the design and service behavior of adhesive joints are dictated by the multitude of condition variables that might affect the constituent parts of the joint, i.e., adhesive and adherend material. These interactions are complex; however, to complicate matters, boundary layers or macrointerfacial phases exist in adhesive joints. These phases may be considered as distinct states of matter having their own physical, chemical, and mechanical behavior. Since this boundary layer is composed of contiguous layers of adhesive and adherend surface layers, it is chemically and rheologically anisotropic. Furthermore, adhesive-adherend couples are usually characterized by a modulus "mismatch" between adhesive and adherend. If the adhesive is being used to bond two different adherends, an even more complicated (three-material) modulus mismatch condition results.

With structural adhesive joints, stress concentration factors are important criteria in structural design planning and specification activities. Engineering designers specifying adhesive joint fastening methods recognize this and also realize the need for more readily adaptable engineering design parameters. Perhaps the application of finite element analysis (FEA) methods [4-6] will provide the key to obtaining this much-needed information. However, at present, only the mechanical strain parameters in adhesively joined macrostructures can, to a degree, be adequately analyzed. FEA methods are not yet adaptable to detailing strains of microscopic (molecular) significance. The influence of other service-life variables such as moisture penetra-

TABLE 7-1 Some Typical Service-life Variables for Adhesive Joints

Intensive—Primary	*Intensive—Specific*
A. Mechanical stress (static, dynamic, impact)	A. General industrial—ambient temperature, CO_2, H_2O, oil vapor, others
B. Temperature	B. Aerospace—fuel, radiation, vacuum, atmosphere (moisture, freeze-thaw cycling), others
C. Physicochemical environment	C. Hydrospace—hydrostatic pressure, hydrodynamic forces, marine corrosion and fouling, others
Extensive	*Random*
A. Adhesive material	A. Careless workmanship
B. Adherend material	B. Voids in bondline
C. Joint design	C. Improper surface preparation
D. Extent bond line	D. Improper cure of adhesive
E. Bond line thickness	E. Poor joint alignment
F. Interfacial considerations	F. Murphy's law variables
G. Others	G. Others

tion (corrosion) and thermal degradation effects of the adhesive materials is far from understood or even capable of being examined by such methods. It is expected that phenomenological approaches will be needed. However, as an initial step, the physicochemical cumulative damage (statistical) theory of mechanical failure of Halpin and Polley [7] and Halpin [8] may be considered. In this approach, the rate of performance degradation of a material is analyzed as a systematic function of such stress variables as temperature, mechanical forces, and environmental (chemical or corrosive) stresses.

As presented in Table 7-1, structural adhesive joints are subject to influence by a great number of external condition variables. For the purpose of clarity, the nature of these variables has been categorized as being *intensive*, *extensive*, and *random*. Figure 7-1 illustrates the interpretive impact that these variables have on the experimental and theoretical approaches that can be applied to predicting the real service-life conditions. This, then, serves to diagram the analytical strategy of this chapter. In this chapter, the theoretical and experimental approaches that have been used to describe the effect of primary intensive variables of mechanical stress, temperature, and physicochemical environment (chemical/corrosion) on adhesive joint permanence will be reviewed. Where appropriate, the interactions and influences of the extensive variables (materials, geometry, design) will be considered. The role of the random variables, although important from a statistical margin of safety viewpoint, will not be detailed since they are not subject to scientific examination. The topic of reliability will not be considered except in the context of defining, on a quantitative basis, the endurance limit of an adhesive joint system based on an identified critical failure mode.

2. STRESSES IN ADHESIVE JOINTS

One of the most important intensive variables in the design performance of an adhesive joint is the stress environment into which the joint is to function. Assuming a proper joint is made interfacially, such factors as materials and mechanical properties (low strain and especially high strain and fracture properties) as well as joint design or geometry are of considerable interest in the scheme of service-life prediction. The stressing of materials can be divided into two categories; response and failure. Both categories relate to the intensity of the stress field imposed on the adhesive material.

2.1. Mechanical Response of Polymer Materials

The response aspects of a polymeric material may be readily discussed in terms of the analysis of Kelley and Williams [9]. Here the deformation response of a polymer through a broad time or temperature scale is interpreted in terms of the relaxation spectrum of the polymer. One form of this relaxation spectrum can be expressed (see Fig. 7-2) in terms of the relaxation modulus $E_{rel}(t)$ as

$$E_{rel}\left(\frac{t}{a_t}\right) = E_e + \frac{E_g - E_e}{[1 + (t/a_t)/\tau_0]^n} \qquad (1)$$

where t/a_t is the reduced time, incorporating the Williams-Landel-Ferry (WLF) [10] shift factor a_t; E_g and E_e are the modulii of the polymer in the glassy and rubbery range, and n and τ_0 are the slope and characteristic relaxation time associated with the midpoint of the glass-to-rubber transition range. From the power law of Eq. (1), one notes that for engineering purposes the viscoelastic response of polymeric material (adhesive) may be characterized by five parameters; viz., E_g, E_e, n, τ_0, and T_g. The glass temperature T_g is contained in the a_t factor from the WLF equation, $\ln a_t = -40(T - T_g)/[52 + (T - T_g)]$. Kelley and Williams [9] have analyzed Eq. (1) in terms of a so-called interaction matrix involving the five response parameters and ten polymer "microstructural

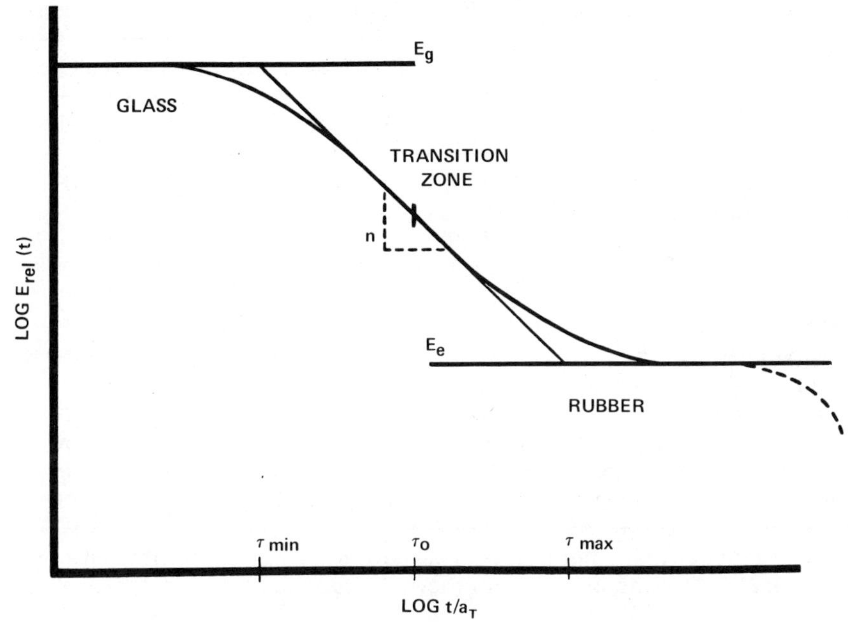

FIG. 7-2 Tensile relaxation modulus as a function of reduced times.

descriptors." An abbreviated version of their interaction matrix is presented in Table 7-2. The interaction of these power law parameters and microstructural descriptors indicate a systematic approach to evaluating the relative importance of each of these parameters in influencing the shape of the relaxation curve. As later stated by Kelley and Trout [11], this interaction matrix may also be useful for sorting out some of the important aging contributions in the material system (as a function of thermooxidative aging, for example).

Basically, the use of the interaction matrix approach provides a systematic tabulation of the factors that are needed to develop useful molecular structure-property relationships; unknowns in the understanding of the effect of molecular parameters on macroscopic response behavior may also be brought to light. From Table 7-2, it is shown that crosslink density, molecular weight, molecular weight between entanglements, degree of crystallinity, and volume fraction of filler and of plasticizer all have a strong effect on the modulus of polymers in the rubbery state, E_e. This information would then provide some guidelines for the laboratory development of adhesive materials with high temperature strength. Furthermore, if one is dealing with the creep behavior of adhesive joints in service, he may wish to learn more

TABLE 7-2 Engineering Properties of Polymers: Interaction Matrix

	Modified power law parameters[a]				
Molecular characteristics	E_g	E_e	τ_0	n	T_g
Crosslink density	N	S	N	M	N
Chain stiffness	N	N	U	M	S
Monomeric friction coefficient	U	N	S	U	S
Solubility parameter	M	N	U	U	S
Molecular weight	N	S	N	N	S
Heterogeneity index	N	N	M	N	M
Molecular weight between entanglements	N	S	N	N	N
Degree of crystallinity	N	S	S	S	N
Volume fraction of filler	N	S	M	S	M
Volume fraction of plasticizer	N	S	S	N	S

[a]Key to response: N = negligible interaction; S = strong interaction; M = moderate interact.on; U = unknown response.

Source: Ref. 9.

about the effect of chain stiffness and solubility parameter on the characteristic relaxation time τ_0. It is felt that more use should be made of the interaction matrix approach to the formulation and development of materials for adhesive joint systems and for the establishment of relevant R & D programs in the field.

2.2. *Fracture of Adhesive Polymers*

The fracture of polymers has been the subject of numerous papers and review articles [12,13]. Some of these publications have been directed toward describing the fracture of glassy adhesives. The foundation of all present-day theories of fracture of materials is still, however, the Griffith crack model of failure [14]. The Griffith model, adapted from the work of Inglis [15], states that the tensile strength of a brittle (glassy) material is said to depend upon the existence of inherent cracks and flaws in the material. The strength is considered to be proportional to the energy required to propagate a crack, flaw, or otherwise a concentrator of stress in the material under an *imposed stress field*. Accordingly, the condition for spontaneous crack growth and therefore separation of the material is

$$\frac{\partial}{\partial a}(4a\gamma - \varepsilon) > 0 \tag{2}$$

where a is the crack length, γ the fracture surface energy, and ε is the stored elastic energy; $\varepsilon = \pi^2\sigma_0 a^2/E$, where σ_0 is the stress applied to the crack and E is the Young's modulus of the material. According to Eq. (2), the inherent crack in a material will propagate if the rate of release of the stored energy in the vicinity of the crack is greater than that energy required to create new surfaces at the crack tip. The final equation for tensile strength σ_t derived by Griffith takes the form

$$\sigma_t = \left(\frac{2E\gamma_c}{\pi a}\right)^{1/2} \tag{3}$$

where γ_c is termed the critical fracture surface energy. A detailed analysis of the Griffith equation has been made by Berry [16].

Further physical significance of the γ_c term has been put forward by Orowan [17], whose modification of the Griffith equation resolved the problem in that the Griffith approach fails to adequately describe the tensile failure of more ductile (than glass) materials. Orowan's modification was to separate the fracture energy terms into two elements: γ_s, the energy to create new surface, and P, the energy for plastic deformation. Orowan then redefined the Griffith equation as

$$\sigma_t = \left[\frac{2E(\gamma_s + P)}{\pi a}\right]^{1/2} \tag{4}$$

In another advance of the Griffith approach, Irwin [18,19] derived the Griffith-type formulation based on a critical stress criterion. Irwin observed that the stress field in the vicinity of the crack tip could be defined by a single stress concentration parameter K, proportional to the stress intensity factor. When the stress at this crack tip reaches a critical value, the crack will begin to propagate abruptly. This critical value of stress concentration K_c then defines the fracture toughness of the material and was considered by Irwin to be a material property. From the Irwin analysis, the important conclusion is reached in that crack opening always occurs at a given value of local stress intensity, regardless of the structural design into which the material is used. There are, however, some characteristic behaviors that should be mentioned. For example, with viscoelastic materials, K_c is found to depend on temperature and strain rate. Furthermore, since crack propagation can occur under conditions of either tension or shear, it is expected that K_c also depends on failure mode.

Experimentally, the critical stress intensity factor K_c can be related to the Griffith fracture theory by a second parameter; a term called the critical energy for crack extension, G_c (also known to be proportional to the fracture toughness) was developed and refers to the energy required to propagate a crack in a material. The fracture toughness or Irwin critical energy for crack propagation term G_c is purported to have important engineering significance. The connection between this term and the Griffith fracture surface energy term is a direct one, namely, $G_c = 2\gamma_c$; as seen, this equation parallels that of the work of cohesion of a liquid, $W_c = 2\gamma_\ell$ where γ_ℓ is the surface tension of the liquid. In each case, the term implies that it is the work necessary to create 2 unit areas of new surface—of solid surface of surface energy γ_c (or of liquid surface of surface energy, γ_ℓ). Of importance, the Irwin interpretation has brought a new status to the meaning of fracture. This approach has paved the way to the present situation that fracture toughness of materials in adhesive joint configurations can now be measured experimentally for various modes of crack openings. The most common mode of measuring fracture toughness of materials and adhesive joints is the cleavage mode using the beam method of Ripling et al. [20,21]. The symbolism used for this parameter is G_{IC}, which is the critical energy for crack propagation in a material in the tension or cleavage mode.

This theoretical and experimental work of Griffith, Orowan, and Irwin has laid the foundation for modern-day fracture mechanics of materials. Concurrently, the interpretation of the fracture of structural adhesive joints is, of course, based on this crack propagation model. The most extensive experimental work on the fracture mechanics of adhesive joints has been carried out by Mostovoy and Ripling [22-25]. The effects of adhesive materials, corrosion effects, and dynamic fatigue behavior have been studied using these methods. The field is becoming well established and is undergoing mature development. Of significance is the fact that this fracture mechanism method of Mostovoy is now being employed in the application study of adhesive materials in engineering laboratories as well as adhesive suppliers laboratories [26,27]. Furthermore, an ASTM (tentative) standard test is presently being established.

2.3. *Static Fatigue of Adhesive Joints*

The response of an adhesive joint to stress in a use environment is complex because the stressing mode is a random experience in both periodicity and intensity. However, for the purpose of analysis, one may categorize these imposed stresses as being static or dynamic. The distinction between the two is relative depending upon the time scale within which the stress load is applied. If a steady (dead) load is applied to a structural material, and the material responds to the load by deforming to the extent that permanent distortion (tensile, shear, or compression set) results, the term "static fatigue" is used to describe such a phenomenon. In the case of oscillatory or stepwise repetitive impulsive stresses caused by vibration during operation or even in random or periodic functional impact, the term "dynamic fatigue" is used. In the case of adhesive joints, both static and dynamic fatigue situations are encountered in practice; and more often, as is the case with most functional engineering materials, dynamic stresses are superimposed on static stresses. Therefore, in order to understand the permanence and endurance behavior of an adhesive joint, one must consider the overall response of adhesive joint materials to stress. To add to this complexity, the behavior of the adhesive joint with respect to the other intensive variables, temperature, and physicochemical environment are also dependent upon the stress environment imposed on the joint.

A model qualitatively describing static fatigue was published by Lewis [28], and in essence summarizes, in qualitative terms, the failure mechanisms in materials. In this work, the behavior of a polymeric adhesive material in a static stress field has been classified according to the performance of the microstructure of the polymer under various environmental (relative) stress levels. Depending upon the stress level imposed on the adhesive material, its response to static stress can be categorized into four regimes of behavior:

1. *Hookean deformation.* When at very low stress levels, reversible deformation of the polymer occurs.
2. *Modulus damage.* At higher stress levels, the polymer undergoes an irreversible (isothermal) deformation, the inherent (Griffith) cracks distort and open, and such localized stresses relax and the crack opening stops by a blunting process. Such a mode results in a lowering of the modulus in the system similar to the "Mullins effect" observed in rubber [29]; it is a form of stress annealing.
3. *Strength damage.* At still higher stress levels, irreversible crack distortion and opening occurs leading to a steady cumulative damaging of the internal defect structure in the adhesive. Such a mechanism will lead to eventual failure of the adhesive joint. Finally, an incipient failure zone is proposed.
4. *Terminal damage.* At stresses at or close to the breaking stress of the adhesive joint, cracks open, grow rapidly, and coalesce; at this stage, failure of the joint is eminent. These various regimes of damage are summarized in Table 7-3.

It should be pointed out here that when an adhesive joint (or structural material in general) is stress-loaded to a condition anywhere

TABLE 7-3 Rheological Response of Adhesive Material under Static Stress

Designation	Description	Effect of damage on ultimate strength
0—Hookean	Reversible elastic response	None
I—Modulus damage	Modulus lowers; self-strengthening	None or slight increase
II—Strength damage	Irreversible crack opening; continue to grow	Lower strength
III—Terminal damage	Catastrophic failure; cracks coalesce	Drastic lowering

in the *strength damage region* and especially near the *terminal damage region*, the material traverses all the "lower-order" rheological damage conditions and rapidly accumulates all these "lower-condition" damage characteristics. Therefore, the rheological damage pattern or configuration of an adhesive material in a joint depends on previous history. One might expect that in viscoelastic materials, molecular relaxations take place. If the viscoelastic materials enter the *strength damage region* rapidly, due to the inhibition of molecular relaxation, the material may reach the terminal or fracture zone at a high-stress level and at a much more rapid rate than if the stress were imposed at an equilibrium-accommodating loading rate. On the other hand, if the viscoelastic material was susceptible to creep or flow under the stress conditions, the material would fail at a lower stress; failure is rate-dependent. These multioperational, time-dependent rheological damage conditions complicate the systematic study of the behavior of viscoelastic materials under large deformation, a subject that has not yet yielded to suitable analysis.

These views of rheological damage in materials present a classically qualitative view of the molecular dynamics of static deformation and fatigue. One can now proceed further by attempting to look at static fatigue in adhesive joint materials analytically by interpreting the deformation/damage event in terms of the creep involved. Here, viscoelastic models can be conveniently employed. Let us assume that the adhesive polymer (in the joint) is under a low shear stress level where the material can respond in a reversible manner; no damage occurs. Interpreting this response in terms of a single Kelvin-Voigt element (parallel spring and dashpot) creep (or more accurately, deformation) in the system will occur according to the following equation:

$$\varepsilon_1 = \frac{\sigma_S}{G_1}(1 - e^{-t/\tau_0}) \tag{5}$$

where ε_1 is the shear deformation (strain), G_1 the shear modulus of the spring element, σ_S is the static load applied to the system, and τ_0 is the relaxation time (the time constant for exponential decay) and t the elapsed time.

Figure 7-3a diagrams the creep response and recovery of a linear viscoelastic material as presented by Eq. (5). As shown, when stress

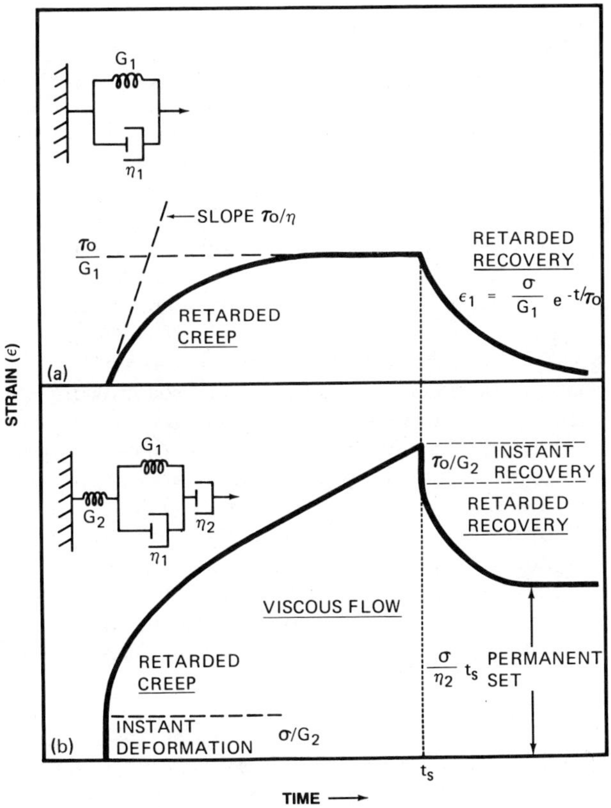

FIG. 7-3 Creep response of Kelvin-Voigt models.

is applied, the material deforms to a steady strain level; when the stress is released, the polymer fully recovers from its excursion into the stressed state. If the stress σ_s is small, and no molecular damage occurs, the system responds in a reversible manner. An adhesive joint material having a stress response demonstrated by Fig. 7-3a would be stressed below its endurance limit. However, at high stress loadings, a mode of steady-state creep may occur which may ultimately lead to separation of the joint. Such irreversible behavior cannot be demonstrated by Eq. (5) (the simple Kelvin-Voigt model). However, a four-parameter (composite) model, combining a Kelvin-Voigt element with a Maxwell element (spring and dashpot in series) must be employed. The creep response of such a four-parameter model is given in Fig. 7-3b, and may be described by the following equation:

$$\varepsilon_c = \varepsilon_1 + \varepsilon_2 = \frac{\sigma_s}{G_1}(1 - e^{-t/\tau_0}) + \frac{\sigma_s}{G_2} + \left(\frac{\sigma_s}{\eta^2}\right)^t \qquad (6)$$

where ε_c is the creep strain of the composite model and is the sum of the strain contributed by the Kelvin-Voigt element ε_1 and the Maxwell element ε_2. The σ_s/G_2 term is a measure of the instant-aneous strain of the spring in the Maxwell element while the $(\sigma_s/\eta_2)^t$ is a measure of the linear flow of the dashpot or viscous component of the Maxwell element.

From Fig. 7-3b, in a mode of deliberate, continuous stress application, the polymer will continue to flow (creep) as long as the stress is applied. Unless the stress is released, in a practical case, an unstable (failure) condition will be reached. While this condition of instability is not predicted by this modeling approach, the composite model is surprisingly good for describing the response behavior of a polymeric material to changes in molecular weight and crosslinking [30]. More refined multiple-element models have been developed to describe more fully viscoelastic behavior of polymers. For creep tests, a generalized Kelvin-Voigt model is used involving the linking of a multitude of Kelvin-Voigt elements in a series. This generalized model takes into consideration that the polymer cannot be accurately described by one characteristic retardation time but a continuous distribution of retardation times. From this, a characteristic "retardation spectrum" for a polymer is derived resulting in an equation describing the low stress-strain response behavior of viscoelastic materials. The analogous derivation is made for a multitude of Maxwell elements arranged in parallel. From this, assuming a continuous distribution of relaxation times, a characteristic "relaxation spectrum" for a polymer is obtained. This approach forms the basis for deriving the relaxation modulus equation defined in Eq. (1) of this chapter. For details on this subject, the reader is referred to books on polymer physics covering the subject of theory of linear viscoelasticity [29,31].

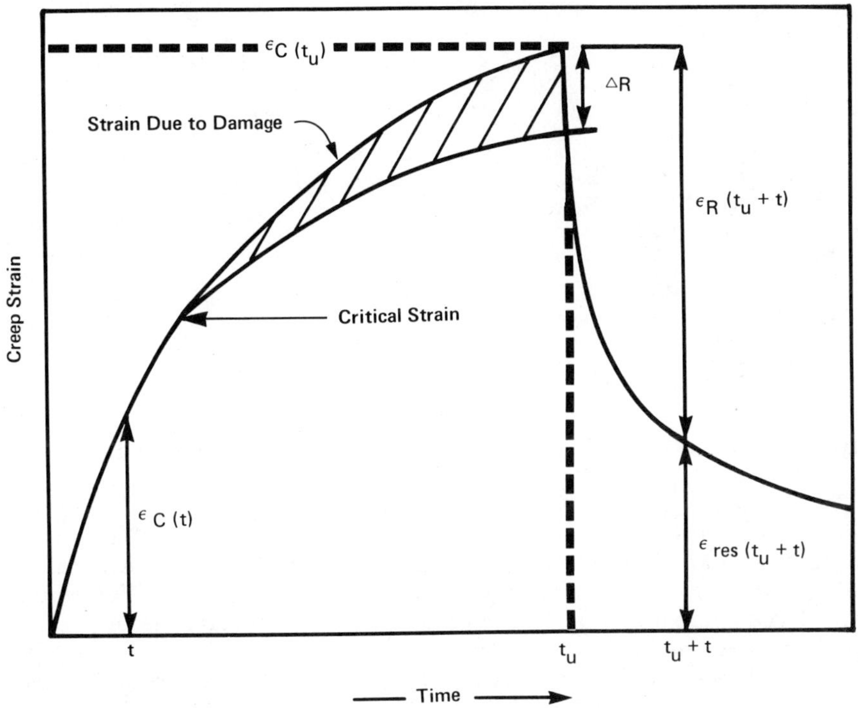

FIG. 7-4 Wright (see Ref. 32) description of creep/craze damage in polymeric materials.

Related to creep mechanism in polymeric materials, Wright has recently employed creep measurements to measure the initiation and growth of microdamage in plastics [32]. As shown in Fig. 7-4, creep curves (creep strain) were obtained along with the recovery creep strain in order to define a damage parameter ΔR. It was observed by Wright that a damaged (surface-crazed, for example) plastic material recovers more rapidly from a creep-strained condition than an undamaged specimen; one can then proceed to define ΔR according to the following equation:

$$\Delta R = \varepsilon_R(t_u + t) - \varepsilon_c(t) \qquad \text{for } t << t_u \tag{7}$$

where $\varepsilon_c(t)$ equals the creep strain at time t, and $\varepsilon_R(t_u + t)$ is the recovery strain at time t after the straining load has been removed. Related to adhesive joint strength, it is presumed that such an experi-

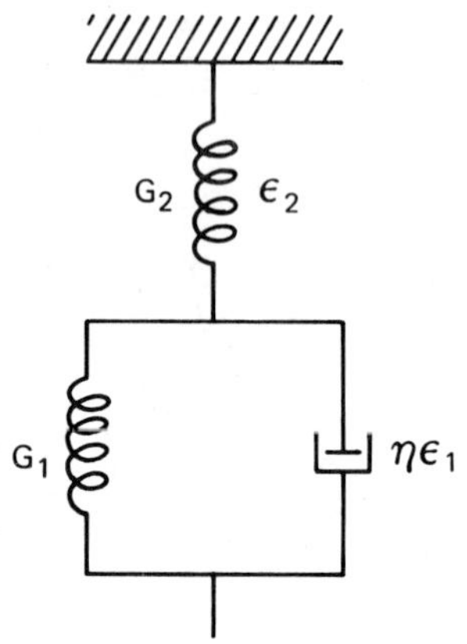

FIG. 7-5 Hata model for mechanical response of adhesive joint.

mental technique can be applied to measuring the damage in adhesive joints due to an imposed static fatigue environment. The ΔR term may indeed serve as a true rheological damage parameter which would reflect the micromechanical damage behavior of adhesive materials under high-stress load conditions.

Viscoelastic models have been used to demonstrate the mechanisms of failure in adhesive joints in a unique way by Hata [33]. In a three-element model, composed of a Kelvin-Voigt element coupled with a spring, the dependence of adhesive strength on the rate of separation, temperature, and type of failure in the joint was analyzed. A useful interpretation of adhesive joint strength resulted. The Hata model, shown in Fig. 7-5, states that the mechanical response of the adhesive material is given by the Kelvin-Voigt model whose modulus and viscosity are given by G_1 and η (strain ε_1), and the interfacial (boundary layer) force is represented by a spring of modulus G_2 (strain ε_2). In the case where it can be assumed that the adherend in the subject adhesive system is rigid, undergoes no deformation during stressing of the joint, stress on the joint can be given as

$$\sigma = \varepsilon_2 G_2 = \varepsilon_1 G_1 + \frac{\eta d\varepsilon_1}{dt} \tag{8}$$

Assuming this stress-strain relation holds up to failure, Hata proposes that a failure criterion can be introduced by stating the failure occurs at the interface when the strain in the surface spring ε_2 reaches a critical value ε_{2_c}, or else, in the adhesive layer, ε_1 reaches a critical value ε_{1_c}. Therefore,

$$\sigma_b^S = \varepsilon_{2_c} G_2 \qquad \text{Interfacial failure} \tag{9}$$

$$\sigma_b^B = \varepsilon_{1_c} G_1 + \eta \left(\frac{d\varepsilon_1}{dt} \right)_{\varepsilon_1 = \varepsilon_{1_c}} \qquad \text{Cohesive failure} \tag{10}$$

with σ_b^S and σ_b^B representing the breaking stress at the surface and in the bulk (adhesive), respectively, the failure mode will depend upon the rate of strain and the values of G_1 and G_2. The Hata model therefore serves to present a simple means of describing the visco-elastic behavior of an adhesive in a structural joint. While qualitatively presenting a good physical picture of adhesive joint strength behavior, continued development of this theory would require the measurement of the modulii and viscosity parameter in the interfacial boundary layer region of the adhesive-adherend couple. This would be an experimentally challenging task, but would certainly lead to a substantial advance in understanding the mechanical aspects of adhesive joint strength.

From this discussion of creep, mechanical response, and adhesive failure of polymeric materials by the use of mechanical models, one can easily picture the time-dependent response of viscoelastic materials with regard to their fracture behavior. Employing the four-element model (see Fig. 7-3b) and the failure criterion assumed by Hata, one can easily demonstrate the two mechanisms of failure that can be exhibited by viscoelastic materials: failure by (a) *a critical stress mechanism* or (b) *a critical strain mechanism*. This can simply be stated, following Hata's simplifying assumption, that if stress is applied at a rapid enough rate so that the dashpot does not respond quickly enough, the spring will take all the load and will stretch out and because of the nature of the spring material, at a critical stress, the "spring" will break. The material model is now "discontinuous"; the material has fractured depending on the attainment of a *critical stress* in spring G_2. Now, if the time for the application of the stress is slow and the dashpot responds, flow of the material ensues. Movement of the piston in the cylinder occurs until the piston becomes "decoupled" at a critical strain ε_{2_c}; the simple model of *critical creep strain failure* is demonstrated.

It is important to point out that although spring-and-dashpot models are useful for conceptualizing, they are gross oversimplifications of viscoelastic behavior. These models are for small deformations where Newtonian flow and/or Hookean molecular response prevails. In this respect, these models only apply to amorphous polymers [31]. The analysis of large deformation and terminal properties of visco-elastic materials by using mechanical models, as was done by Hata and is done here, must be considered a convenience; it's a way of qualitatively illustrating complex, time-dependent, mechanical response phenomena.

The relevance of this elementary review of creep response (deformation) of polymers to adhesive joint studies may not be readily apparent. However, considering the engineering applications for adhesives, creep of these material systems is of universal importance. While agreeing about the importance of creep in structural adhesives, too little basic research has been done on the subject; although some work has been published [34,35]. The problem with such studies lies in the difficult experimental technique that is required to properly measure microstrains in these composite structures, the complications in the modulus mismatch between adhesive and adherend, the minute strain levels that would characterize the properly stressed joint, and the complex strain patterns that result from geometric constraints. The application and development of holographic [36] and/or Moiré [37] photoelastic techniques to functionally stressed adhesive joints may provide some leads. A better understanding of micromechanical damage in structural adhesives would more than likely pave the way to the establishment of badly needed adhesive joint test methods that will provide information of significance to engineering design.

A systematic experimental study of the static strength and endurance limit of lap shear adhesive joints has been reported by Lewis et al. [38]. In this work, a modification of the Prot method [39] was employed as an accelerated test method for determining the endurance limit of adhesive joints. In this technique, the load on a material specimen is increased at a constant (programmed) rate until the specimen fails. This can be done using the pivot-beam-type apparatus or, if available for long-term testing, universal tensile test machines (Instron or MTS). In his original analysis, Prot assumed that sustained load test results were representative of a hyperbolic law expressed as

$$t_b = \frac{K^1}{\sigma_b - EL} \tag{11}$$

where t_b = time to break
σ_b = stress at break
EL = endurance limit
K^1 = material constant

Prot considered a variably applied stress starting at an arbitrary value ("preload") and increasing linearly with time. Thus, the stress σ at any given time t may be expressed in terms of the initial stress σ_i and the rate of increase of stress α in the equation

$$\sigma = \sigma_i + \alpha t \tag{12}$$

If σ_i is set equal to EL, the theory predicts that the stress at break, σ_b, depends linearly on the half-power of the rate of increase of stress, such that

$$\sigma_b = EL + K\alpha^{1/2} \tag{13}$$

A general theory, related to the Prot method, has been proposed by Loveless et al. [40]. In their work, it was assumed that under stress level conditions which are higher than the endurance limit, a specimen will suffer damage. This damage was considered to be proportional to the amount by which the stress exceeds the endurance limit and also to the time for which the stress acts. They derive the following equation for the endurance limit:

$$\frac{\alpha t_b^2}{2} = K - (\sigma_i - EL)t_b \tag{14}$$

where t_b is the time to break a test specimen that is stressed to failure at a rate α, K is a material constant, and σ_i is a so-called preload term or the stress at t = 0. The endurance limit can then be determined by plotting αt_b^2 vs. t_b and determining the slope of the straight line obtained which is equal to 2(EL - σ_i). The preload term σ_i is an experimentally adjustable parameter. This analysis was found by Loveless [40,41] and others [42] to present a useful interpretation of the endurance limit concept.

In full consideration of the assumptions that were made in the derivation of Eq. (14), endurance limit measurements were made under controlled temperature and humidity conditions (23°C and 50% RH) on aluminum lap shear joints (ASTM type 1002) prepared with a series of different adhesive materials. The results of these experiments are presented in Table 7-4.

TABLE 7-4 Comparison of Endurance Limit and Short-term Strength of Lap Shear Joints

Type	LSS (psi)	EL (psi)	EL/LSS
Nylon-epoxy	6258	2826	0.45
Nitrile-phenolic	4164	1728	0.42
Epoxy-novolac	5052	2181	0.43
Nitrile-epoxy	5482	2455	0.45

Source: ASTM D-1002-AL/AL (see Ref. 38).

As shown, the endurance limit of the structural adhesives tested are between 0.42 and 0.45 of the short-term strength. Furthermore, there is no obvious trend between the measured endurance limit and the modulus of the adhesive material. From additional experiments, estimates of the statistical variability of some of the determined values were made with the result that the endurance limit/lap shear strength (EL/LSS) ratio can vary between ±0.05 to ±0.18. Considering this, one can then assume the lower range of the endurance limit to be about 0.25 of the short-term strength. Such information may be of great interest to a design engineer who wishes to incorporate a realistic margin of safety into the specific adhesive joint structure he is designing. In this consideration, however, one must keep in mind that the data presented here are for an ASTM type D1002 lap shear joint, a structural joint arrangement very rarely used by engineers. Furthermore, the data were obtained in a standard, controlled environment, 23°C and 50% RH. However, the measured endurance limit as determined by the described accelerated method (Prot) does provide some idea of what one might expect. One is able to get an order-of-magnitude estimate of what the endurance limit of such joints might be.

As mentioned in the quoted work [38], the modified Prot approach was found, through sustained loading tests, to provide a reasonable estimate of the long-term strength of an adhesive joint. However, some ambiguities in the theory indeed exist. Criteria for damage are ill defined. Furthermore, for the adhesive joints, the sustained load tests did not wholly support the basic assumption of the Prot equation, namely that plots of stress level to break vs. time should follow a hyperbolic law. In the reported study and in some recent unpublished work [43], the sustained load vs. time-to-break data appear to more adequately fit a semilog function. This behavior is presented in the following equations (Table 7-5) derived from a least-squares fit of the data presented in Ref. 38. The correspondence of adhesive joint strength data, under static stress fatigue conditions, to a semilogarithmic law is not unusual. Zhurkov and coworkers (see Ref. 44, p. 27) found that the time dependence of tensile strength for a great variety of materials (metals, plastics, and polymer fibers) followed an exponential law,

$$t_b = A \exp - (a'\sigma_0) \qquad (15)$$

where A and a' are constants and σ_0 the applied stress. This equation was also found to be valid over a wide range of temperatures; the coefficients A and a', however, change with temperature.

The fitting of static fatigue data (relating stress level vs. time to break) on a log time scale does not support the concept of the stress endurance limit as defined in earlier discussions. This difficulty of

TABLE 7-5 Least-squares Fit* of Static Fatigue Data for Several Structural Adhesives

Adhesive type	K	M	R*R
Nylon-epoxy	0.71	-0.025	0.59
Nitrile-phenolic	0.68	-0.047	0.85
Epoxy-novolac	0.63	-0.026	0.70

*Data fit to equation $\sigma/LSS = K + Mt_b$, where LSS is the short-term lap shear strength (ASTM D1002), and σ is the static load applied, and t_b is the observed time to break of the specimen. Aluminum adherends were used.

Source: Adapted from Ref. 38.

determining a meaningful time-independent stress endurance limit for materials exists with other schemes for predicting the long-term strength of materials. For example, a widely used method is the stress-time-temperature approach presented by Larson and Miller [45] describing the rupture and creep behavior of metals. This method was later used by Goldfein [46] to describe the long-term rupture and impact stresses in reinforced plastics. In this approach, a time-temperature superposition parameter was devised by defining long-term strengths at ambient temperatures in terms of short-time static tests at elevated temperatures. Related to this, a superposition method for predicting the ultimate strength of polymers has also been proposed by Smith [47] and Kaelble [48]. A review of the application of the time-temperature superposition principle to the long-term engineering properties of plastics has been published [49]. These approaches are based on the viscoelastic equations of state introduced by Ferry and colleagues [50,51]. While these superposition methods have proven to be useful for empirically predicting the high-strain mechanical behavior of plastic materials, the interpretation is characterized by its prediction of strength or stress endured through a finite time scale.

Referring back to Table 7-5 and the fit of the static fatigue data to an equation of the form,

$$\frac{\sigma}{LSS} = K + M \log t_b \tag{16}$$

we observe that the K term represents a finite condition; it is the relative stress level at $\log t_b = 0$ (when $t_b = 1$ hour). This condition is empirically fixed by the units of t_b. It is possible then that the static fatigue behavior of adhesive joints deviates from this linear relationship at breaking times approaching zero. On the other hand, one cannot be certain, from the data analyzed, that at very long times the relative stress level becomes a constant giving rise to a time-independent endurance level. Such general questions remain to be answered. If the log t_b plots do not level off within a reasonable experimentation time, one may impose the assumption that the fatigue loading will fail after 10^6 years (an exaggerated example) has been reached. This is, of course, philosophically (and physiologically) acceptable since, in practice, few engineering structures are planned to endure for periods of time that will span multiple civilizations. However, for the present day-to-day decisions an engineer has to make in carrying out his duties as a structural designer, he may be more willing to specify materials that have a measured (finite-time) endurance limit. Hence, more work remains in this field where measured endurance limits can be used for establishing engineering design parameters for adhesive joints and other materials.

2.4. Dynamic Fatigue Response of Adhesive Joint Materials

Classically, a dynamic test has been defined as any test which subjects a volume element of material to a stress which varies with time. More specifically, one relates a dynamic mode of stress application as one which varies periodically in magnitude. However, the feature of periodicity need not be maintained in a strict definition of the term, but its existence indeed simplifies mathematical treatment of the phenomenon. The importance of cyclic stresses and strains (especially sinusoidal) has been increasingly recognized in the failure of material components through fatigue [52].

Dynamic fatigue in adhesive joint materials manifests itself by damage and failure of the material in ways similar to static fatigue. However, in dynamic situations, materials undergo a time-dependent response caused by the driving force, i.e., sinusoidal, square wave, random wave, or a high-velocity impact mode of stress application. In service, these dynamic working stresses are often superimposed upon an existing static load. The main point of dynamic fatigue conditions in polymeric materials is that the rate of stress application causes the polymeric material being stressed to respond viscoelastically, that is, to exhibit nonequilibrium, time-dependent modulus behavior. More specifically, if a polymer is stressed at a rapid enough rate such that nonequilibrium stress conditions are periodically induced into the material, at any point in time during stressing, various amounts of elastic energy may be stored or lost as heat in the material. The basic (single-cycle) response of the polymer can still be described by the creep curves of Fig. 7-3. The time scale or sequence of stress application and release would change the detailed shape of the curve. The basic features of the response (stress application and creep recovery) will remain the same. There will also be complicating overlapping effects. Mathematically, the case of the Kelvin-Voigt model for a dynamic mode of stress application is more complex than for static loading. To illustrate, the stress pattern of a Kelvin-Voigt element subjected to a sinusoidally applied strain, $\varepsilon = \varepsilon^1 \sin \omega t$, of frequency ω and of maximum amplitude ε^1, can be expressed as

$$\sigma_{(\omega)} = G_1 \varepsilon^1 \sin \omega t + \eta_1 \omega \varepsilon^1 \cos \omega t \tag{17}$$

where the sine term is the stress component in phase with the strain and the cosine term is 90° out of phase with the applied strain. At a continuously applied dynamic stress σ_0, the cyclic strain on the specimen can be shown to follow the expression [29,31]

$$\varepsilon = J'_{(\omega)} \sigma_0 \cos (\omega t) + J''_{(\omega)} \sigma_0 \sin (\omega) \tag{18}$$

where $J'_{(\omega)}$ is the storage compliance equal to

$$J'_{(\omega)} = \frac{1}{E(1 + \omega^2 \tau^2)} \tag{19}$$

and $J''_{(\omega)}$ is the loss compliance of the system, or

$$J''_{(\omega)} = \frac{\omega \tau}{E(1 + \omega^2 \tau^2)} \tag{20}$$

The loss compliance term is based on the fact that part of the deformation gives rise to nonrecoverable work done by the applied force; this nonrecoverable work manifests itself by heating the strained sample [53]. For linear viscoelastic media at low strain levels, the energy loss per cycle is given by

$$\text{Loss per cycle} = \pi \sigma_0^2 J''_{(\omega)}$$

$$= \pi \sigma_0^2 \frac{\omega \tau}{E(1 + \omega^2 \tau^2)} \tag{21}$$

The loss per cycle, being proportional to $J''_{(\omega)}$, is therefore a function of frequency. It increases as the frequency is increased, reaches a maximum at $\omega = \tau$, and decreases at higher frequencies. Thus, a viscoelastic material being sinusoidally strained at or near its characteristic retardation time will be in its maximum "lossy" state. It will absorb, into its molecular structure, a maximum of mechanical energy. Such a coupling, if carried out at the proper long-term fatigue conditions, can cause chemical changes to occur in the polymeric media leading to rheological damage and ultimate failure of the polymer. Hence, we have defined a fundamental basis for fatigue damage in polymers; the situation becomes much more complex as dynamic fatigue conditions are imposed at higher strain levels where nonlinear viscoelastic phenomena predominate. In this case, mechanochemical reactions may take place in the polymer [54]. In severe straining conditions, a massive heating of the material occurs, as can be observed with dynamically flexed elastomers [55]. In the service life of an adhesive material, such dynamic fatigue conditions can exist, causing damage to the joint material. These conditions, with a superimposed static stress and/or thermal and chemical service-life environment, are all possible. The effects are interrelated and must be considered as part of the envirodynamic system the adhesively bonded component is subjected to during its lifetime.

Experimentally, the fatigue of materials is normally represented by the so-called S-N curves. (Plots of peak stress level, S, on the material vs. the number of cycles, N, to failure.) A schematic diagram of such curves is presented in Fig. 7-6. Here, an adaptation of the curve of Manson and Hertzberg [13] is made showing the various regimes of damage in the materials (prior to ultimate failure). Regarding this typical fatigue behavior diagram, Fig. 7-6 shows that in the case of strength damage in static fatigue, crack initiation and crack propagation precede the final failure of the material. Regarding fatigue crack propagation rates, Paris [56,57] has concluded that crack propagation rates da/dn can be related to the stress intensity factor range ΔK by the equation

$$\frac{da}{dn} = C\,\Delta K^n \qquad (22)$$

where C and n are the material constants. With this equation, a controversy exists in the fracture mechanics field with regard to the exponent n—Paris originally predicted n = 4, where n is related to the volume of the plastic zone at the crack tip. However, others [58] have considered n = 2. Regardless of the correct magnitude of n, it is found to be a function of frequency, environment, mean stress, temperature, and material variables [13]. Therefore, n may be said to be related to the quality of the crack that is being opened by the cyclically applied stress. However, the true meaning of n is still a subject of much discussion.

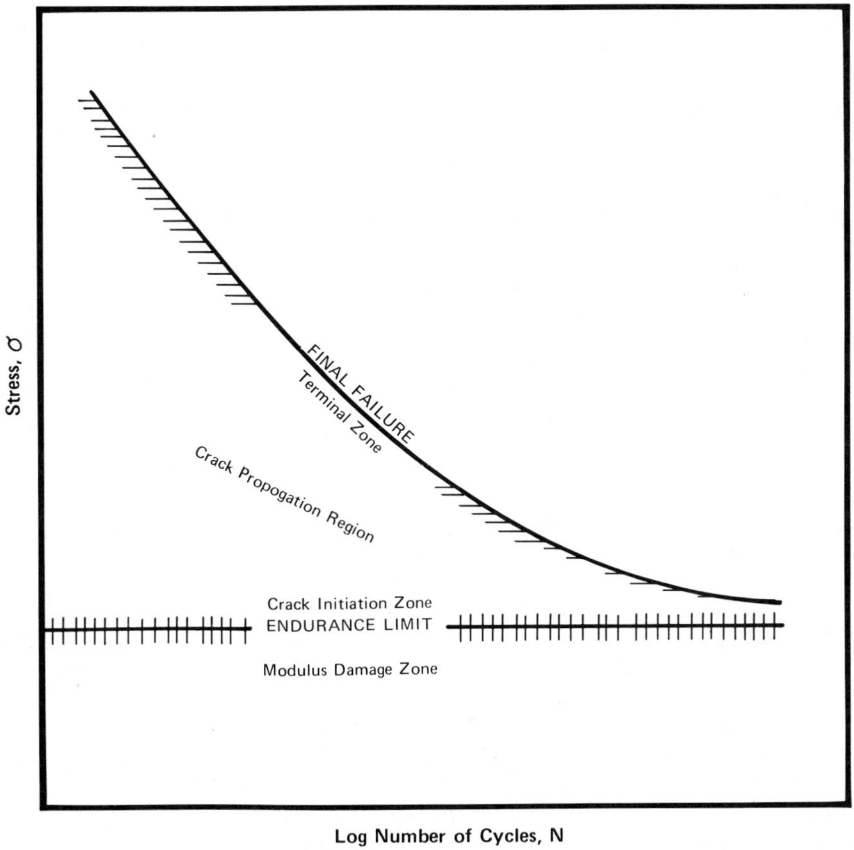

FIG. 7-6 Schematic S-N curve for the dynamic fatigue of polymeric materials.

From a phenomenological viewpoint, one can envision a material system undergoing sinusoidal deformation accumulating damage during each cycle or each time it is stressed or strained beyond its damage endurance limit. This is shown in Fig. 7-7, which describes the material being deformed sinusoidally where the upper ranges of the deformation traverse the endurance limit in the material. During each cycle, the material makes an excursion into the damage zone and accumulates an increment of damage. It can readily be seen that if the peak amplitude of the applied deformation, A_0, is below the endurance limit, little or no damage will accumulate (assuming that the frequency of the oscillation is low enough such that isothermal conditions

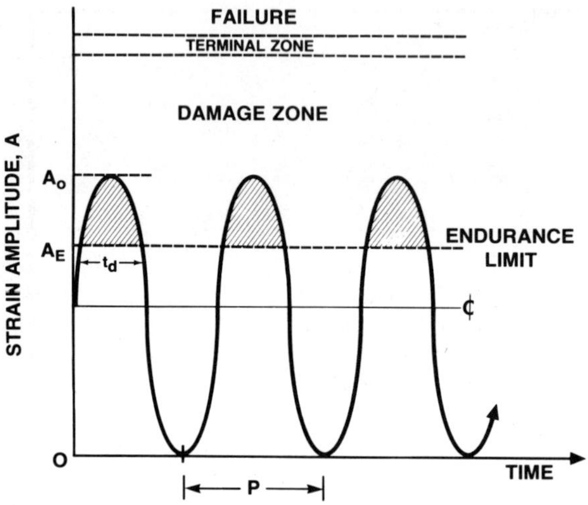

FIG. 7-7 Schematic diagram of the accumulation of damage during the sinusoidal deformation of a material system.

prevail). On the other hand, if the maximum amplitude of the vibration is high, and enters (or goes beyond) the failure zone, the material will fail in the first cycle.

As shown in Fig. 7-7, the accumulation of time that the material spends in the damage zone is shorter than the total elapsed time. As an example, if a material is fatigued by a strain forcing it into the damage region, it can be shown that the residence time per cycle in the damage zone, t_d, is given by the equation

$$t_d = \frac{1}{2}\left(P - 4 \text{ arcsine } \frac{A_E}{A_0}\right) \quad \text{for } 0 < A_E \text{ and } A_E < A_0 \tag{23}$$

where P is the period of the vibration, and A_E and A_0 are the amplitude level at the endurance limit and the maximum amplitude of the symmetrical sinusoidal stressing force, respectively (see Fig. 7-7).

From Eq. (23), one can make a first-approximation conversion of dynamic fatigue test data to the static fatigue elapsed time interval $|T_d|$ by

$$|T_d| = Nt_d = \frac{N}{2}\left(P - 4 \text{ arcsine } \frac{A_E}{A_0}\right) \tag{24}$$

This expression is, of course, a rough approximation of the accumulated damage time in a dynamic (sinusoidal mode) fatigue test. It should hold for low-frequency test. The equation does not, however, account for the fact that damage occurs at a more rapid rate when the amplitude of the stressing cycle is close to A_0 (the rate of damage is highest at A_0) and also the closer A_0 is to the breaking stress of the material. Furthermore, damage may not accumulate linearly with time. The rheological damage reaction may be autocatalytic, accumulating at a faster rate during the later cycles to failure. In addition, one is not considering the viscoelastic response in the material during the dynamic stressing process. As stated by Lepie and Adicoff [59], and as was previously discussed, in a dynamic test, the total energy input E_t during deformation is the sum of an elastic (recoverable) part E' and a loss (nonrecoverable) part E". The nonrecoverable energy is lost as heat (internal friction) or else results in damage to the viscoelastic material, especially if the stressing amplitude is high. In either case, mechanical loss in the material manifests itself by a heating of the material. The dynamic mechanical effects of viscoelastic materials used in adhesive joints has been studied by Mostovoy [60]. Related to these heating effects, Manson and Hertzberg [13] state and Mostovoy [60] reasserts that in order to properly carry out dynamic fracture fatigue tests in polymers (and adhesive joints) the testing frequency must be kept below 3 Hz to avoid nonlinear effects due to nonequilibrium, nonisothermal crack growth conditions.

The subject of the accumulation of damage in a dynamic fatigue test is one of general interest, and so far little has been done to define an adequate experimentally measurable damage function. All work employing the linear cumulative damage analysis approach can be subject to criticism because the damage function term is a purely statistical identity. No assessment or explanation of the nature of the damage is given. Recently, Lepie and Adicoff [59] have analyzed the dynamic mechanical behavior of highly filled elastomeric polymers (solid rocket propellant) in terms of energy balance and damage. Their approach essentially employs the phenomenon of the difference in the stress-strain behavior in a filled elastomer before and after an applied strain history. This damage process resembles the phenomenon referred to earlier in this chapter (p. 323) as the "Mullins effect" [29]. The internal damage in these filled systems may be attributed to dewetting effects between the polymer matrix and the particles and bulk rupture in the matrix polymer. In the present case, it was assumed, from a molecular primary bond rupture calculation, that the major contribution to damage in the material system was that of the rupture of the load-bearing molecular chains. Citing this work by Lepie and Adicoff is encouraging, since it outlines and attempts to explain damage in terms of rheological parameters. More work of this type should be done. It will lead to a much more basic understanding of structural failure in viscoelastic materials in general and adhesive joint systems in particular.

3. EFFECT OF TEMPERATURE ON POLYMERIC ADHESIVE MATERIALS

The effect of thermal energy on polymeric material systems is well known, principally from the point of view of causing first- and second-order phase transitions in the materials and, of course, degradation. The phenomenon of phase transitions in polymers is a well-studied subject [61,62] in polymer physics. Thermal energy and its impact on materials is measured by the intensity scale of kT energy units, where k is Boltzman's constant (1.37×10^{-16} erg/°C) and T is the absolute temperature. In the case of polymers, an increase in thermal energy causes an increase in molecular motion which can induce many changes to occur in the polymer. All of these changes are manifested by the thermal energy being absorbed by the polymer molecules and causing them to vibrate at higher amplitudes.

3.1. Thermal-Mechanical Response of Polymeric Materials

A change in temperature can change the rheological state of a polymer. If it is a glassy amorphous solid, an increase in temperature can cause the polymer to soften to a rubber by passing through its glass transition temperature T_g. The material, in passing through its glass transition temperature, goes from a cohesively strong solid (may be brittle or tough, depending on the composition of the polymer) to a cohesively weak, creep-prone, deformable rubber. Then at still higher temperatures, if the polymer is amorphous, it may liquify and/or degrade. If the polymer is crystalline, meaning it possesses regions of order between numbers of polymer molecules and segments, the polymer may go through its glass transition condition with no noticeable softening effect until it reaches its crystalline melting temperature T_m. Above this melting temperature, the polymer becomes a molten fluid; again at still higher temperatures the polymer will degrade. Some useful relationships between T_g and T_m have been observed; namely, for polymers with symmetrical repeating units $T_g/T_m = 1/2$, while for polymers with unsymmetrical repeating units $T_g/T_m = 2/3$. In all cases the glass transition temperature of polymers is always less than its crystalline melting temperature. Crystalline polymers, because of their sharp transition between being a cohesively strong solid below T_m and a viscous fluid above T_m, are used as hot melt adhesive systems. Generally, in hot melt adhesives technology, one would like to have as sharp a transition between the solid and molten state as reasonable. Therefore, crystalline polymers whose T_g values are close to T_m make the best hot melt adhesives.

Crosslinked polymer systems are much less mobile under thermal stress than are thermoplastic polymers. Due to their 3-D network, thermoset polymers will soften slightly at high temperatures through a chemical degradation mechanism. This condition is usually irreversible except with not so highly crosslinked systems such as certain epoxy systems that may exhibit an observable glass transition condition. In other specifics, rubber-modified epoxy adhesives may exhibit low-temperature transitions due to the fact that such adhesives are two-phase systems; the low-temperature transition reflects the glass transition temperature of the second rubbery phase [63]. The strength of polymeric adhesive materials at various temperatures then depends on a number of factors: the T_g of the base polymer and/or resin modifiers, the degree of crystallinity (if any), and the degree of crosslinking (if any).

Basically, the glass transition temperature of an adhesive polymer is an important factor to consider when predicting the permanence and endurance of adhesive joints. The glass transition temperature depends on composition in a great variety of ways (stiffness or geometry of backbone polymer, amount of plasticization, degree of crosslinking, degree of crystallinity). The strength-temperature profile of adhesive materials, therefore, is a broad subject and is a specific characteristic of a given adhesive system. However, to summarize, generalizations can be made regarding the strength states of various polymer types. Figure 7-8 reviews the strength-temperature behavior that adhesive materials and joints may go through depending upon composition. Some formulated adhesive materials systems may exhibit a thermal response character that is a combination of the three basic types of polymers shown. The general durability performance of adhesive joints depends upon the rheological state (environmental temperature) of the polymeric adhesive material. For example, amorphous polymers in their rubbery state are structurally deficient. Under stress, they would be subject to creep and flow, as was discussed in Sec. 2 (see Fig. 7-3). Therefore, the behaviors in strength illustrated in Fig. 7-8 suggest that adhesive joints are structurally durable only if the adhesive polymer exists in its glassy state (below T_g) under its service condition. Only highly crosslinked polymers, where the T_g is obscured by the stiff network structure, are presently used in the formulation of high-utility-temperature adhesives (phenolic and polyimide polymers are some examples of this type of adhesive polymer system).

3.2. Thermal Degradation of Adhesive Polymers

The temperature dependence of durability (at constant nominal stress) for solids and polymers can be expressed by an Arrhenius expression,

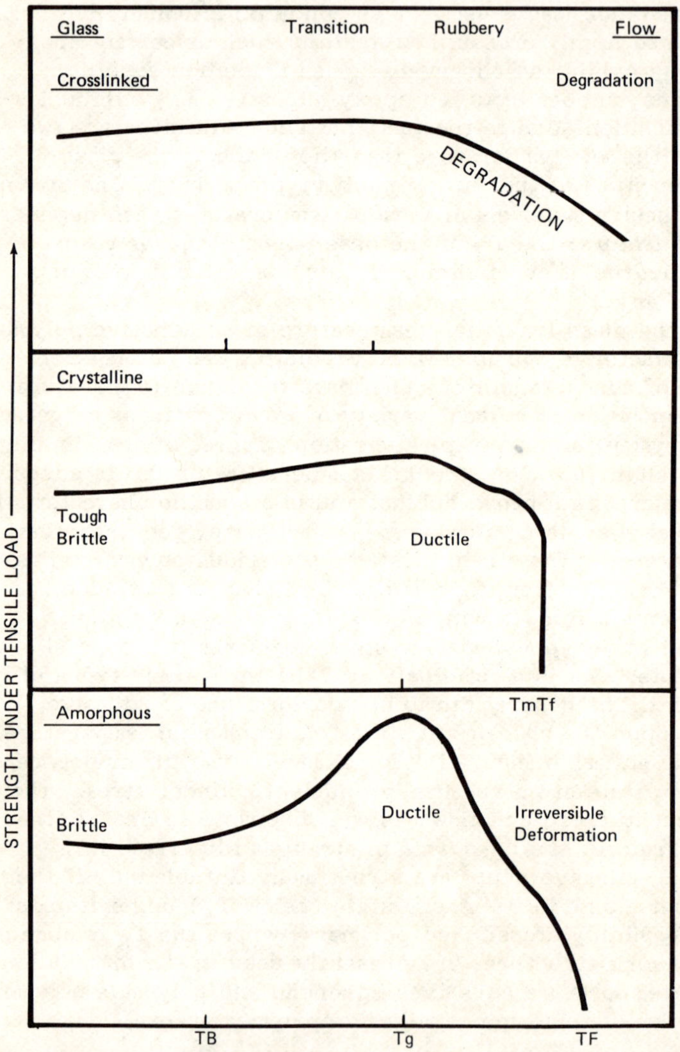

FIG. 7-8 Diagram of thermal temperature/strength states of various polymer types.

$$D = D_0 \exp \frac{U_0}{kT} \tag{25}$$

where D is the failure time, D_0 a constant numerically similar to the period of thermal oscillation of the atoms, and U_0 is the activation energy of the failure process. It is found that the activation energy U depends upon the applied static stress σ_s and decreases according to the formula

$$U = U_0 - c\sigma_s \tag{26}$$

where U is the intrinsic activation energy related to the value of activation energy at a zero stress condition. It is found that U is of the same magnitude as chemical bond energies for polymers and c is a coefficient which depends on the nature of the materials. The durability of a polymeric material under stress can then be given by the Tobolsky-Eyring [64] expression,

$$D = D_0 \exp \frac{U_0 - c\sigma_s}{kT} \tag{27}$$

This equation states that stress and temperature are the basic factors which cause the failure of polymeric solids; the stress decreases the activation energy, and higher temperature increases the thermal motion leading to the chemical breakage of bonds. Equation (27) is observed to be valid over a wide range of material types and stressing times (10^3 sec up to 10^7 sec) [44].

Although the environment has an influence on the time dependence of strength, it is not usually found to be critical. In any case, a surface-active environment is known to influence the values of U_0 and c and can cause deviations from this equation of durability. Deviation in linearity from Eq. (27) can also occur if the polymeric structure changes under stress, i.e., the polymer stress anneals or else becomes more oriented (crystalline) under stress. One might also expect complications if the polymer system undergoes a chemical "postcuring" reaction before degradation occurs.

Regarding the degradation behavior of epoxy resins, Neiman and coworkers [65] have shown that such resins degrade at a relatively low temperature, 200 to 250°C, whether they are cured by a polyamide (amine) or by anhydride curative. The maximum degradation rate is reported to increase with temperature according to the Arrhenius law expressed by Eq. (25). An apparent activation energy for decomposition was calculated to be 18 to 35 kcal/mol. This value compares with U_0 = 55 kcal/mol for polymethyl methacrylate [44]. Such activation energies are in the range of chemical bond dissociation energies.

Referring to this brief review of the temperature durability of polymeric materials as a basis for understanding the behavior of the thermal degradation of homogeneous materials, one can now turn to the case of structural adhesive joints. The problems of thermal durability are multiplied here since one is now dealing with a complex material system containing at least two homogeneous material phases, base adhesive material and adherend, and at least one interfacial phase. Furthermore, if the adherend is nonporous, i.e., impervious to atmospheric oxygen diffusion, the geometry of the adhesive joint may influence the thermal durability of the joint. Much subjective data exists in the literature on the thermal durability of polymers [66], including various structural adhesive material systems; it is unfortunate that at present, no systematic studies have been carried out. Some qualitative results showing the degradation of structural adhesives under various thermal environments have been presented by Krieger and Politi [67]. In their work, the dramatic influence of the type of surface metal (in metal to metal joints) on the high-temperature durability of adhesive joints (epoxy phenolic, polyimide adhesives) has been demonstrated. Obviously, the metal surface catalyzes the degradation of the adhesive material. Furthermore, the detrimental influence of atmospheric oxygen (vs. nitrogen atmosphere) on adhesive joint thermal durability is also demonstrated. Overall, the thermal degradation of a structural adhesive joint is a complex chemical kinetic phenomena that will not yield easily to a simplistic approach.

4. PHYSICOCHEMICAL ENVIRONMENT ASPECTS OF ADHESIVE JOINTS

Another important intensive variable controlling performance of an adhesive joint is the chemical or corrosive environment in which the joint is to function. The effects of various chemical environments, especially humidity, on the durability of adhesively bonded interfaces is a subject of broad interest [68-73]. However, few theoretical approaches exist that suitably describe the failure mode of joints under the influence of chemical environments. Here, we shall briefly review the few models that exist to predict the performance of adhesive joints under the influence of chemical environment and then proceed to develop arguments leading to a new model based on the attachment site theory of adhesive joint strength [63,74,75].

4.1. *Existing Models of Adhesive Joint Performance*

Many of the existing theories describing the deleterious effects of chemical environments on the durability of adhesive joints are modifications of the Griffith theory of critical fracture energy previously discussed [14]. Some take into account the effect of the presence of a liquid phase on the thermodynamic work of adhesion. In earlier work, Fowkes [76] suggested that the free energy of a liquid γ_L may be considered to be made up of two components, dispersion (London) parameter γ_L^d and polar (Keesom) contribution γ_L^p. Thus, the surface free energy of a liquid is given by the following expression:

$$\gamma_L = \gamma_L^d + \gamma_L^p \tag{28}$$

He further proposed that the attractive forces across the interface between two immiscible liquids was due to the interaction of like forces only. He then developed an expression for the interfacial free energy between two liquids based on dispersion forces alone.

Owens and coworkers [77,78] and more recently Gent and Schultz [79] expanded Fowkes' equation to include the polar contributions as well. Thus, the interfacial free energy γ_{AB} between two liquids A and B is given by

$$\gamma_{AB} = \gamma_A + \gamma_B - 2\left[\left(\gamma_A^d\gamma_B^d\right)^{1/2} + \left(\gamma_A^p\gamma_B^p\right)^{1/2}\right] \tag{29}$$

The reversible work of adhesion, W_{AB}, which is a measure of attraction across the interface, is given by

$$\begin{aligned} W_{AB} &= \gamma_A + \gamma_B - \gamma_{AB} \\ &= 2\left[\left(\gamma_A^d\gamma_B^d\right)^{1/2} + \left(\gamma_A^p\gamma_B^p\right)^{1/2}\right] \end{aligned} \tag{30}$$

Due to difficulties in measuring directly the surface and interfacial free energies of solid systems, the above concepts developed for liquid systems could not be directly applied to solid-solid interfaces. However, the development of a simple contact angle method was later developed by Owens and Wendt [77] to measure the components of surface free energy of a solid due to dispersion and polar contributions. Based on these surface free energy concepts, Owens [78] concluded that the condition for spontaneous separation of an adhesive-adherend interface is a negative thermodynamic work of adhesion. The condition for spontaneous separation may also be expressed as

$$\gamma_{AB} > \gamma_A + \gamma_B \tag{31}$$

If the work of adhesion is positive, it means simply that work must be done on the system to separate the solids, and separation will not be spontaneous. Gent and Schultz [79] discussed the inequality presented by Eq. (31) in terms of the work of detachment W_d, where in an inert atmosphere

$$W_d = W_{AB} = \gamma_A + \gamma_B - \gamma_{AB} \tag{32}$$

where γ_A, γ_B, and γ_{AB} are now the surface free energies of the adhesive, substrate, and adhesive substrate interface. For detachment in the presence of a wetting liquid, the detachment equation becomes

$$W_{LAB} = \gamma_{LA} + \gamma_{LB} - \gamma_{AB} \tag{33}$$

In terms of contact angle,

$$\gamma_{LA} = \gamma_A - \gamma_L \cos \theta_{LA} \tag{34}$$

and

$$\gamma_{LB} = \gamma_B - \gamma_L \cos \theta_{LB} \tag{35}$$

Substituting in Eq. (33) gives

$$W_{LAB} = W_{AB} - \Delta \tag{36}$$

where $\Delta = \gamma_L(\cos \theta_{LA} + \cos \theta_{LB}) = W_{AB} - W_{LAB}$; this quantity Δ is important because it expresses, in experimentally measurable quantities, the reduction in work of detachment in the presence of a liquid compared to the work of detachment in the absence of a liquid. In terms of surface free energy,

$$\Delta = \gamma_A + \gamma_B - \gamma_{LA} - \gamma_{LB} \tag{37}$$

As described by Owens [78], if Δ is greater than the work of detachment, W_d, in the absence of the liquid, then W_{LAB} is negative and the adhesive and substrate will spontaneously separate in the presence of the liquid.

Schonhorn [80] and Schonhorn and Frisch [81] used the example shown in Fig. 7-9 to illustrate the above concepts. Furthermore, Table 7-6 gives the dispersion and polar contributions to the total surface tension of the various components of the adhesive joint from this work. From these surface tension values, the thermodynamic work of adhesion may be calculated. For the aluminum-oxide-

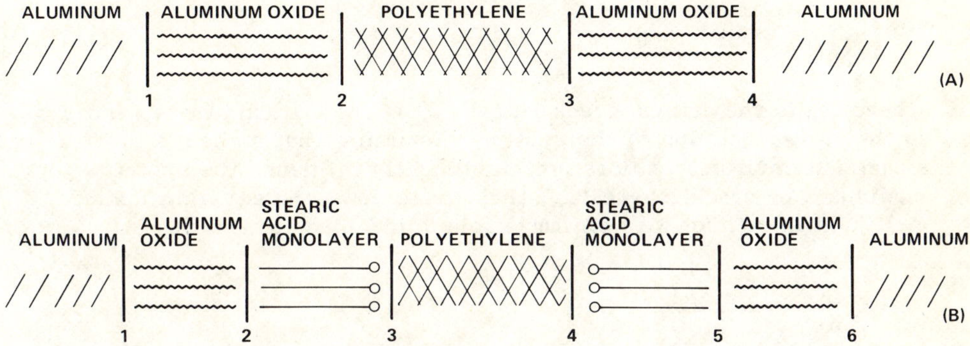

FIG. 7-9 Schematic representation of polyethylene to aluminum adhesive joints. (Source: Ref. 81.)

polyethylene system in the absence of water the work of adhesion, W_{AB}, is calculated as 120 erg/cm^2. However, in the presence of water, the thermodynamic work of adhesion W_{LAB}, takes a negative value of -312 erg/cm^2 predicting spontaneous separation of polyethylene from aluminum oxide in the presence of water. The interaction of the environment with the interface may be prevented by the introduction of a monolayer of stearic acid, for example. In this case, the thermodynamic work of adhesion for aluminum oxide-stearic-acid-polyethylene system in the absence of water is 56 erg/cm^2. In the presence of water, W_{LAB} for this system is still positive, 101.8 erg/cm^2, and the joint is durable under this moist environment.

Schonhorn and Frisch [81], using the above concepts of thermodynamic work of adhesion, modified the Griffith theory of critical surface energy for fracture and arrived at the following equation:

TABLE 7-6 Dispersion and Polar Contributions to the Total Surface Tension

Substance	γ_S^d (dyn/cm)	γ_S^p (dyn/cm)	γ_S (dyn/cm)
Aluminum oxide	100	900	1000
Stearic acid monolayer	23	0	23
Polyethylene	36	0	36
Water	21.8	51.0	72.8

Source: Adapted from Ref. 81.

$$2W_{AB} - \frac{2\pi a_c \sigma_c^2}{E} = 0 \tag{38}$$

where $2a_c$ is the critical crack length, σ_c is the critical stress, and E is the elastic modulus of the system. Assuming that a change in the immersion environment does not significantly influence the system modulus E or crack length $2a_c$, they wrote down the equation for an A-B joint in contact with an immiscible liquid L as

$$2W_{AB}^{L} - \frac{2\pi a_c (\sigma_c^L)^2}{E} = 0 \tag{39}$$

It then follows that

$$\left(\frac{\sigma_c^L}{\sigma_c}\right)^2 = \frac{W_{LAB}}{W_{AB}} \tag{40}$$

The above equation gives a quantitative relationship between the critical stress to failure in the presence and absence of a liquid environment.

Kaelble and coworkers [82,83] applied the above concepts to predict the environmental degradation in the interlaminar shear strength of two types of graphite-epoxy composites. Both these two composite types used high-tensile-strength graphite fiber as the reinforcement material. They differed only in the type of epoxy resin matrix employed. The two types of epoxy systems used are respectively designated as modified epoxy (American Cyanamid Co., Bloomingdale Division Type BP-907) and cycloaliphatic epoxy (Hercules 3002). Table 7-7 summarizes the surface properties of the fiber reinforcement and the two types of adhesive matrix. Kaelble and coworkers obtained the following expression to predict the degradation in the interlaminar shear strength of the composite due to immersion in water for an infinite time, e.g.,

$$\frac{\lambda_{b\infty}^{L}}{\lambda_{b0}} = 1 - f_I + f_T \left(\frac{\sigma_c^L}{\sigma_c}\right) \tag{41}$$

where λ_{b0} is the interlaminar shear strength before water immersion, $\lambda_{b\infty}^{L}$ is the interlaminar shear strength after water immersion for infinite time, and f_I is the fractional area for interfacial failure in the composite. From the surface properties of the components of the

TABLE 7-7 Surface Energy Properties of Components in Graphite-Epoxy Composite Laminates at 23 ± 1°C

Material	γ_S^d (dyn/cm)	γ_S^p (dyn/cm)	γ_S (dyn/cm)
High-tensile strength graphite fiber	25.9 ± 1.5	25.7 ± 3.3	51.6 ± 2.3
Modified epoxy (cured under dry N)	37.2 ± 3.1	8.3 ± 2.0	45.5 ± 1.2
Cycloaliphatic epoxy (cured under dry N)	28.9 ± 3.2	25.0 ± 4.7	53.9 ± 2.0
Water	21.8	51.0	72.8

composite, the degradation in the interlaminar shear strength of the laminate was estimated. In the case of modified epoxy type laminates $\lambda^L_{b\infty}/\lambda_{b0} = 0.59$ and cycloaliphatic epoxy type laminates $\lambda^L_{b\infty}/\lambda_{b0} = 0.47$. These values were in fair agreement with the experimentally measured values of 0.56 and 0.51 for the two types of laminates, respectively.

In other work, Halpin and coworkers have applied the fundamentals of fracture mechanics to describe the effect of environment on the fatigue properties of adhesive bonded joints [7,84-86]. Their model assumes:

1. Joint contains preexisting flaws.
2. Flaws develop in a manner determined by material properties and environment.
3. The failure load is a decreasing function of flaw size.

The fatigue lifetime, then, is simply the time required to produce a sufficient strength degradation via flaw growth to permit failure. The rate of flaw growth is governed by the Paris law [56] given previously by Eq. (22). Davis and coworkers [87] and others [88], based on their experimental studies, arrive at the conclusion that the environment has relatively little effect on the exponent n in the Paris law, but has a strong effect on the preexponential term C. These authors use an Arrhenius-type relation to describe the effect of environment on the fatigue crack growth. If the chemical environment is humidity, the following equation applies:

$$C = A' \exp \frac{D'}{RW} \tag{42}$$

where RW is the ratio of weight after humidity exposure to weight before. By proper choice of the constants A' and D', these authors are able to fit the experimentally measured values of C into equations of the type given above.

Kaelble and coworkers [89] summarize the Halpin approach to describe the kinetics of the cumulative damage due to mechanical, thermal, and environmental factors by the following equation:

$$\phi = \exp \frac{-Kt^b}{A_t A_m A_c} \tag{43}$$

where ϕ is the cumulative degradation function, t is exposure time, K is a system constant, b is a time exponent, and A_t, A_m, and A_c are time shift factors relating to respective thermal, mechanical, or chemical effects. If one measures the interlaminar shear strength of the adhesive joint as a function of exposure time to follow degradation in the joint, the above equation takes the form

$$\phi = \frac{\lambda_b(t) - \lambda_{b\infty}}{\lambda_{b0} - \lambda_{b\infty}} = \exp \frac{-t}{\tau} \tag{44}$$

where the time exponent b is assumed to be 1.0 and relaxation time $\tau = A_t A_m A_c / K$. Rearranging the above equation, the interlaminar shear strength $\lambda_b(t)$ at exposure time t is obtained as

$$\lambda_b(t) = \lambda_{b\infty} + (\lambda_{b0} - \lambda_{b\infty}) \exp \frac{-t}{\tau} \tag{45}$$

Kaelble and coworkers [82] from their experimental measurements of $\lambda_b(t)$ on modified epoxy and cycloaliphatic-epoxy-type graphite fiber composites estimated the respective environmental relaxation times for the two composites to be 61 ± 18 and 92 ± 59 hr for degradation of the joint by immersion in water at 100°C. While the approach by Kaelble and coworkers was based on the interlaminar shear adhesion in graphite fiber-epoxy resin advanced composites, its applicability to the interfacial zones of structural adhesive joints is readily apparent.

All of the various theories on the chemical/corrosion aspects of adhesive joints reviewed thus far fall into two classes: (1) theories based on the thermodynamic work of adhesion and (2) models based on the kinetics of flaw growth. In the work of Schonhorn and Frisch, the thermodynamic model correctly predicted the probability for the spontaneous separation of the adhesive from the adherend in the presence of a hostile environment, and in some cases, even provided some insight into procedures to preclude premature failure. However, the intensity of the attractive forces between the adhesive and adherend is not a measure of the adhesive joint strength. The adhesive joint strength will be dictated by a multitude of factors such as the physical and mechanical properties of the adhesive system, adherend material, and the macrointerfacial phases existing at the boundary layer between the two constituent parts of the joint. As a result, the thermodynamic work of adhesion may be considered only as one of the variables controlling the strength and endurance of adhesive joints. The kinetics approach, on the other hand, by proper choice of the constants can be used to correctly predict the chemical degradation of the adhesive joint strengths. However, one cannot yet apply any physical significance to the constants used and relaxation times calculated.

4.2. Attachment Site Theory of Adhesive Joint Strength

It is our intention here to develop a model for the chemical degradation of the adhesive joint strengths based on the attachment site theory [63,74,75]. The elements of an adhesive joint are identified in Fig. 7-10 as (a) bulk adhesive; (b) adhesive-adherend boundary layer; and (c) bulk adherend. The theory examines the physical nature of the boundary layer from a different viewpoint. Lewis and Forrestal [63] found that when various controlled amounts of an adhesion-promoting compound [bis(2-chloroethyl)vinyl phosphonate, BCVP] was radiation-grafted onto the surface of polypropylene, the existence of these conditions of interfacial attachment was encountered. In this study, the surface-grafted polypropylene was used as a hot melt adhesive for various metals. When very small quantities (less than 0.1%) of the BCVP were grafted onto polypropylene, low-adhesive joints were observed. This condition was characterized by complete boundary (apparent interfacial) failure, signifying weak surface attachment. At somewhat higher grafting levels, 0.2 to 0.4%, mixed-type failures

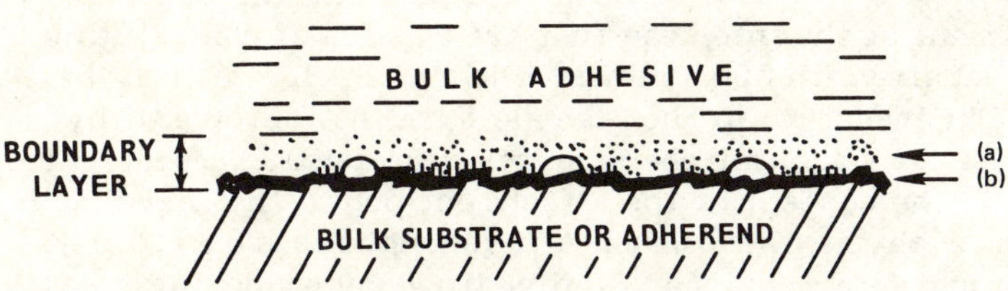

FIG. 7-10 Diagram of boundary layer at adhesive/adherent interface: A—contiguous polymer phase; B—adherend surface layer.

were observed, accompanied by higher joint strengths. Finally, at a particular grafting level (0.2-0.5%, depending on metal type), maximum joint strengths exhibiting cohesive-type failures were observed. A condition had been reached wherein the strength of the attachment of the grafted polypropylene adhesive to the surface was not greater than the cohesive strength of the bulk adhesive material.

These experiments, as well as others, enable one to generalize on the influence of the degree of interfacial surface attachment on adhesive-joint strength. The different interfacial states of adhesion are summarized in Fig. 7-11. In the *boundary failure region*, designated by B, the strength of the polymer-substrate adhesive bond is controlled by physical and mechanical forces active at the interface. In this region, the strength of interfacial contact is much less than the cohesive strength of the adhesive material. At a certain level, the degree of surface attachment begins to influence the strength of the joint. This critical degree of surface attachment may be referred to as the threshold value. For example, the modified polypropylene-to-metal adhesive joint systems previously described had a threshold value at 0.15% BCVP graft. In the *transition zone*, the strength of the adhesive joint is very sensitive to the degree of surface attachment. In this region, mixed failure is observed (designated by B/C or C/B). Now at a critical degree of surface attachment, which can be referred to as a minimal or saturation value, the adhesive joint will rupture, showing totally cohesive failure (designated by C). The attainment of this state of surface attachment would represent the ultimate strength of the joint. Above this so-called saturation degree of surface attachment, the strength of the joint levels off into a cohesive plateau. Under these conditions, further increase in the degree of surface attachment does not result in an increase in the mechanical strength of the adhesive joint. The influence of the adherend surface on the quality of surface attachment in joints has recently been emphasized by Brockman [90].

The nature of these proposed "attachment sites" cannot yet be described in quantitative terms other than that they are believed to be of morphological size rather than molecular size. These boundary layer attachment sites can best be described as being microscopic areas of attachment at the interface that are capable of contributing to the mechanical strength of an adhesive joint. Such sites may be considered similar in nature to the "tensile ligaments" proposed by Krafft [91,92] in his model for the strength of metals. A more appropriate analogy is the concept of the attachment points that have been proposed to describe the phenomenon of frictional adhesion [93,94].

The attachment site model may be quantified by employing a semi-empirical curve fitting procedure to the joint strength behavior shown in Fig. 7-11. This may be approximated by the equation

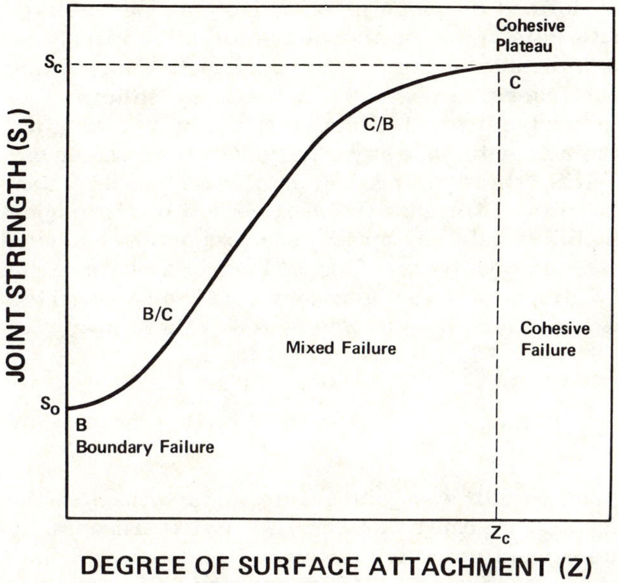

FIG. 7-11 Schematic diagram of the interfacial states encountered in adhesion and adhesive joint strength as a function of the strength of boundary layer.

$$S_j = S_c(1 - \exp - \Sigma\ \alpha_i\ Z_i^{Q_i}) \tag{46}$$

where Z_i represents the number of a particular type of site (per unit area) and α_i and Q_i are corresponding characteristic parameters. For example, in Ref. 63, a summation of two types of sites was proposed, those with chemisorbed character and those with physisorbed character. For the sake of simplicity, if one groups the sites into two types—namely, Z_e, the number of effective sites, and Z_0, the number of ineffective or inoperative sites—we arrive at the expression

$$\Sigma\ \alpha_i Z_i^{Q_i} = \alpha_0 Z_0^{Q_0} + \alpha_e Z_e^{Q_e} \tag{47}$$

Substituting Eq. (46) into Eq. (47) and normalizing the Z_i term by Z_0 to satisfy the boundary conditions, we arrive at a final expression:

$$S_j = S_c \left\{1 - \exp - \left[\alpha_0 + \alpha_e\left(\frac{Z_e}{Z_0}\right)^{Q_e}\right]\right\} \tag{48}$$

where S_j and S_c were defined previously; α_e and Q_e are parameters believed to be characteristic of the nature of the effective interfacial attachment sites. As indicated in Fig. 7-11, the effective site density at the cohesive plateau transition point is Z_c, which by definition equals the sum of the ineffective and effective sites, or rearranging, $Z_0 = Z_c - Z_e$. It is implicit here that any addition to the number of effective attachment sites, through chemical or physical modification or surface treatments, occurs through the replacement of the inoperative sites. As we shall illustrate, Z_c and Z_e are proportional to experimentally measured variable quantities. The α_0 term is a characteristic constant that can be defined from the boundary condition that when no effective sites are operative, $Z_e = 0$, and $S_j = S_0$; therefore,

$$\alpha_0 = -\ln \frac{S_c - S_0}{S_c} \tag{49}$$

Now, from suitably planned adhesive joint strength experiments, the parameters α_0, α_e, Q_e, S_c, S_0 can be determined by experimental measurements and subsequent computer analysis. Equation (48) is unique in that it contains parameters of thermodynamic, kinetic, and rheological significance. The characteristic terms Q_i and α_i may be assigned to be identified with the termodynamic interfacial free energy of the adhesive-adherend system. The ultimate strength S_c, on the other hand, has a rheological significance, in that it is determined by the rheological state (cohesive strength) of the polymer at a given temperature.

In some recent studies, a derivation of the attachment site theory equation has been accomplished from first principles employing the concept of the development of adhesive joint strength via the formation of discrete boundary layer morphologies [95]. In this work, it was assumed that the nucleation and growth of attachment site morphologies follows the same mechanism as for classical phase transformations. However, in the case of adhesive joint formation, the time variable is replaced by other variables such as the concentration of adhesion-promoting species in the boundary layer. The kinetics of attachment site formation and hence the development of adhesive joint strength can then be developed following the Avrami [96-98], Johnson and Mehl [99], and Natarajan and Davidson [100] approach for phase transformation kinetics. By relating the volume of attachment sites formed to the boundary layer strength through a boundary layer ultimate strength parameter K*, we arrive at an equation for the strength of the boundary layer:

$$S_b = K^*[1 - \exp - (\alpha^* Z^{Q^*})] \tag{50}$$

Correcting for any inherent residual boundary layer strength through a correction parameter S_0, Eq. (50), the strength of the boundary layer is then expressed as

$$S_b = K^* \{1 - \exp - [\alpha_0^* + \alpha^*(Z)^{Q^*}]\} \tag{51}$$

where

$$\alpha_0^* = \ln \frac{K - S_0}{K}$$

Equation (51) is then the general form of Eq. (48) presented in the earlier discussion with the exception that in Eq. (51), α^* is now a derived rate constant factor (kinetic term) dependent upon the particular geometric shape of the formed attachment site, growth rate effects as a function of attachment site concentration, and the potential nucleation site density. The Q^* term, from crystallization theory, is the effective dimension of the attachment site growth mode. K^* now represents the ultimate theoretical strength of the boundary layer. With this derivation, the interpretation of the attachment site model has been refined to present a more meaningful and generalized representation of the boundary layer.

As thus far developed, the attachment site theory describes a model of adhesive joint strength which is essentially a "weak link" model. The assumption made is that

$$S_j = \min[S_b, S_c, S_d] \tag{52}$$

where the joint strength S_j is a minimum value of the three components of a simple joint: S_c, the cohesive strength of the bulk adhesive; S_d, the cohesive strength of the bulk adherend; and S_b, the strength of the interfacial boundary layer. The proposed morphological theory thus describes the control of boundary layer strength in terms of surface attachments. This new refined interpretation of the attachment site theory is schematically presented in Fig. 7-12. Here, using a particular illustrative condition, as the attachment site density Z increases from its inherent joint strength condition, (S_0), the joint strength, characteristically increases. Between points X and Y (see Fig. 7-12) the boundary layer is the weakest link in the joint; boundary layer failure predominates. When the vicinity of point Y is reached, again in this illustrative case, the weakest link converts to being the cohesive strength of the adhesive polymer. Now S_c determines the ultimate strength of the joint; the observed "cohesive plateau" has been reached. Yet as determined by the function described in Eq. (50), the boundary layer strength S_b can still increase, but its

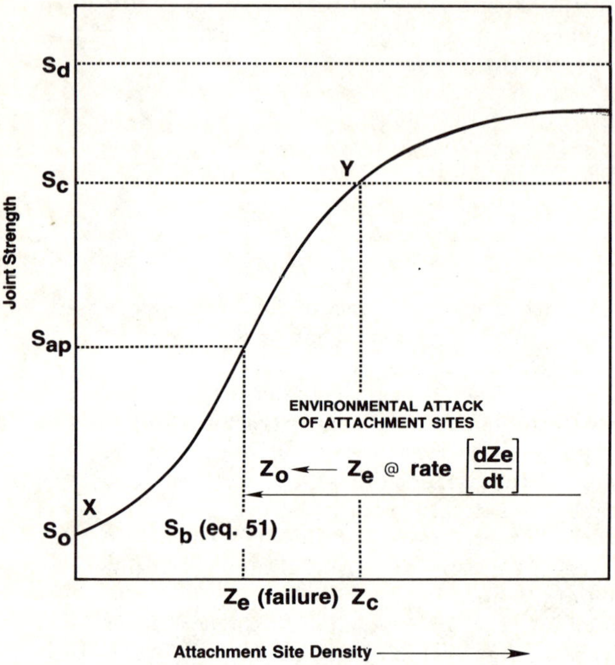

FIG. 7-12 Attachment site theory as described by boundary layer morphology/weakest link approach.

increased strength is not noticed because S_c is the weakest link. To further illustrate, the adherend strength S_d does not enter into the picture because in this particular illustrative case it is stronger than the adhesive material.

In studying Fig. 7-12, one can easily recognize the innate flexibility of this newly refined interpretation of the attachment site theory of adhesive joint strength. For example, specific adhesive joint systems will have characteristic S_c, S_d, and S_b values. Knowing these parameters then enables one to predict the ultimate strength of the particular joint system. In practice, then, one might be able to predict the failure mode of the joint if these strength parameters are known. It is felt that the theoretical interpretation of the boundary layer in terms of characteristic morphological entities will have a significant impact on the future development of the attachment site theory of adhesive joint strength.

4.3. Effects of Deleterious Environments

In the presence of a deleterious environment, it can be suggested that the strength of the adhesive joint is deteriorated through depletion of the number of attachment sites Z due to reaction with the attacking environment. The kinetics of depletion of Z will depend on the diffusion rate of the environmental species through the attachment sites and/or the reaction rate between the attachment sites and the environmental species. It is proposed then that in the presence of a corrosive environment the attachment sites are attacked by the corrosive atmosphere and the rate of depletion of the number of effective attachment sites is a kinetically controlled interaction. The kinetics of the interaction between the environmental, attacking species and the attachment site may be considered to be governed by a whole sequence of subsidiary rate-dependent steps. The slowest of these rate-dependent steps controls the depletion kinetics of the attachment sites. A typical sequence of these steps can be: (1) diffusion rate of the environmental species into the adhesive joint; (2) chemical reaction rate of the environmental species and the attachment sites. The chemical reaction between the environmental species and the attachment sites are thermally activated. When an externally applied static stress is superimposed on the adhesive joint, the chemical interaction may be further enhanced by a stress corrosion mechanism induced by the increase in physicochemical diffusion rate in the dilationally stressed media. To illustrate, suppose that the following equation represents the reaction between the environmental species and the bonds of the attachment site:

$$\text{--} \; B \; \text{--} + \eta[M] \rightleftarrows B[M]^{\eta} \tag{53}$$

The value of η determines the order of reaction; [M] is the concentration of environmental species. If the diffusion of the environmental species into the boundary layer is the rate-controlling step, then the depletion rate of the attachment sites is given by

$$\frac{dZ_e}{dt} = \frac{1}{\eta}\frac{d[M]}{dt} \tag{54}$$

where d[M]/dt is the steady-state molecular flow rate. For the diffusing species of gaseous environments, the equation developed by Knudsen [101] and by Snowden [102,103] for the diffusion rate based on the kinetic theory of gases can be modified to give

$$\frac{d[M]}{dt} = \frac{(\rho_0 - \rho_c)Ka}{(2\pi MkT)^{1/2}} \tag{55}$$

where ρ_0 is the pressure of the gaseous environment at the mouth of the diffusion path, ρ_c is the pressure at the attachment site being attacked, a is the cross-sectional area of the diffusion path at the attachment bond under attack, and K is an impedance factor. Also, M is the molecular weight of the diffusing species, k is the Boltzman constant, and T is the absolute temperature. The impedance factor K depends on the geometry of the diffusion path, the Young's modulus of the boundary layer and the stress intensity at the attachment site being attacked by the environmental species.

If the thermally activated reaction rate is the rate-controlling step, on the other hand, the depletion rate of the attachment site is given as [104]

$$\frac{dZ_e}{dt} = Z_w \beta \tag{56}$$

where Z_w is the number of attachment sites per unit width of the diffusion front and β is the reaction rate for one site. In the presence of an applied load, Wiederhorn [105] expressed the reaction rate β as

$$\beta = C(T)\rho_c^{\eta} \exp \frac{-U^* + bS_{appl}}{kT} \tag{57}$$

where C(T) is a slowly varying temperature-dependent term, U^* is an activation energy, and bS_{appl} is a term arising from the free energy reduction as the external loading is applied.

The relative importance of the diffusion and reaction steps in the depletion of the number of attachment sites in the boundary layer depends on the stress and thermal conditions operative in a joint. At high levels of stress or temperature, the diffusion of the environmental species may be expected to be the determining factor in the kinetics of an adhesive joint deterioration.

In summary, the important feature of the described attachment site model of adhesive joint strength is its ability to describe, in theory at least, the deterioration of joints under adverse environments. To do this, one regards the deterioration of joint strength as a conversion from Z_e-type sites to Z_0-type attachment sites. From a kinetic viewpoint, the rate of disappearance of interfacial attachment sites, dZ/dt, can be quantitatively described from properly designed laboratory experiments and the proper use of Eqs. (54) and (56). To illustrate the environmental response of adhesive joint systems under an applied mechanical (in-service) stress, S_{appl} can be qualitatively described by Fig. 7-12. Here, the conversion of effective attachment sites into ineffective sites is shown by an arrow running counter to the effective-number-of-sites baseline. In this case, the joint strength would be dropping according to the $S_c \rightarrow S_{appl}$ part of the curve. When the remaining number of effective sites equals a critical value Z_e (failure), the applied stress on the joint equals the joint strength and incipient joint failure occurs. While this simplified picture does not take into consideration geometric effects, it does present, to a first approximation, the deterioration in adhesive joint strength in a degrading environment. Experimental and analytical studies of adhesive joint strength deterioration using the above approach should be considered so that a full picture as to the applicability of the attachment site model to describing the environmental behavior of structural adhesive joints may be developed. Perhaps of some significance in interpreting this approach is the fact that Eqs. (48) and (51) are of the same form as the Weibull distribution function [106] where the α term is a scalar factor and the Q term is a shape factor.

5. SCHEMES OF SERVICE-LIFE PREDICTION OF ADHESIVE JOINTS

Up to this point, a general review of the most important influential factors that characterize the endurance and permanence behavior of structural adhesive joint materials has been presented. A broad picture of the nature of the important intensive variables of stress, temperature, and chemical environment and their effect on material systems was presented from a mechanistic standpoint. The treatment has been from a more or less isolated view, in that each variable effect was treated independently. It is known that there is a strong interrelation between the influences of these variables and the behavior of the atoms and molecules that constitute the enduring material system [107-109]. To give a specific example, heat history can have an effect on the mechanical strength of polymers. Matsuoka [110] has shown that the ultimate properties of certain engineering plastics (ABS) depend on the annealing conditions. Local brittle-to-ductile transitions can be induced by stress concentrations; if the plastic is annealed, the brittle-to-ductile transition is more difficult to induce and the material breaks at a lower strain energy. Another coupling effect is found when plastics absorb moisture and become plasticized. In nylon and other polar polymeric materials (epoxy and polyurethanes) the effect of moisture absorption on mechanical properties and dimensional stability is dramatic [111,112]; modulus, hardness, impact strength, and flexural strength are affected. Developing this point further, the application of stress to a polymer can produce a volumetric (dilatational) strain on the material that will increase its free volume leading to a more rapid interdiffusion of moisture (or other externally migrating reacting species) [113]. This may lead to time-dependent fracture of the polymeric material phenomenologically similar to stress corrosion [114-116]. Such "stress corrosion" effects have been observed by Minford [71] and others [72,90] on adhesive joint systems. Engineering-wise, one must consider the variety of service condition variables collectively. For example, the role of engineering design of adhesive joints on the mechanical stress and environmental durability of such joints in service is important. The extensive variable of joint geometry must always be considered a dominant factor in the practical use of adhesive joints. To smooth out stress concentrations and decrease the amount of joint edge exposure are perhaps two obvious engineering design considerations when one is designing for permanence and durability. An important condition to attain is that of the lowest stress concentrations at joint edges. Such points or zones would be most vulnerable to environmental attack. The subject of stress concentration and design of adhesive joints is one of active engineering interest. For more information, the reader is referred to the many publications on the subject [117-119].

5.1. *Generalized Scientific Approach*

In a basic sense, pertaining to structural adhesives, one can examine the permanence and endurance of adhesive joint materials in the same way that one examines all material systems. In doing this, one can define the environment into which a functional material exists in terms of a multidimensional "environmental space." Primarily, a coordinate system made up of mechanical stress, temperature, radiation flux, and chemical environment is involved. The elements of such a coordinate system (assuming the radiation level to be constant) are depicted in Fig. 7-13. Here material systems are acted upon by external forces which under certain conditions will change the molecular composition and/or arrangement of the materials. The imposed external forces of mechanical service stresses, temperature, and external chemical environment are the three important intensive variables in which all functioning engineering materials systems must exist. These variables may be referred to as "open" external energy fluxes. For completeness, the coordinate of radiation energy must be included in the realm of a materials habitation in "environmental space." This is important for materials exposed to the sun's radiation in outdoor exposure, and is especially important for engineering materials to be used in outer space. However, since in many ordinary circumstances the radiation energy flux is a constant and/or steady-state phenomena (as is the bombardment of materials by cosmic rays), for simplicity it is convenient to exclude this intensive variable parameter. This facet of the external environment will not be included in these present considerations.

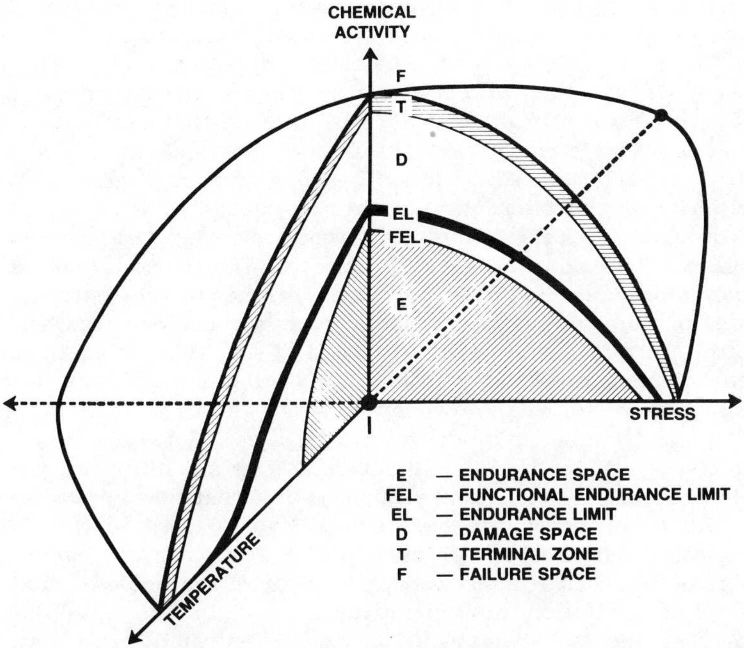

FIG. 7-13 Schematic diagram of energetic response of materials in environmental space: E—endurance space; FEL—functional endurance limit; EL—endurance limit; D—damage space; T—terminal zone; F—failure space.

Having listed the primary forces that impinge on a material during its existence in environmental space, let us now turn to the more complex situation of how materials react or respond to these external environmental forces. One viewpoint that can be described is that the response of materials to a mechanostatic (or mechanodynamic) energy field is a change in its stored mechanical strain energy; the response of materials to a thermostatic (or thermodynamic) energy field is a change in its stored thermal (kT) energy. Finally, the response of materials to an external chemical energy field may involve a change in the material's stored chemical energy—its chemical potential, free energy accumulation. Chemical energy storage or loss will occur relative to the nature and magnitude of the outside chemical driving force (vapor pressure or atmospheric gases, concentration of a reactive species, etc.). All these conditions are capable of inducing mechanical, physical, and chemical responses in the material, macroscopic- as well as molecular-level responses. Basically, these interactions must all occur with the specification of upholding the law of energy conservation; i.e., the energy may be transformed and/or absorbed, but overall, the total energy change between the system and its open surroundings must be zero. Overall, this interpretation can present the foundation for the establishment of a branch of physical science that can be called *enviroenergetics*. In a strict sense, the field of classical thermodynamics (or thermostatics, as it should be more correctly called [120]) is based primarily on the interaction of thermal energy on material behavior. Chemical (and electrochemical) reactions are considered as classical chemical thermodynamics. Furthermore, the thermodynamics of mechanical deformation is also a subject of interest in the realm of the energetics of material systems; such a topic is considered in the realm of continuum mechanics. It is proposed, then, that by employing the term "enviroenergetics" to the study of materials (as outlined in Fig. 7-13) one can overcome the overlap and ambiguities in terminology that presently exist in the fields of thermodynamics. The establishment of this view of the energetics of materials goes much beyond the scope of this present chapter. However, it is felt that new approaches are needed and the proposed approach should be consulted further.

From the coordinate system depicted in Fig. 7-13, one can easily display the principle theme of enviroenergetics from the point of view of the survival of a functioning engineering material in environmental space. Here the concept of endurance space is illustrated. In service, a material can first exist (molecularly and compositionally unchanged) within certain limits of endurance with regard to an imposed, external combined force field of stress, temperature, and chemical environment. In order to serve an engineering function, the limits of endurance of the material may be less than what is required for molecular reversibility. To illustrate, a material may soften thermally, and not degrade; yet it is not able to perform its specified engineering function in this thermally activated (yet reversible) state. There is, however, a lower temperature (a limiting temperature) below which the material can perform its function. This temperature can be referred to as its functional endurance temperature. Therefore, engineeringwise, functional endurance space must be defined relative to the service-life variables that the engineering materials or structures (material system) is to endure; interactions between the variables are also considered. Functional endurance space may then be considered a characteristic "volume" in environmental space wherein the material or material system can permanently serve to perform its function without failure. Of course, each material will have its own characteristic environmental space diagram (characteristic volume in environmental space). Furthermore, in material and material systems, the shape of the functional endurance space (volume) region depends upon the molecular properties and behavior of the materials involved and, or course, the intended service-life application of the material.

Relative to endurance space, the concept of damage space and failure space can easily be illustrated. These conditions, defined somewhat by the discussions and arguments about rheological damage used in the context of Table 7-3 and Fig. 7-7 of this chapter, are shown in Fig. 7-13 as a volume envelope surrounding the outer regions of endurance space. The principle can be expressed here that any time the material "ventures" into (exposed to extreme external environments) and resides at any length of time in the damage space volume, irreversible "damage" occurs to the material. Damage may be accumulated at a rate and degree depending upon its residence time in this damage space region and the intensity of the environmental condition—how close to the terminal space or failure space region the material is residing. From this proposed description, it can logically be assumed that the venture of a material into any region of damage space and beyond will cause irreversible changes in the shape of the endurance space volume. To illustrate, if a material is partially degraded thermally, it may be degraded in such a manner that it will not be able to respond to the same ultimate stress as the undegraded material; it becomes mechanically weaker. In addition, thermal degradation may have changed the morphology and/or chemical composition of the subject material in such a way that will render it more susceptible to attack by an adverse chemical environment. With this in mind, one can again readily project on Fig. 7-13 how a change in shape of the endurance space volume will occur. In the case of thermal effects on materials, in some cases annealing or additional crosslinking can occur which may cause what we can refer to as "constructive" changes in a material. Nevertheless, like the case of damage by thermal degradation, irreversible changes occur in the endurance space volume profile. Under this particular treatment, a new material structure is formed.

The concern here is for damage or partial destruction of a material material's structural integrity by the exposure of the material to an environmental condition of stress, temperature, or chemical adversity where the intensity level of one or more of the coordinate fluxes are higher than the endurance limit for the given material. Damage may thus be described as a condition where the cohesive elements of the material are systematically destroyed by either a mechanical, thermal, physical, and/or chemical process or a complex combination of these. In any event, the result is that molecular and/or morphological cohesive elements in the materials are irreversibly changed—permanently destroyed in the sense that the damaged material can no longer behave and function in exactly the same way it did before. In an attempt to quantify this, one can consider all materials that reside in environmental space to have a characteristic number of cohesivity endurance elements, N_e, and that each element occupies a volume V_n in this proposed environmental endurance space. A material then has an extent of N_eV_n as the total volume it occupies in environmental endurance space, V_e. From the standpoint of damage, then, environmental destruction of a material can then be quantitatively described in terms of the destruction and depletion of these cohesive endurance elements. Damage can then be manifested by a shrinking of the endurance space volume—a "shortening" of the endurance limit reaches at the various coordinates. In the case of "constructive" damage where the material may cure, crystallize, or become annealed by the imposed environment, the situation becomes more complex; the cohesivity elements in such a "changed" material cannot be simply and uniformly defined.

Basically, the concept of N_e introduced above is arbitrary. N_e is a scalar quantity conceived to depict the endurance capacity of a material. The destruction of these cohesive endurance units is also indiscriminate; that is, elements destroyed by stress, temperature, chemical, or a radiation environment have identical final states that are indistinguishable. The assumption then, from a functional standpoint, is that the element can no longer perform its function no matter how it was destroyed.

The coordinates of stress, temperature, and chemical environment are represented in Fig. 7-13 as singular scalar quantities. This, however, is true only for the temperature coordinate. The stress coordinate is a tensor quantity composed of the three principle axes of normal stress on a material and six shear stresses. This stress tensor can be conveniently written in a nine-element matrix form which can be broken down into an isotropic (or hydrostatic) pressure tensor and a deviatoric stress tensor. Mechanical failure of a material is complex and is indeed a function of the stress concentrations as they build up in the stressed material. The stress pattern in the deformed material depends upon the magnitude of the externally applied force, its mode, and its direction relative to the geometrical shape of the stressed body. Stress patterns in mechanically deformed solids are within the realm of engineering micromechanics and analysis and will not be herein discussed. The reader is referred to the many texts that exist on the subject [121,122]. Under real service conditions (outdoor exposure, for example), the chemical environment coordinate is also complex. It may be composed of oxygen, water (liquid and vapor), carbon dioxide, and sulfur dioxide (atmospheric pollutants), to name a few of the components of the contemporary atmosphere. Furthermore, the material may also be subject to biochemical attack. Therefore, the chemical environment coordinate is complex. Unlike the mechanical stress coordinate, however, it can be separated as a set of multiple scalar quantities, chemical energy fluxes, impinging on the material. In many cases, only one species in the chemical environment functions as the critical destructive agent. In many cases, then, this coordinate may be considered a single scalar coordinate;

however, one must be alert for synergistic effects of the combined chemical environment.

Having pointed out the complexities of the coordinate system shown in Fig. 7-13, another complicating factor about real engineering structures is that they are usually multiple material systems; adhesive joints, for example, are characterized by at least two material components plus the interfacial phase (or phases). This multiple-materials composition feature, impacted by the multicoordinate environmental space, indeed results in a situation of compounded multiplicity; the scope of studies of permanence and endurance of material systems is thus clearly illustrated.

5.2. *Engineering Approach*

For the most part, the material specifications of present engineering design structures comes from knowledge about the system collected from previous experience. This information is obtained from (1) published handbooks of tabulated data and documented literature; (2) performing special laboratory tests or simulation experiments; (3) experience with previous functioning structures; or else (4) exists in the undocumented, state-of-the-art experience and knowledge of the design engineer himself. Operationally, engineers design with conservative safety margins. This conservatism is a result of an engineer's ultimate responsibility for designing a successful structure. In many cases, he has to take direct responsibility should the structure fail.

As a result, many engineering structures and components are overdesigned. This is not a serious situation in most cases. However, it does reflect a lack of confidence and general understanding of materials engineering properties. In spite of this lack of understanding, some predictive schemes are emerging which should help to point the science and engineering materials researchers in the right direction. One such scheme is the approach of Kelley and Trout [11]. While their scheme relates to the elements of service-life prediction of solid rockets, it has general engineering applicability, including structural adhesive joints. Their predictive scheme is presented in Fig. 7-14.

The outline in Fig. 7-14 describes the impact of *load definition* (intensive variable) and *geometry* (extensive variable) on the stress analysis and material properties in the structural system. The material (in its specific geometry) reacts to the external load by either mechanical responses (low strain) or else fracture (high strain). The interaction of these parameters in defining the overall service life of a structure is clearly illustrated. Load definition can be obtained from previous knowledge of the stress environment the structure is expected to encounter in service. The geometric factors are dictated by the constraints imposed on the structure, such as size and weight (materi-

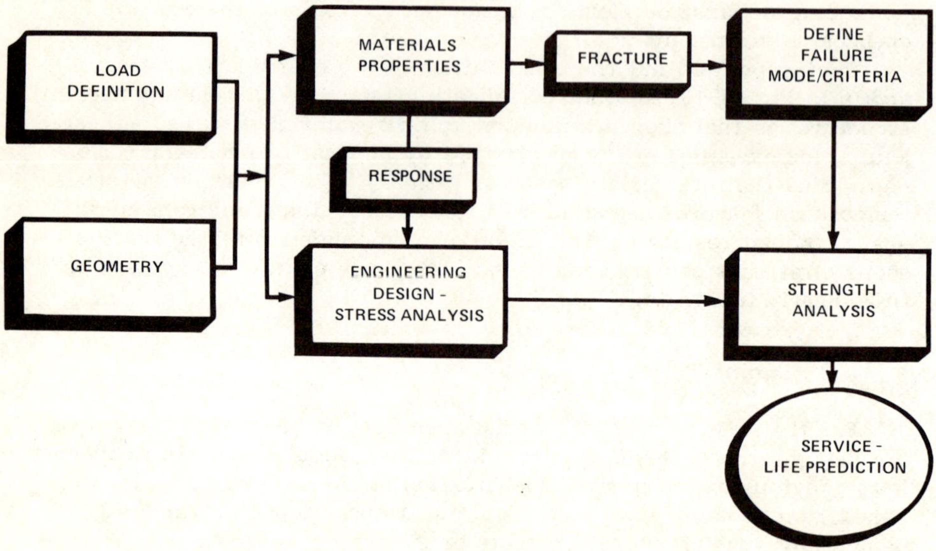

FIG. 7-14 Elements of service life prediction. (Source: adapted from Ref. 11.)

als considerations and cost may also be important). This example therefore illustrates the practical importance of understanding the nature of the intensive and extensive variables that go into establishing the endurance and permanence of engineered materials systems.

From the approach of Kelley and Trout, one is able to outline a sequence of activities needed to qualify and specify an adhesive joint system for an engineering application. The outline in Table 7-8 reviews these steps; more correctly, it outlines a series of thought processes which will help to guide the materials engineer and researcher toward systematically getting the desired information. The first step is the obvious definition of the service-life conditions. These are defined from previous experience or else are estimated by extrapolation methods. Most important would be an estimate of the upper limits of the intensive variables the adhesively bonded structure is expected to encounter (stress-temperature-chemical environment). Next, from a materials research and testing point of view, specific failure modes in the adhesive joint system should be reviewed and identified. Here, a specific adhesive material system may be specified. The rate of damage of the particular joint system under the various extreme service conditions should be determined by experimental testing procedures; or else if data already exists, a damage mode analysis should be made. This

TABLE 7-8 Sequence for Considering P/E/R (Permanence/Endurance/Reliability) in Engineering Design of Structures

1. Define extent of service condition variables.
2. Identify specific failure mode(s).
3. Determine rate of damage for each failure mode at the extreme service conditions.
4. Define critical failure mode.
5. Establish endurance limit of system.
6. Determine reliability of determined endurance limit value(s).
7. Plan margin of safety for engineering design from established endurance limit.

is the most critical and may be the most time-consuming (experimentally) step. Failure can be defined as having the component or structure fall short of meeting a performance specification for either properties or service [123]. Failure by a mode, such as yielding, fracture, fatigue, or corrosion, are often easily identified and/or diagnosed. When failure is by other means, e.g., thermal oxidation/pyrolysis, biological attack, or radiation damage, the problem is not so easily identified. Overall, identifying failure damage in adhesively bonded structures can sometimes be especially difficult. Adhesive joint damage may remain undetected because of the proximity of the adherends, which may mask any visible signs of incipient failure.

A diagram illustrating a possible set of experimental data that one would compile to understand the failure modes of an adhesive joint system is presented in Fig. 7-15. When the data are closely analyzed and the extent of ultimate service life is specified, the critical failure mode and time can be defined. If this ultimate service lifetime is longer than (within the limits of statistical variability) the expected service life of the engineering structure, the material specified for the adhesive joint system may now be qualified for use in the structure. The endurance limit of the whole structure is, of course, determined from the limits of durability of its "weakest link." However, from an adhesive joint standpoint, one has the option of modifying the joint geometrically or materially to accommodate the required service-life endurance limit. If the critical failure mode is mechanical yielding, a geometric change in the joint may suffice as an extender of the ultimate service life. Failure by a chemical corrosion mode may require a surface coating of anticorrosion primer for the adhesively bonded surfaces. However, if the critical failure mode is pyrolytic degradation, changing the geometry of the joint or priming the bonded surfaces may have little effect, unless in the case of priming the failure is a catalytic thermal degradation of the adhesive in the boundary layer.

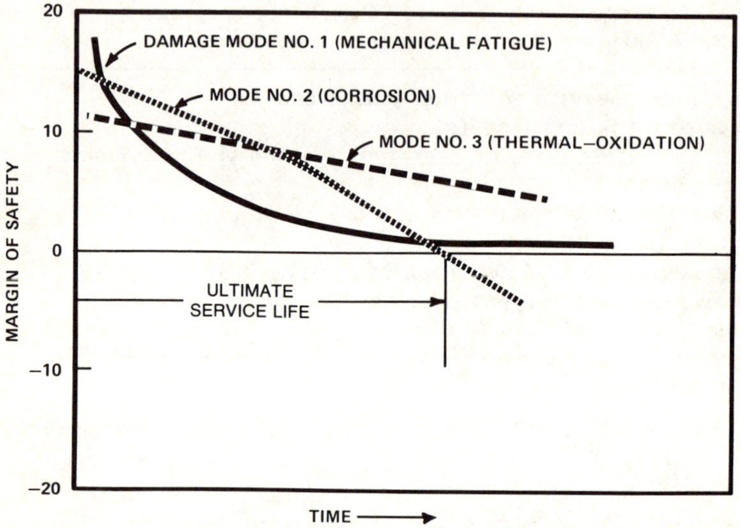

FIG. 7-15 Determination of critical failure mode (Source: adapted from Ref. 11.)

Nevertheless, in this case, the specified adhesive material system must be modified or changed to circumvent the observed premature failure of the adhesively bonded structure. This example, then, points out the importance of knowing the chemical behavior of the adhesive material before it is employed in the structure.

As stated previously, the most important step in the service-life prediction of adhesive joint structures is the definition of the critical failure mode and the assignment of a "margin of safety" to the functioning component. Often the engineer must resort to prototype construction (or trial manufacture) of a part of a structure and subject it to extreme and/or accelerated service testing in order to confidently assign margins of safety to his materials and component design specifications. The testing and characterization of engineering materials is a key to their successful application. Often, too little attention is paid to the development, understanding, and relevance of these so-called standard tests. In practice, much work lies ahead in establishing the credibility and utility of various test methods as they apply to defining engineering parameters.

From an engineering viewpoint, the National Materials Advisory Board of the National Research Council, has issued a report on the testing of components and structures for the prediction of performance

[120] that sheds some light on the subject of service-life testing. In this work, the use of the "Mathes ladder" approach [124] to establish a testing hierarchy was described. To illustrate, the Mathes ladder, presented in Fig. 7-16, proceeds from simple tests of basic materials where well-defined test specifications are available to increasingly complex tests. Although more complex tests are often necessary, the correct interpretation of information obtained from the lower "rungs" of the ladder may reduce or even eliminate the need for more complex tests. However, the need to understand the limits of the simpler tests is paramount to their proper use. Very often, the "surprise" failure in service is associated with a complete lack of understanding of the effects of the service environment on the basic materials used in the construction of the structure. To be certain, however, prototype testing under simulated service conditions may uncover some unexpected failure conditions that will call for corrective design adjustments and sometimes redesign.

The most difficult-to-predict failure situations are the coupling effects: interactions between the stress condition, temperature, and chemical or external environment which converge unpredictably to cause premature failure. In recent years, pertaining to adhesive joints, problems with biological attack of polyurethane-based encapsulating materials in moist environments and water vapor diffusion through adhesively bonded aircraft structures in freeze-thaw cyclic service conditions have been discovered in service [125,126]. Unexpected serious failures in the aircraft structures and functional components

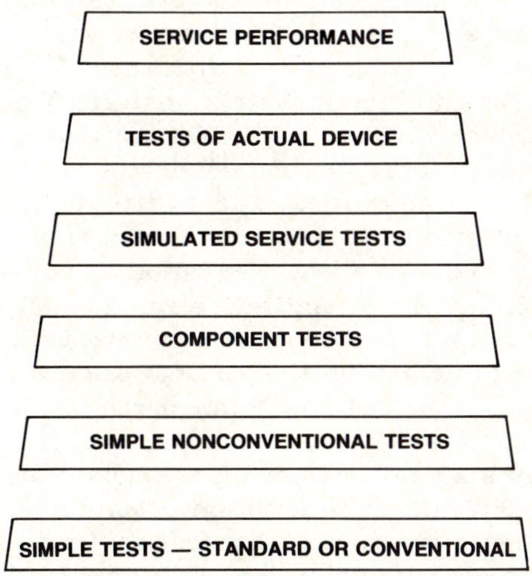

FIG. 7-16 Testing hierarchy as illustrated by the "Mathes ladder."

were the result. In other cases, long-term creep of nylon-epoxy adhesively bonded joints (under moist tropical environments) has been observed in helicopter blade applications; here, thus far, the functional performance of the aircraft has not been too adversely affected. To the materials scientist, prototype testing of components for the most part may at first appear redundant; to the engineer who may be unfamiliar with the molecular architecture of the materials he has employed in the structure he has created, complex testing is necessary so that he may gain confidence in the materials usage and assurance in his design. He is now able to reexamine or recheck the nature and intensity of the variables that act on his structural systems; sometimes, he knows too well, some new, unexpected critical performance conditions may enter into the picture.

5.3. *Adhesive Joint Interaction Matrix*

In order to construct a complete picture of the multiplicity of environmental space and material response parameters that characterize the functioning of an adhesive joint system, the significant expression of the attachment site joint strength model, Eq. (47), can be subjected to an interaction matrix analysis employing the methodology of Kelley and Williams [9]; a similar review of the structure-property relationships in polymers is presented in Table 7-2 of this chapter. Assuming Eq. (47) to be broadly applicable, one can consider it to be an "equation of surface state" for the adhesive joint system. This equation, through the proposed definition of the described parameters, is capable of defining the joint strength in terms of the surface states of the ruptured and/or environmentally deteriorating joint. With these assumptions, the basic elements of an interaction matrix analysis for adhesive joints are presented in Tables 7-9 and 7-10. The various joint materials and interfacial boundary layer characteristics as related to the influence of mechanical stress, temperature, and hydrolytic and/or chemical environment are outlined with reference to the engineering parameters derived from the adhesive joint "equation of surface state." To complete this interaction matrix, quantitative relationships between the materials characteristics and the proposed adhesive joint engineering parameters are needed to determine the strong, moderate, and weak interactions between the exponential power law parameters in Eq. (47) and the materials descriptors. This, of course, cannot be done quantitatively at this early state. Considering the conditions of forming a strong joint, Table 7-9 shows an interaction possibility between 90 different elements. When a thermal or chemical environment is superimposed on a stressed adhesive joint, an additional 120 elements is added to the ensemble, as shown in Table 7-10. A general lack of knowledge about the nature of these interacting parameters

TABLE 7-9 Adhesive Joint Interaction Matrix: Formation of Adhesive Joints (Equilibrium Conditions)[a]

Joint material characteristics	Phase(s)[b]	State parameters					
		S_c	S_0	α_0	α_e	Q_e	(Z_c)
Relaxation modulus	B	S	M	M	U	U	M
Shear modulus	B/A	S/S	M/N	M/U	U/U	U/U	M/N
Ultimate strength	B	S	N	N	U	U	S
Poisson's ratio	B/A	M/M	N/N	N/N	U/U	U/U	M/M
Glass temperature	B	S	M	M	U	U	M
Crosslink density	B	S	S	S	U	U	S
Self-diffusion constant	B	M	N	N	U	U	N
Fracture toughness I	B/I	S/S	N/M	N/M	U/U	U/U	S/S
Surface free energy	I	M	S	S	U	U	M
Surface roughness	A	M	S	S	U	U	M
Site area	I	U	U	U	U	U	S
Boundary layer thickness	I	S	M	M	U	U	S

[a]Key: B = bulk material characteristic; I = interfacial characteristic; A = adherend characteristic; S = strong interaction; M = moderate; N = negligible. U = unknown.

[b]Interactions rated according to specific phase.

TABLE 7-10 Adhesive Joint Interaction Matrix: Superimposed Thermal/Chemical Environment[a]

Joint material characteristics	Phase(s)[b]	State parameters					
		S_c	S_0	α_0	α_e	Q_e	(Z_c)
Thermal environment							
Glass temperature	B	S	M	M	U	U	N
Crosslink density	B	S	M	M	U	U	N
Coeff. thermal expansion	B/A	M	N	N	U	U	N
Thermal conductivity	B/A	N	N	N	U	U	N
Oxygen diffusivity	B	S	M	M	U	U	M
Activation energy—thermal	B/I	S	M	M	U	U	S
Surface catalysis	I/A	S	M	M	U	U	S
Chemical environment							
Activation energy—chemical	B/I	S	M	M	U	U	S
Vapor diffusivity	B	S	M	M	U	U	M
Surface catalysis	I/A	S	M	M	U	U	S
Solubility product constant	I	S	M	M	U	U	S
Electrochemical potential	I	S	M	M	U	U	M
Electrical conductivity	B/A	N	N	N	U	U	M

[a]Interaction rating refers to the influence that certain materials characteristics have in preventing changes in the state parameters when the joint is subjected to a deteriorating environment. Key: B = bulk material characteristic; I = interfacial characteristic; A = adherend characteristic; S = strong; M = moderate; N = negligible; U = unknown.

[b]Interactions rated on collective phases.

exists. However, some qualitative estimates as to the degree of interaction among the various parameters can be made; these are presented in Tables 7-9 and 7-10. Strong, moderate, and negligible interactions are noted along with the stated unknown interactions. Regardless of the present correctness of the stated qualitative opinions expressed in Tables 7-9 and 7-10, the status of the materials structure-property relationships in the field of adhesive joint strength materials technology has been surveyed in an orderly fashion. This, in itself, as pointed out by Kelley and Williams [9], makes such an interaction-matrix exercise worthwhile. Assuming a proper matrix has been constructed, such an outline should help to guide future work. In addition, this methodology should serve to point out where uncertainties exist and ultimately lead to the development of a unified theory of adhesive joint strength of significance to engineering design applications. Meanwhile, much work remains to be done in verifying the magnitude of the interactions between the parameters displayed in the proposed interaction matrix. This is in addition to the need to more fully substantiate the proposed equation of state and to establish the proper choice of materials characteristics to be used in setting up the matrix.

6. SUMMARIZING REMARKS

In this chapter, a systematic and selective review of the field of adhesive joint endurance and permanence has been presented—a review relative to the state of the art. Rather than being an exhaustive search of the literature and a compilation of research results and subsequent analysis, an attempt has been made to discuss what is considered to be the most relevant technical factors that contribute to the durability of structural adhesive joints. Where necessary, new interpretations have been conceived. Of most notable significance is the methodology and formulations expressed by the proposed attachment site theory of adhesive joint strength. This theory presents a means of describing, in engineering terms, the development of adhesive joint strength. Basically, it assumes that weak boundary layers are rendered strong by a progressive increase in the number of so-called effective attachment sites in the boundary layer. Conversely, the environmental deterioration (thermal and/or chemical, etc.) of a joint involves the steady-state random destruction of these effective attachment sites at the interface. Overall, this attachment site interpretation is a convenient approach to modeling the boundary layer in adhesive joints as well as other situations where adhesively bonded interfaces are present. Attachment sites are pictured to be mechanically functioning ligaments, patches, or zones which characterize the boundary layer and are viewed to be of a morphological or surface roughness size. These "macrosites" exist and are affixed between an adherend and an adhesive by chemical and/or physical attractive forces, mechanical and/or diffusional intermixing, or electrostatic attraction forces or a combination of all the above-mentioned mechanisms. This array of macrosites then functions as the intermediate zone (boundary layer) that functionally attaches the adhesive to the adherend. While being presented as a hypothetical model of the boundary layer in an adhesive joint, such a topographical model of the interface probably exists in some form in real adhesive joint interfaces. Supportive experimental evidence for this view of adhesive-adherend interfaces may perhaps be obtained through careful morphological studies. Though at present no quantification of the attachment site morphologies has been carried out, it is felt that this new methodology for the study of adhesive joints will indeed lead to new insights and interpretations of the basic mechanisms that control adhesive joint strength and functional durability.

Summarizing in the broad sense of an overview of this chapter, the topic of permanence and endurance of adhesive joints was first discussed in terms of the important (primary) service variables that an adhesive joint (material) encounters while performing its engineering function. The primary intensive variables of stress, temperature, and chemical environment were treated in some detail. In the review of the stress variable, the use of viscoelastic models for adhesive joint materials was employed to describe the response of adhesive joints to stress. A special emphasis was made on the subject of creep of polymeric materials and static fatigue of adhesive joints. Rudimentary discussions of creep were presented in view of recent interpretations of the response of polymeric materials to static stress. The regimes of response were presented in terms of the types of rheological damage imparted to viscoelastic materials while under different stress levels. These regimes were specifically outlined as in Table 7-3 and designated as 0—*Hookean response*, where at low stresses no damage to the polymeric material occurs; I—*modulus damage*, where at higher stress levels a polymeric material suffers a loss in modulus but no loss in ultimate tensile strength; II—*strength damage*, where at still higher stress levels cracks and/or crazes occur in the material; these cracks propagate and/or form continuously, so long as the stress is applied. The final regime of rheological damage is III—*terminal damage*, where at stress levels at or near the failure stress, the failure cracks or crazes coalesce and incipient, catastrophic failure of the material occurs. Further studies of the static stress response of polymeric materials in the context of adhesive joints should be encouraged in an attempt to bridge the gap between the understanding of the creep properties of polymeric materials in relation to the static stress response and endurance limit properties of adhesive joints. It is felt that a better understanding of the static stress response of adhesive joints must be obtained before much scientific progress can be made as to understanding the dynamic response or fatigue behavior of adhesive joints. The collective response of static, dynamic, impact, and random stresses, however, are important engineering considerations, especially in the field of adhesives for aerospace applications. In reference to applications, it is worthy to remark here about the overwhelming need for more meaningful test methods for adhesive joints. Existing test procedures (peel, lap shear, butt, impact, torque shear, etc.) are all employed and represent the foundation for the present "tragicomedy" of qualification testing. There is some hope for the future, however. The work emerging from the studies of the fracture mechanics of adhesive joints shows promise. The application of this fracture mechanics information by engineers to the design of adhesively bonded structures, while in its infancy, is making steady progress.

Probably the best-understood area in the study of durability of joints is the thermal response of adhesive joint materials, that is, the thermal softening behavior of polymers. In most systems, adhesive joint strengths parallel the thermal softening behavior (glass temperature, crystalline melting temperatures) of the adhesive polymer. The thermal aging behavior of adhesive joints, however, is quite a different consideration. While the thermal softening behavior of the adhesive polymer also enters the picture, the controlling factors lie in understanding the nature of the specific chemical reactivity of the adhesive material under the thermal environment. Factors such as the physical organic chemistry aspects of thermal pyrolysis and thermal oxidation, the catalytic effects of surfaces, surface chemistry, surface oxidation, effects of superimposed mechanical stress, and complex chemical kinetic degradation schemes must be considered. Much literature exists on these subjects as they relate to polymeric materials. The application of this vast amount of information and the approaches used here have not been too effectively applied to studies of the thermal degradation of adhesive joints. More systematic studies are needed.

Probably the most complex area of adhesive joint strength durability is the area of chemical environment effects. One reason for this is that a real service environment is characterized by a multitude of chemicals and chemical effects. Temperature, diffusion of gases, mechanical stress, surface chemistry, and electrochemical corrosion all enter into the picture and add interactive complications. The combined influences of these factors lead to a further superimposed or overlay of complexity. In this regard, the field of permanence and endurance of adhesive joints can, to an advantage, be analyzed in terms of an interaction matrix scheme, where the different molecular characteristics of adhesive joint materials can be analyzed in terms of factors of engineering significance. Using this approach, a sorting out of the more important and less important factors is possible. Hence lies the profound significance of the proposed attachment site model of adhesive joint strength. A methodology for studying adhesive joint strength and durability has been formulated; an equation of surface state has been proposed. It is predicted that a systematic, experimental followthrough on this proposed methodology will contribute greatly to the further understanding and development of the science and technology of adhesive joints.

Another new interpretive concept presented in this chapter is felt worthy of further comment—that is, the model of "environmental space." This view has been proposed as a generalized scheme for describing the response of materials to existing and functional environmental stresses. While serving as a basic, more organized model for studying or understanding the thermodynamic properties of materials, the environmental space model has the important unifying feature of relating scientific knowledge and understanding about the behavior of materials into information of engineering significance. It serves as an interpretive bridge between the science of materials behavior and the engineering application of materials. More about this approach should be forthcoming as the scope of the field of enviroenergetics develops and becomes more clearly understood.

ACKNOWLEDGMENT

The authors wish to thank Dr. Frank N. Kelley, Head, Polymer Institute, University of Akron, Akron, Ohio, for his comments which served to help us sort out the right characteristic parameters to be included in the compiled interaction matrix.

REFERENCES

1. H. Liebowitz, ed. *Fracture: An Advanced Treatise*, 7 vols., Academic Press, New York, 1969–1972; see especially volume, *Fracture of Non-Metals and Composites*.
2. H. T. Corten, J. Adhesion 3:103 (1971).
3. E. J. Ripling, S. Mostovoy, and H. T. Corten, J. Adhesion 3:107 (1971).
4. N. L. Harrison and W. J. Harrison, J. Adhesion 3:195 (1972).
5. R. D. Adams and N. A. Peppiatt, University of Bristol Paper: Stress analysis of adhesive joints, presented at the 12th An. Conf. Adhesion and Adhesives, City University, London, April 9 and 10, 1974.

6. E. E. House and J. A. White, Bond structural modifications to operational aircraft, paper presented at 19th Nat. SAMPE Symp., Buena Park, Calif., April 23-25, 1974.
7. J. C. Halpin and H. W. Polley, J. Comp. Mater. 1:64 (1967).
8. J. C. Halpin, *Composite Materials Workshop*, Technomic Publisher, Westport, Conn., 1968, p. 87.
9. F. N. Kelley and M. L. Williams, Rubber Chem. and Tech. 42, No. 4:1175 (1969).
10. J. D. Ferry, *Viscoelastic Properties of Polymers*, Wiley, New York, 1960.
11. F. N. Kelley and J. L. Trout, AIAA Paper No. 72-1085, AIAA/SAE 8th Joint Propulsion Specialist Conf., New Orleans, November 29-December 1, 1972.
12. E. H. Andrews, J. Mater. Sci. 9:887 (1974).
13. J. A. Manson and R. W. Hertzberg, Crit. Rev. Macromol. Sci. 1, No. 4:433 (1973).
14. A. A. Griffith, Phil. Trans. Royal Soc. 4227:180 (1921).
15. C. E. Inglis, Proc. Inst. Naval Architects, March 14, 1913.
16. J. P. Berry, J. Appl. Phys. 34:62 (1963).
17. E. Orowan, Phys. Soc. Rep. Prog. Phys. 12:186 (1948).
18. G. R. Irwin, Handbuch Phys. 6:551 (1958).
19. G. R. Irwin, Trans. ASME, Ser. D 82, No. 2:417 (1960).
20. E. J. Ripling, S. Mostovoy, and R. L. Patrick, Mat. Res. Stand. 65, No. 3:129 (1964).
21. E. J. Ripling, S. Mostovoy, and R. L. Patrick, ASTM Spec. Tech. Pub. 360:5, 1963.
22. S. Mostovoy and E. J. Ripling, J. Appl. Polym. Sci. 13:1083 (1969).
23. S. Mostovoy and E. J. Ripling, J. Appl. Polym. Sci. 15:641 (1971).
24. S. Mostovoy and E. J. Ripling, J. Appl. Polym. Sci. 15:661 (1971).
25. S. Mostovoy and E. J. Ripling, Fracturing characteristics of adhesive joints, Contract #N00019-73-C-0163, Materials Research Laboratory, Glenwood, Ill., October 1974.
26. W. D. Sell, Proc. 19th Nat. SAMPE Symp. and Exhib., Vol. 19, April 23-25, 1974, p. 196.
27. L. C. Hopper, SAMPE Nat. Tech. Conf., Vol. 20, San Diego, Calif., April 1975, p. 513.
28. A. F. Lewis, Adhesives Age 15, No. 6:38 (1972).
29. F. Bueche, *Physical Properties of Polymers*, Interscience, New York, 1962, p. 49.
30. S. L. Rosen, *Fundamental Principles of Polymeric Materials*, Barnes and Noble, New York, 1971, p. 204.
31. R. S. Lenk, *Plastics Rheology*, Wiley, New York, 1968, p. 125.
32. D. C. Wright, RAPRA Members J., p. 153 (June 1974).

33. T. Hata, in *Recent Advances in Adhesion* (L. H. Lee, ed.), Gordon and Breach, London, 1973, p. 269.
34. J. P. Thomas, J. Appl. Polym. Sci., Appl. Polym. Symp. 3:109 (1966).
35. M. E. R. Shanahan and W. C. Wake, Creep phenomena in adhesive joints, Synopsis of Lectures, 12th An. Conf. Adhesion and Adhesives, April 9-10, 1974, City University, London.
36. E. Camatini, *Optical and Acoustical Holography*, Plenum Press, New York, 1972.
37. A. J. Durelli and V. J. Parks, *Moire Analysis of Strain*, Prentice-Hall, Englewood Cliffs, N.J., 1970.
38. A. F. Lewis, R. A. Kinmonth, and R. P. Kreahling, J. Adhesion 3:249 (1972).
39. E. M. Prot, Rev. Metallurg., 45, No. 12 (1948); English trans. E. J. Ward, WADC Tech. Rep., Sept. 1952, pp. 52-148.
40. H. S. Loveless, C. W. Deeley, and D. L. Swanson, SPE Trans. 2, No. 2 (April 1962).
41. H. S. Loveless and D. E. McWilliams, Polym. Eng. Sci. 10, No. 3: 139 (1970).
42. G. Schumacker, Materialpruf. 10, No. 3:88 (1968).
43. J. R. McDowell, unpublished results, Lord Corporation, Hughson Chemicals Division, Erie, Pa., June 1974.
44. G. M. Bartenev and Y. S. Zuyev, *Strength and Failure of Viscoelastic Materials*, Pergamon Press, Oxford and London, 1968, pp. 31, 189.
45. F. R. Larson and J. Miller, Trans. ASME, p. 765 (July 1952).
46. S. Goldfein, ASTM Bull. No. 224:36 (1947).
47. T. L. Smith, SPE J. 16:1 (1960).
48. D. H. Kaelble, *Physical Chemistry of Adhesion*, Wiley-Interscience, New York, 1971, p. 295.
49. J. T. Seitz and C. F. Balazs, Polym. Eng. Sci. 8, No. 2:151 (1968).
50. M. L. Williams, R. F. Landel, and J. D. Ferry, J. Amer. Chem. Soc. 77:3701 (1955).
51. J. D. Ferry, *Viscoelastic Properties of Polymers*, Interscience, New York, 1969.
52. S. Rabinowitz and P. Beardmore, J. Mater. Sci. 9, No. 81 (1974).
53. P. Zoller and H. Bent, Polym. 15, No. 4:239 (1974).
54. W. F. Watson, in *High Polymers*, Vol. 19 (E. M. Fattes, ed.), Wiley, New York, 1964.
55. G. G. Winspear, ed., *Vanderbilt Rubber Handbook*, R. T. Vanderbilt Company, New York, 1968, p. 316 (refer to ASTM Test D623).
56. P. C. Paris and F. Erdogan, Trans. ASME 85:528 (1963).
57. P. C. Paris, Proc. 10th Sagamore Army Mater. Res. Conf., p. 107, 1964.

58. J. M. Barsom, Fatigue-crack propagation in steels of various yield strength, paper presented at 1st Nat. Cong. Pressure Vessels and Piping, San Francisco, May 10, 1971; see Ref. 13, p. 461.
59. A. H. Lepie and A. Adicoff, J. Appl. Polym. Sci. 18:2165 (1974).
60. S. Mostovoy and E. J. Ripling, Fracturing characteristics of adhesive joints, Final Report, Contract # N00019-73-C-0113, Materials Research Laboratory, Glenwood, Ill., January 1974.
61. R. F. Boyer, Rubber Chem. Tech. 36:1303 (1963).
62. M. B. Bever and M. Shen, Mater. Sci. Eng. 15:145 (1974).
63. A. F. Lewis and R. Saxon, in *Epoxy Resins* (C. May and Y. Tanaka, eds.), Marcel Dekker, New York, 1973.
64. A. V. Tobolsky and H. Eyring, J. Chem. Phys. 11:125 (1943).
65. M. B. Neiman, B. M. Kovarskaya, L. I. Golubenkova, A. S. Striznkova, I. I. Laventovskaya, and M. S. Akutin, J. Polym. Sci. 56:383 (1962).
66. M. B. Neiman, ed., *Aging and Stabilization of Polymers*, Consultants Bureau, New York, 1965.
67. R. B. Krieger and R. E. Politi, in *Aspects of Adhesion*, Vol. 3 (D. J. Alner, ed.), Chemical Rubber Co., Cleveland, 1969, p. 46.
68. Graphite composites for ocean ordnance, Final Report prepared by Fiberite, #FTR-73-1, NOC Contract #N00017-73-C-4305, October 1973.
69. R. F. Wegman, H. C. Hamilton, and M. J. Bodnar, SAMPE Nat. Tech. Conf., Palo Alto, Calif., October 1972, p. 425.
70. Durability of adhesive bonded joints, Report prepared by Bell Helicopter, #299-099-495, USAF Contract #F33615-71-C-1668.
71. J. D. Minford, in *Treatise on Adhesion and Adhesives*, Vol. 3 (R. L. Patrick, ed.), Marcel Dekker, New York, 1973, pp. 79-122.
72. W. B. Jones, SAMPE Nat. Tech. Conf., Buena Park, Calif., April 1974, p. 63.
73. P. M. Stifel, SAMPE Nat. Tech. Conf., Buena Park, Calif., April 1974, p. 75.
74. A. F. Lewis and L. J. Forrestal, Symp. Rec. Dev. Adhesion Science, ASTM Special Technical Publication No. 360, 1963, p. 59.
75. A. F. Lewis and R. T. Natarajan, ACS Preprint Booklet, Organic Coatings and Plastics Division, Spring meeting, Philadelphia, April 1975.
76. F. M. Fowkes, Ind. Eng. Chem. 56:40 (1964).
77. D. K. Owens and R. C. Wendt, J. Appl. Polym. Sci., 13:1741 (1969).
78. D. K. Owens, J. Appl. Polym. Sci. 14:1725 (1970).
79. A. N. Gent and J. Schultz, paper presented at 162nd meeting, American Chemical Society, Organic Coatings and Plastics Chemistry, September 1971; Preprints, 31:113 (1971).

80. H. Schonhorn, paper presented at the 12th An. Conf. Adhesion and Adhesives, City University, London, April 1974.
81. H. Schonhorn and H. L. Frisch, J. Polym. Sci., Polym. Phys. Ed. 11:1005 (1973).
82. D. H. Kaelble, P. J. Dynes, L. H. Crane, and L. Hans, ASTM Composite Reliability Conf., Las Vegas, Nevada, April 1974.
83. D. H. Kaelble, P. J. Dynes, and L. H. Crane, J. Adhesion 7:25 (1975).
84. J. C. Halpin, M. E. Waddoups, and T. A. Johnson, Int. J. Frac. Mech. 8:465 (1972).
85. J. C. Halpin, J. R. Kopf, and W. Goldberg, J. Comp. Mater. 4: 462 (1970).
86. J. C. Halpin, J. Comp. Mater. 6:208 (1972).
87. A. W. Davis and T. A. Johnson, ASTM Composite Reliability Conf., Las Vegas, Nevada, April 1974.
88. M. E. Waddoups, R. V. Wolff, and D. J. Wilkins, Reliability of complex large scale composite structure: Proof of concept, AFML Report #TR-73-160, July 1973.
89. D. H. Kaelble, P. J. Dynes, and E. H. Cirlin, J. Adhesion 6:23 (1974).
90. W. Brockman, SAMPE Nat. Tech. Conf., Buena Park, Calif., April 1974, p. 198.
91. J. M. Krafft and J. H. Mulhern, ASME Trans. 62, No. 64:64 (1969).
92. J. M. Krafft, Trans. ASME, Paper No. 74—Mat 13, Meeting, Miami Beach, June 24-28, 1974.
93. D. Bulgin, G. D. Hubbard, and M. H. Walters, Proc. 4th Rubber Technology Conf., London, 1962, p. 173.
94. D. F. Moore and W. Geyer, Wear 22:113 (1972).
95. R. T. Natarajan, ACS—Organic Coatings and Plastics Div. Preprints, March 1977—Borden Award Symposium.
96. M. Avrami, J. Chem. Phys. 7:1103 (1939).
97. M. Avrami, J. Chem. Phys. 8:212 (1940).
98. M. Avrami, J. Chem. Phys. 9:177 (1941).
99. W. A. Johnson and F. R. Mehl, Trans. AIME 135:416 (1939).
100. R. T. Natarajan and T. Davidson, J. Polym. Sci., Polym. Phys. Ed. 10:2209 (1972).
101. M. Knudsen, *The Kinetic Theory of Gases*, Methuen, London, 1950.
102. K. U. Snowden, Acta Met. 12:295 (1964).
103. B. R. Lawn, Mater. Sci. Eng. 13:277 (1974).
104. S. M. Wiederhorn and L. H. Bolz. J. Amer. Ceram. Soc. 53:543 (1970).
105. S. M. Wiederhorn, J. Amer. Ceram. Soc. 50:407 (1967).

106. W. Wiebull, J. Appl. Mech. 18:293 (1951); *Fatigue Testing and Analysis of Results*, Pergamon Press, New York, 1961.
107. A. J. Kinloch, Environmental failure of structural adhesive joints: A literature survey, August 1973, AD784890 or ERDE TN95.
108. R. A. Gledhill and A. J. Kinloch, J. Adhesion 6:315 (1974).
109. A. J. Kinloch, W. A. Dukes, and R. A. Gledhill, ACS—Coatings and Plastics Preprints 35, No. 1:546 (1975).
110. S. Matsuoka, Polym. Eng. Sci. 14, No. 3:162 (1974).
111. Gulf Oil Corp., Effects of moisture on physical properties of Gulf nylon, Technical Bulletin.
112. R. D. Deanin, *Polymer Structure, Properties, and Applications*, Cahners Books, Boston, 1972, pp. 75ff.
113. D. H. Kaelble, *Physical Chemistry and Adhesion*, Wiley-Interscience, New York, 1971, pp. 295, 360.
114. L. J. Broutman, SPE J., p. 283 (March 1965).
115. P. Hittmar and R. Ullman, J. Appl. Polym. Sci. 6:191 (1962).
116. A. R. C. Westwood, J. Mater. Sci. 9:1871 (1974).
117. G. Anderson, Ph.D. dissertation, University of Utah, 1973; University Microfilms, Ann Arbor, Mich.
118. W. J. Renton and J. R. Vinson, Analysis and design of anistropic bonded joints, AFOSR, TR75-0125 (MAE TR 171), August 1974.
119. Army Materials and Mechanics Research Center, The role of mechanics in design: Structural joints, Proc. Army Symp. Solid Mech., 1974, AMMRC—MS 74-8, September 1974.
120. I. M. Klotz, *Chemical Thermodynamics*, W. A. Benjamin, New York, 1964, p. 3.
121. A. S. Tetelman and A. J. McEvily, *Fracture of Structural Materials*, Wiley, New York, 1967.
122. O. C. Zienkiewicz, *Finite Element Method in Engineering Science*, McGraw-Hill, London, 1971.
123. Testing for prediction of material performance in components and structures, NMAB—#288, National Research Council, Washington, D.C., May 1972.
124. K. N. Mathes, Trans. AIEE 67:1236 (1948).
125. Aerospace structural adhesives, NMAB—#300, National Research Council, Washington, D.C., July 1974, pp. 34, 78.
126. J. S. Noland, Amer. Chem. Soc. Preprint Booklet OCP—35, No. 1, April 1975, p. 525.

Author Index

Numbers in parentheses indicate that an author's work is referred to. Underlined numbers give the page on which the complete reference is listed.

Subject Index